Origins of Pest, Parasite, Disease and Weed Problems

Origins of Pest, Parasite, Disease and Weed Problems

The 18th Symposium of
The British Ecological Society
Bangor, 12–14 April 1976

edited by

J. M. Cherrett

Department of Applied Zoology,
University College of North Wales, Bangor

and

G. R. Sagar

School of Plant Biology,
University College of North Wales, Bangor

Blackwell Scientific Publications

OXFORD LONDON EDINBURGH MELBOURNE

First published 1977

British Library Cataloguing in Publication Data

Origins of pest, parasite, disease and weed
    problems : the 17th Symposium of the British
    Ecological Society, Bangor, 12–14 April 1976.
    —(British Ecological Society. Symposia; 17th).
    Bibl.—Index.
    ISBN 0–632–00226–3
    1. Cherrett, John Malcolm   2.   Sagar, Geoffrey Roger
    3.   Series
    632   SB599
    636.089'6   SF605
    Field crops—Diseases and pests—Congresses
    Weeds—Congresses
    Veterinary medicine—Congresses

Distributed in the U.S.A. by
Halsted Press,
a division of
John Wiley & Sons Inc
New York
and in Canada by
J.B. Lippincott Company of Canada Ltd, Toronto

Printed in Great Britain by
Western Printing Services Ltd,
Bristol
and bound by Kemp Hall Bindery, Osney Mead, Oxford

Contents

Problems past, present and future

Preface

The book comprises most of the papers presented at a meeting held in Bangor from 12 to 14 April 1976 together with short reports of four discussion meetings which took place during one evening.

Our aim as organizers of this, the eighteenth symposium of the British Ecological Society was to examine the ways in which pest, parasite, disease and weed problems have originated and to explore the ways in which future problems might arise. This topic has not, to our knowledge been the subject of a symposium. We wished to emphasize the origins of problems and not solely the origins of the causative organisms. From this examination it is suggested that problems of the future might first be predicted and secondly (and hopefully) avoided.

In inviting our contributors we deliberately sought to bring together many diverse interests. We realized the risks inherent in this approach, the danger that specialists might not find common ground or speak a common language. As far as the meeting was concerned these fears proved groundless.

In the first paper Pimentel emphasizes that problems caused by pests, parasites, diseases and weeds arise from a wide range of causes many of which are examined in greater detail by later authors. A knowledge of the population dynamics of potential pest, parasite, pathogen and weed organisms is of value in understanding and perhaps predicting the type of problem that may arise in given conditions. Southwood reviews the relevance of population dynamic theory to pest status.

The evolution of injurious organisms and their biogeography are important causes of problems some of which are dealt with specifically by Garnham, Lupton, Shattock and Simmonds and Greathead as well as by several other authors. It has frequently been claimed that 'monoculture', one of the most characteristic features of modern plant and animal husbandry is a major cause of upsurges in pest, parasite, disease and weed problems. After an introduction by Way, monocultures of forestry (Gibson and Jones), marine fisheries (McVicar and MacKenzie) and temperate cereals (Potts) are reviewed. The four appear to agree that there is no simple correlation between the degree of monoculture and the incidence of specific problems.

Apart from the foregoing obviously ecological (biological) origins of problems, mans' attitudes reflected in economic and commercial objectives may be crucial. Norton and Conway, Bundy and Lewis explore some of these influences; the interdependence of ecology and economics is exposed.

As Parker emphasizes, the use of our understanding of the origins of problems for prediction is notoriously difficult as the most successful predictions are the ones that eventually prove false because they are acted upon in time. However, against backgrounds of experience, Freeman on storage pests, Sachs on game animals, Coaker on crop pests and Bradley on pests of man have also ventured into this field, extrapolating current trends. In the final paper however Surtees warns that scientific understanding is not sufficient. Political considerations coupled with inadequate systems of communication (editors are given some stick!) can prevent known solutions being put into practice.

During the course of the symposium there were four open discussions on ways of preventing problems arising. These were stimulating and the conclusions are briefly reported. We thank J. L. Harper, R. Johnson, G. A. Norton and D. Pimentel for leading these discussions and J. Grant, E. R. B. Oxley, R. Whitbread and J. R. Witcombe for reporting.

Introduction

The ecological basis of insect pest, pathogen and weed problems*

DAVID PIMENTEL *Department of Entomology and Section of Ecology and Systematics, College of Agriculture and Life Sciences, Cornell University, Ithaca, N.Y. 14853, U.S.A.*

Introduction

At present the world population of four thousand million humans is having difficulty feeding itself; an estimated five hundred million humans are malnourished and at least one thousand million more are hungry (UN 1974, USDA 1974). As the world population continues to grow rapidly toward seven thousand million projected for the next twenty-five years, more acute food shortages are expected (Brown & Eckholm 1974, Pimentel *et al.* 1975a).

For several thousand years about fifteen crops have provided about 90% of the world's plant food (Harrar 1961, Mangelsdorf 1966, Thurston 1969). These crops include rice, wheat, maize, sorghum, millet, rye, barley, cassava, sweet potato, potato, coconut, banana, common bean, soybean, and peanut. Only about eight livestock types provide more than 90% of the animal protein; these include cattle, buffalo, goats, sheep, hogs, chickens, ducks, and turkeys.

All crops and livestock are attacked by pests and on a worldwide basis the losses to pests are high. At present world crop losses to pests (insects, pathogens, weeds, mammals, and birds) are estimated to be about 35% (Cramer 1967). Mammal and bird losses appear to be more severe in the tropics and subtropics than in the temperate region, but these losses are still low compared with the three major pest groups insects, pathogens, and weeds.

Further, available evidence tends to suggest that the 'green revolution' technology has intensified losses to pests (Pradhan 1971, Ida Oka, personal communication). One reason for this increased crop loss is the increased

* A publication of the Cornell University Agricultural Experiment Station, New York State College of Agriculture and Life Sciences, A Statutory College of the State University of New York.

susceptibility of the new high-yield varieties to insects, pathogens, and weeds. In the past farmers most often selected seeds from individual plants that survived best under native cultural conditions. The plants that yielded best under these conditions are genotypes that also naturally harboured alleles resistant to insects and pathogens. They were also successfully competitive with weeds (Ida Oka, personal communication).

Post-harvest losses to pests in the world are estimated to be about 20% (Pimentel *et al.* 1975a) and range from an estimated 9% in the U.S. (USDA 1965) to 40% and 50% in some of the developing nations especially those in the tropics. The prime pests of harvested foods are microorganisms, insects, and rodents. When post-harvest losses are added to pre-harvest losses, worldwide food losses due to pests are estimated to be about 48% (35% preharvest +20% loss of the harvested crops).

In the United States pre-harvest losses to pests are estimated to be about 33% in spite of modern pest control technology. If losses from birds and mammals (about 1%) are included in the estimate, total losses are 34%. This is about the same as the 35% estimate for the entire world. However, with post-harvest losses of about 9% (USDA 1965), total losses in the U.S. are estimated to be about 40% (34% pre-harvest + 9% of the harvested crop). This 40% is less than the estimated world average of 48%. The loss of 48% of world food crops occurs in spite of all types of pest controls used. This is a significant loss of valuable food.

In the world today an estimated 4·1 thousand million pounds of pesticides are applied annually (W. Turtle, personal communication). About 34% of the total is applied in North America, about 45% in western and eastern Europe, and the remaining 21% in the rest of the world (primarily developing countries). It is estimated that by 1985, the total quantity of pesticide applied annually in the world will total more than 5 thousand million pounds (Pimentel 1974).

Of the 4·1 thousand million pounds of pesticide used in the world about 25% or 1·2 thousand million pounds are used in the United States (Fowler & Mahan 1975) (Fig. 1). Pesticide applications to crop and farm lands are estimated at 700 million pounds, of which 54% are insecticides, 36% herbicides, and 10% fungicides. In addition about 500 million pounds of pesticides are used by government agencies, industries, and home-owners for various pest problems.

Of particular significance is the fact that pesticides used in U.S. agriculture are not evenly distributed over all crops (USDA 1975). For example, about half the insecticide used in agriculture is applied to the non-food crops of cotton and tobacco (Table 1). Of the food crops, corn, fruit, and vegetables receive the largest amounts of insecticides. Forty-one per cent of herbicides is used on corn and the remaining 59% is distributed among the other crops

(Table 1). Most of the fungicidal material is applied on fruit and vegetables, with only a small amount used on field crops (Table 1).

According to the latest data, cropland (including pastures) in 1971 totalled about one thousand million acres, of which only 65% was treated with insecticides, 17% with herbicides, and 1% with fungicides (USDA 1975). If cropland devoted to pastures is removed from the total acreage, then the percentage of cropland treated with insecticides, herbicides, and fungicides increases to 15%, 41%, and 2%, respectively.

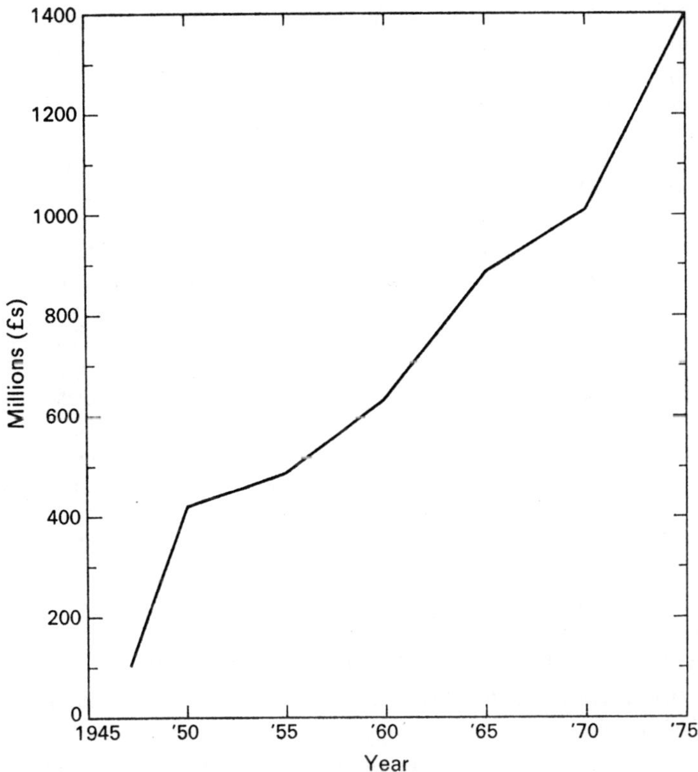

Figure 1. Estimated amount of pesticide produced in the United States (USDA 1971; Fowler & Mahan 1975).

Although cotton receives 47% of all the insecticide used in agriculture, about 40% of the total cotton acreage receives *no* insecticide treatments at all (Table 1). The largest percentage (79%) of acres treated is located in the Southeast and Delta states; whereas the smallest percentage (37%) treated is in the southern Plains (Texas, Oklahoma). Of the major food crops, only citrus fruits and apples have more than 85% of their acreage treated with insecticides (Table 1). Many acres of small grains and pastures receive little or no treatment with insecticides. Therefore, it can be concluded that for some crops and in some geographic areas insecticide use can be minimal, yet crop yields are presumably satisfactory.

Table 1. Some examples of percentages of crop acres treated, of pesticide amounts used on crops, and of acres planted to this crop (USDA, 1968, 1971, 1975).

Crops	Insecticides		Herbicides		Fungicides		% of total crop acres
	% acres	% amount	% acres	% amount	% acres	% amount	
Non-Food	NA	50	NA	NA	<0·5	NA	1·26
Cotton	61	47	82	6	4	1	1·11
Tobacco	77	3	7	NA	7	NA	0·11
Food	NA	NA	NA	NA	<0·5	NA	98·74
Field crops	NA	NA	NA	NA	NA	19	NA
Corn	35	17	79	41	1	NA	7·43
Peanuts	87	NA	92	3	85	4	0·16
Rice	35	NA	95	2	0	NA	0·22
Wheat	7	NA	41	7	0	NA	6·11
Soybeans	8	2	68	9	2	NA	4·19
Pasture hay and range	0·5	3	1	9	0	NA	68·40
Vegetables	NA	8	NA	5	NA	25	NA
Potatoes	77	NA	51	NA	49	12	0·16
Fruit	NA	13	NA	NA	NA	NA	NA
Apples	91	6	35	NA	67	28	0·07
Citrus	88	2	22	NA	58	13	0·08
All crops	6	54	17	36	0·9	10	NA

NA—not available.

In addition to pesticide controls numerous bioenvironmental pest controls are employed both in the U.S. and the world. Bioenvironmental control is described as any method utilized to reduce pest populations 'by manipulation of the pest's environment and ecology or by altering the pest's physiology, genetics, and behaviour or by a combination of these' (PSAC 1965). Although pesticides are often considered to be the most important control of pests, bioenvironmental controls are probably more important than pesticides worldwide. In the U.S., bioenvironmental controls are employed on more cropland than pesticide controls (Pimentel 1976). No published estimate is available on the extent of bioenvironmental pest controls employed in world agriculture, but it is believed to be substantial.

In spite of the combined bioenvironmental and pesticidal controls, pest populations are consuming and/or destroying almost one-half of the world's food supply. Surely this is a loss that we cannot afford as we face the need for even more food to feed the ever increasing world population.

No one factor is *the cause* of pest problems in either crops or livestock; usually a combination of several factors form the ecological basis of most pest problems. In this discussion, I will consider the ecology of each factor as it contributes to pest problems, but this is done with the recognition that most pest problems are the result of the interactions of a complex of factors.

Monocultures

Natural ecosystems tend to evolve toward stable climax communities for each particular habitat in the world. The natural plant community that exists in climax communities is most often not composed of the plant(s) that man utilizes in agriculture. For agriculture, the natural plant community is removed, destroyed and usually replaced by a single crop species.

As soon as the land is cleared of the natural vegetation, man's battle with what he terms pests begins. The seeds that are planted germinate but so do hundreds of seeds of other plant species that lay in the soil, some of which may have been dormant for many years. Many of these species are annual and biennial plants that are the normal, early successional plants of the seral or pioneer stages of ecosystem development.

All the plants that germinate except the crop plant are considered to be weeds. If the weeds or unwanted plants were not removed, they would either eliminate or seriously reduce the growth and stand of the crop by competing for soil nutrients, moisture, and light.

Many of the annual and biennial plants that emerge, as mentioned, are from seeds that lie dormant in the soil. Removal of the climax vegetation provides ideal temperature, moisture, light, and chemical conditions for the dormant weed seeds to germinate. Other weed invaders are carried in by the wind, water, and animals (NAS 1968, Crafts 1975).

Alteration of the ecosystem and restricting the macro-plant community to a single crop plant has other ramifications. Some microorganisms that remain in the altered ecosystem containing only the crop plant now utilize the crop plant as their food host. These relatively non-specific microorganisms may be highly pathogenic to crop plants. In other cases specific pathogens may be present, having survived previously on a related species, and these infect the new introduction.

Insects are usually less of a problem than weeds in newly altered ecosystems. However, because some insects are broadly adapted to feeding on a wide variety of plant hosts, they too become a problem to some crop plants in newly opened ecosystems (e.g., grasshoppers, crickets, cutworms). In addition, insects have excellent mobility and are likely to find new crop plantings.

Population densities and economics

Included in the ecological parameters of pest populations is the factor of density (Norton & Conway 1977). The usual description of a pest population is one of high densities. Pest populations do not have to exist at such high levels to decrease crop yields. In fact, in some cases the insect, pathogen, weed, mammal, or bird pest population may not be any more abundantly associated with the cultivated crop or livestock than it had been under natural conditions. However, because the numbers of these organisms are sufficiently abundant to damage crops or livestock they are labelled as pests.

For example, a species population may actually exist at relatively low densities in crop ecosystems but it is identified as a pest because it feeds on or causes observable damage to a valuable part of the crop. This is the case with birds feeding on cherries and mice girdling fruit trees. The apple maggot fly (*Rhagoletis pomonella*) attacks individual apples and hence, can be a serious pest even at relatively low densities (J. Brann, personal communication).

The relative effectiveness of biological control is also related to economic pest densities. A parasite or predator population attacking a pest population may keep the pest population stable and at relatively low numbers (e.g., apple maggot fly). However, in terms of economic pest densities, the biological control provided by parasite and predator may be less than successful. Again, this emphasizes the fact that economics determine whether any control method is considered successful in reducing pest population densities.

The terms, biological and natural control, are sometimes misunderstood. In successful biological control, prey or host (pest) population numbers are depressed by natural enemies and stabilized below 'economic density' levels. In contrast, natural control encompasses the numerous mechanisms in nature that limit the size of populations. In natural prey or host populations, numbers are limited and stabilized at levels that result in balanced supply-demand economies in predator–prey and parasite–host population systems (Pimentel 1961a, Pimentel *et al.* 1975b). Thus, the difference between biological and natural control is prey or host density levels at stability; with biological control, prey or host (pest) numbers must be below the 'economic density' level.

In both biological control and natural control, co-evolution occurs in predator–prey and parasite–host systems via genetic feedback, which contributes to the stability of these population systems. Although most ecologists recognize that co-evolution often takes place in nature in interacting parasite-host and predator–prey population systems (including parasites and predators that have been introduced specifically for biological control), a few ecologists deny it (Huffaker 1971, Huffaker 1974). The primary concern of

Huffaker and his colleagues in biological control studies is the reduced chance of success of introduced natural enemies if co-evolution occurs. They apparently are unaware that when co-evolution occurs among natural enemy and pest population systems in biological control, the effectiveness of biological control does not necessarily have to be lost or diminished. This is well documented by the relatively rare instances when biological control deteriorates after it has been successful (DeBach 1964, Huffaker 1971). In addition, in laboratory predator–prey and parasite–host population studies it has been demonstrated that co-evolution generally does not lead to outbreaks of the prey or host populations (Pimentel & Stone 1968, Pimentel & Soans 1971, Pimentel *et al.* 1975b).

In fact, understanding the role of the genetic feedback mechanism in natural population control can broaden our use of biological control (Pimentel 1963, 1973). For example, applying genetic feedback principles can help us in selecting natural enemies for effective biological control. In this regard, it is interesting to note that although biological control agents of a pest species are almost always sought in the same habitat from which the pest came, about 40% of the cases of successful biological control have involved an unrelated natural enemy (Pimentel 1963). Thus, a large portion of the successes in biological control were the result of new associations between parasite or predator and its host. This confirms the ecological principle that a host newly associated with its parasite or predator often lacks any resistance to the natural enemy. Therefore, the new enemy can effectively control the pest.

Introduction of crops and livestock into new biotic communities

Some pest problems occur when crops or livestock are introduced into new biotic communities. For instance, a new crop plant or animal species introduced into a community may attract a new predator or parasite species in the community to feed on it. This happens because the newly introduced species, never having been exposed to the organisms in the biotic community, lacks any natural resistance to any potential parasite or predator. Thus, these new parasites or predators may become serious pests.

For example, when the potato (*Solanum tuberosum*) which originated in Bolivia and Peru (Hawkes 1944), was introduced into the southwestern United States, it acquired a serious pest, the Colorado potato beetle (*Leptinotarsa decemlineata*). Native to the U.S., the Colorado potato beetle had originally co-evolved with and fed on wild sand bur (*Solanum rostratum*) (Elton 1958). When the potato was introduced into the southwest for production, the beetle spread onto the potato plants. Since the potato had never been exposed to this beetle, it lacked any natural resistance to it. Since then, this insect has become the most serious pest of the potato in the world, because it has spread along with the potato.

Another crop that acquired a new pest when it was introduced into a new biotic community was the apple (*Malus* sp.). Originally, the apple maggot fly (*Rhagoletis pomonella*) fed on wild hawthorn (*Crataegus* spp.) in the north-eastern United States. When the apple was introduced from Europe (Bush 1969), the apple maggot fly spread from hawthorn to the apple and is now considered the most serious pest of apple in the northeastern United States.

Serious pest problems have also developed with the introduction of live-stock into new biotic communities. For example, European cattle introduced into South Africa were found to be more susceptible to the 'bont' tick (*Amblyomma hebraeum*) and to 'heartwater' disease than Afrikander (Zebu) cattle (Bonsma 1944). The number of ticks on the European cattle were about four times more abundant than on the native Africander (Table 2).

Table 2.   The number of ticks on Africander and European cattle (Bonsma 1944)

	Totals for 12 cows of each kind		Ratio European: Africander
	Africander	European	
Number of ticks:			
12 counts per cow:			
On the body (800 cm²)	237	1,775	7·5:1
On the escutcheon (200 cm²)	1,529	4,397	2·9:1
Under the tail	2,140	4,789	2·2:1

The mortality in Africander cattle due to heartwater disease averaged only 5·3% over a thirty-month period whereas for European cattle the average was 60·7% mortality. Thus, the European cattle that had never been exposed to either the tick or heartwater disease were highly susceptible to the attack of both pests. Similar results occurred in Australia with Zebu cattle and European cattle and another tick species (*Boophilus microplus*) (Wharton *et al.* 1969). These results further substantiate the importance of genetic feed-back co-evolution in parasite–host systems, i.e., new associations between parasite and host often result in serious parasite outbreaks.

Introduction of pest species

Unfortunately, insects, pathogens, plants, mammals, or birds may be intro-duced into new biotic communities and become pests (Simmonds & Greathead 1977). The ecological principle involved is again the new associations between exploiting species and victims (prey or hosts). Because the victim lacks any natural resistance, the exploiter population can increase to outbreak levels (Pimentel 1961a, Pimentel *et al.* 1975b).

A few examples of newly introduced species that became pests include the European rabbit (*Oryctolagus cuniculus*), Japanese beetle (*Popillia japonica*), Dutch Elm disease (*Ceratocystis ulmi*), and water hyacinth (*Eichhornia crassipes*). In 1859 the European rabbit was introduced into Australia and increased rapidly during the next twenty years (Stead 1935). The association between the European rabbit and the Australian biotic community was a new one. Because the natural plant community lacked any resistance or tolerance to feeding by the rabbit, and few natural enemies of the rabbit existed, rabbits increased rapidly resulting in a severe impact upon the natural vegetation.

The European gypsy moth (*Porthetria dispar*) introduced into the eastern United States is another example of a newly introduced species reaching outbreak levels on its new plant hosts (primarily oaks and related hardwoods) (Forbush & Fernald 1896). The Japanese beetle (*Popillia japonica*) is another introduced insect that became a pest in the eastern U.S. (Smith & Hadley 1926).

Dutch elm disease in the U.S. is the result of introducing (by chance) a fungal pathogen from European elms (*Ulmus* spp.) to the American elm (*Ulmus americana*). The American elm lacked resistance to the new pathogen, which had caused serious destruction of American elms. The spread of the disease was aided by the introduction of the European bark beetle (*Scolytus multistriatus*) which also increased to outbreak levels (Matthysse 1959).

Water hyacinth (*Eichhornia crassipes*) is a plant that was introduced from South America into the U.S. and many other parts of the world (Crafts 1975). It was part of a horticultural exhibit at the Cotton Centennial Exposition at New Orleans in 1884. The plant was so attractive that it was taken home and planted in ponds and gardens (Penfound & Earle 1948). Soon the water hyacinth adapted nicely to its new tropical water habitat and spread so rapidly and formed such thick mats that in some cases people could walk on the plants. Obviously such luxuriant plant growth has had a deleterious effect on navigation and fishing.

The starling (*Sturnus vulgaris*) was introduced into the eastern United States by Eugene Scheifflin, a New York drug manufacturer whose hobby was the study of birds and Shakespeare. He was determined to bring to the United States all the birds mentioned by Shakespeare. The starling was on his list because Shakespeare in *Henry IV* wrote, 'Nay, I'll have a starling shall be taught to Speak nothing but "Mortimer" ' (Laycock 1966).

Many introductions of pests, such as Dutch elm disease, have been accidental but some (the starling) have been intentional introductions. Regardless of the reason for the introduction, there are many instances of newly introduced species becoming serious pests.

Movement of crops and livestock into different climatic regions

Some plants and animals are able to escape severe attack from their usual parasites and predators by the ability to exist in a climatic region in which their natural enemies have difficulty surviving. Thus, we find situations where the distribution of some plants and animals in nature is the result of the differential survival of parasite, predator and host (Elton 1927, Andrewartha & Birch 1954).

In agriculture, climatic conditions influence the relative degree of pest attack on the same crop or animal. For example, potatoes grown in northern Maine or in North Dakota have fewer insect pests and these pests cause less damage than on potatoes grown in warmer regions of the country. The potato aphid (*Macrosiphum euphorbiae*), potato stalk borer (*Trichobaris trinotata*), and potato tuberworm (*Phthorimaea operculella*) are all more serious pests (especially the borer and tuberworm) in the warmer southeast than in the cooler northern and mountain regions. The more serious pest problem is also reflected in the fact that almost 100% of the potato acreage in the southeast is treated with insecticides, compared with 65% in the cooler mountain region (USDA 1975).

A similar pattern is observed with the codling moth (*Cydia pomonella*) which is also a more serious pest of apples in the South than of apples grown in Nova Scotia, Canada. In Nova Scotia, there is usually one generation per year (Putnam 1963). However, in the southern U.S. there are usually two full generations and a partial third (Putnam 1963). With nearly three full generations and a less severe overwintering climate, the codling moth is obviously a much more severe pest in the southern U.S. than in Nova Scotia.

The interaction of the screwworm fly (*Cochliomyia hominivorax*) and livestock is another example. The fly was a pest only in the southern United States (Eddy & Bushland 1956). Now the application of the sterile-male technique has controlled the fly in the southern U.S. and its distribution is restricted to the southern part of Texas (Anon. 1973).

In addition to temperature, moisture can also influence the distribution of pests relative to crops. Corn, for example, in most areas of the United States is seldom attacked by mites. However, in the irrigated moist areas of Nebraska and Kansas, mites are a serious problem on grain corn (H. L. Brooks, D. Gates & Z. B. Mayo, personal communications).

Crop and livestock breeding

One of the most important ecological factors involved in pest problems is the breeding of susceptible crop and livestock genotypes (Lupton 1977). Most texts on plant breeding emphasize that the prime aim of the plant breeder is to increase crop yields. When altering the genetic makeup of the crop plant to increase yield with little or no attention given to pest attack, natural resistance can be lost or greatly reduced. Of importance then is breeding plants that not only have high yields but are resistant to their major pests.

The differences in levels of resistance that may exist in crop plants and their effectiveness is well illustrated with pea aphids (*Acythosiphon pisum*) associated with alfalfa (*Medicago sativa*). Five young pea aphids placed on a common crop variety of alfalfa produced a total of 290 offspring in ten days, whereas the same number of aphids for a similar period on a resistant variety of alfalfa produced a total of only two offspring per aphid (Dahms & Painter 1940). Obviously, a pest population with a 145-fold greater rate of increase on a host plant would inflict greater damage on the host plant than one with an extremely low rate of increase.

Sorghum provides another example. On a susceptible strain of commercial sorghum (*Sorghum vulgare*) the mean rate of oviposition (eggs per generation) of the chinch bug (*Blissus leucopterus*) was about 100. On a resistant strain of sorghum, however, the mean oviposition was less than one (Dahms 1948). In this instance, the animal feeding was reduced by 99% on the resistant plants and this had dramatic effects on the population dynamics of the feeding pest.

Sometimes in an effort to breed crops more palatable to human taste, the plants become more susceptible to pests. In particular, selecting and breeding plants to eliminate such factors as bitterness, hairiness, or other characters that are disliked by man may eliminate the exact factor that helps the plant resist pest attack. For example, cabbage, broccoli, or collard plants grown in wild areas were consumed rapidly and were selectively preferred by slugs and mice (Pimentel 1961b). Wild cruciferae, however, growing in the same wild area were hardly touched at all. The leaves of the young, tame cole plants were palatable, but the leaves of the wild cruciferan plants were bitter and indeed, were repulsive. Whether the observed bitterness of the wild plants affected the mice and slugs remains an unproven possibility.

Another factor in plant breeding that may influence the relative resistance of crop plants is partitioning or reallocating resources within the plant for different purposes. For example, one of the reasons for increased yield in corn has been due to breeding to partition the relative amounts of energy going to tassel development and corn grain development (C. Grogan, personal communication). The size of the tassel was reduced and this energy reallocated to the corn ear. A smaller tassel was found to be quite satisfactory for pollination when corn plants are grown close to one another.

When plant breeders artificially protect plants from pests through the use of pesticides or various other means, natural resistance in the plant may be lost unintentionally during the breeding and selection. With the plant protected from pests, the energy formerly going toward pest resistance may be partitioned to increase yield.

Crop and livestock breeding is indeed complex and difficult and may well alter the resistance to pest attack. Therefore, it is encouraging to note that recent efforts by plant breeders include attention to increasing plant resistance to pests.

Species diversity

Much has been written about diversity and its influence on stability in ecological systems (Brookhaven Symposium in Biology 1969). Frequently, outbreaks of insect pests in agriculture have been attributed to the practice of crop monocultures. For example, Marchal (1908) wrote that when man plants a vast extent of the country with certain crops while excluding others, he offers to the insect pests feeding on these plants favourable conditions for their explosive increase.

On the same subject, Graham (1915) wrote, 'that pines [white, *Pinus strobus*] growing in mixture with other species such as Norway pine [*Pinus resinosa*] are somewhat less subject to attack' from the white-pine weevil (*Pissodes strobi*). Further support for this observation has come from the works of Pierson (1922), Graham (1926), and MacAloney (1930) who noted that the white-pine weevil was excessively abundant in pure stands of white pine, but when the pine grew with mixed hardwoods, infestations were insignificant. According to von Hassel (1925), insect outbreaks were common in the monocultured forests of Germany but seldom occurred in the virgin forests of South America. Trägårdh (1925), comparing outbreaks in the forests of Germany and Sweden, concluded that fewer outbreaks took place in Sweden because about two-thirds of its forests are mixed pine–spruce–birch. Rohrl (1928) and Friederichs (1928) observed that catastrophes occur in forests where man has changed the environment by monoculture, and they suggested that the lack of diversity characteristic of mixed stands was responsible for the outbreaks noted. Damage from two moths was found to be negligible to spruce trees in mixed woods, but in pure stands severe damage often resulted (Barbey 1931). A study of the mango beetle (*Crypto-*

rhynchus gravis) in Java showed that although it causes little damage to mango trees growing in natural forests, it severely attacks the fruit on trees grown alone or when cultured near man's habitation (Voûte 1935). In Sumatra, Schneider (1939) reported great numbers of a lepidopteran (*Oreta carnea*) on cultivated gambir; in the virgin forests where gambir grows naturally no outbreaks of the pest had ever been observed. Working in Turkey, Schimitschek (1940) reported that the beetle (*Ips sexdentatus*) is found primarily in pure woods planted by man. Eidmann (1942, 1943) observed that the African virgin forests are relatively free of insect pests in contrast to areas in which man has altered the environment.

The evidence supporting the proposition that more outbreaks occur in single-species stands under man's cultivation than in natural-mixed-species communities, has led some workers to propose that in virgin forests outbreaks never occur (Schimitschek 1937 in Voûte 1946). Schneider (1939) wrote further that the virgin forests of the tropics are unsuited for insect plagues because of species diversity. Podhorsky (1933) reported that forests not affected by man seldom suffer from insect pests. In contrast, however, both Schenk (1924) and Eidmann (1942) have pointed out that catastrophes are not restricted to cultivated areas, but that damage occurs occasionally in the virgin forests.

Aside from these general observations, no experimental proof exists to support this hypothesis except for the experiments conducted with the biological community associated with the plant *Brassica oleracea* (*Cruciferae*) (Pimentel 1961b, Tahvanainen & Root 1972, Cromartie 1975, Root 1973, Root 1975). In the study by Pimentel, animal numbers on plants grown in monoculture were contrasted with animal numbers on plants grown in the mixed vegetation of a 15-year fallow field. Approximately 300 species of plants and 3,000 species of animals existed in the fallow field and provided the species diversity.

During the first year, 27 taxa were associated with the *B. oleracea* plant grown in the mixed-species planting compared with 50 taxa associated with the plant when grown in monoculture (Pimentel 1961b). Also, three to four times as many parasitic and predaceous taxa were present in the monoculture planting than were present on similar plants grown in the mixed planting. Aphid, flea beetle, and caterpillar populations reached outbreak levels in the monoculture planting, but no such outbreaks occurred in the mixed planting (Figs. 2, 3 and 4). Similar population trends were observed during the following summer (Pimentel 1961b). On the basis of these data, it seemed logical to conclude that the surrounding plant and animal species diversity associated with *B. oleracea* played an important role in preventing population outbreaks of the insect populations directly interacting with *B. oleracea*.

Perhaps another, equally important factor was affecting the stability of the populations in these two communities: the long-term genetic integration of the populations making up the organization of this Cruciferea community (Pimentel 1961b). By 'genetic integration' is meant the long-term association and presumed evolution of species complexes in natural biotic communities.

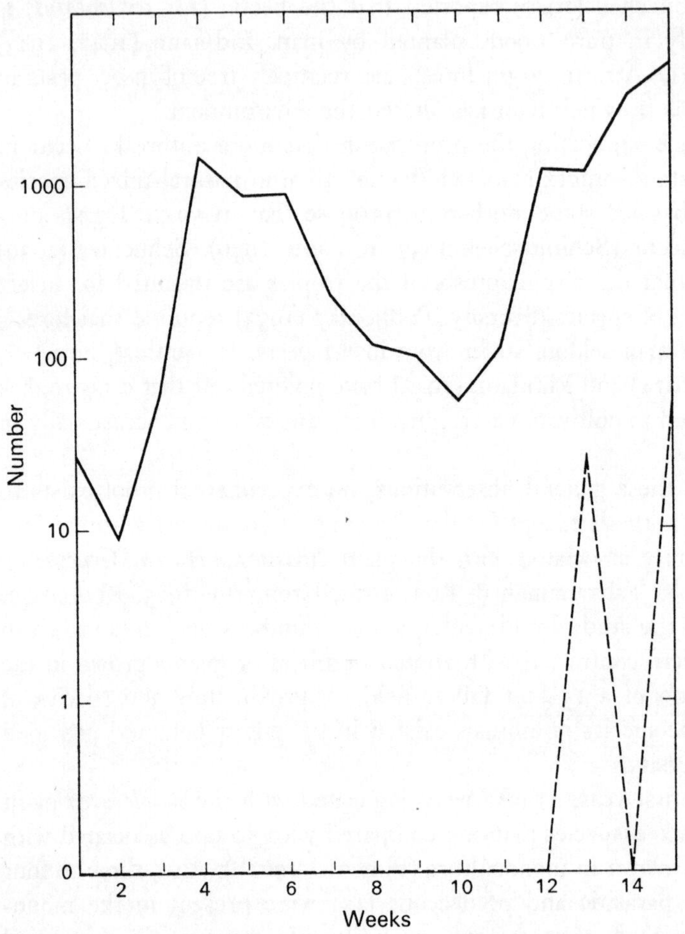

Figure 2.   The log number of aphids per unit plant area in the mixed (– – – – –) and monoculture (———) plantings. (After Pimentel 1961b.)

Then if a subunit (monoculture) is separated from the natural community, the genetic integration is affected and population outbreaks may occur. The balanced supply–demand economy that exists among herbivore–plant, predator–prey, and parasite–host systems as proposed by the 'genetic feedback mechanism' has a direct relationship to this concept (Pimentel 1961a,

Pimentel *et al.* 1975b). Obviously, we need more field data before we will know the complete answer.

Little is known concerning the exact nature of interspecific integration in communities, but as Emerson (in Allee *et al.* 1949) pointed out and

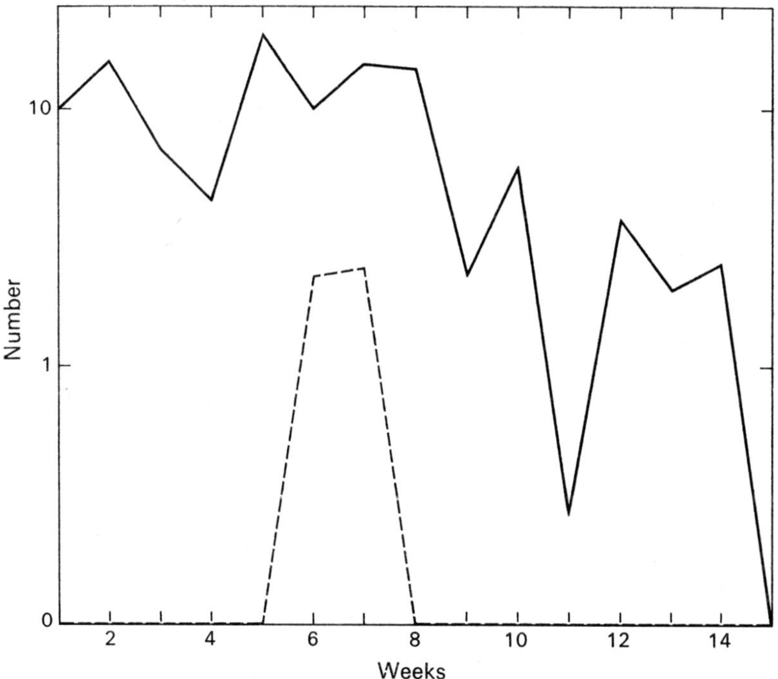

Figure 3. The log number of flea beetles per unit plant area in the mixed (– – – – –) and monoculture (———) plantings. (After Pimentel 1961b.)

experiments have documented, genetic integration is an important factor in maintaining community stability. Species members have evolved together to form an integrated and functional system, with the organized community having a balanced supply–demand economy. The removal of a subunit, as in the case of the monoculture, resulted in isolating an unbalanced, unstable subunit.

Certainly the lack of species diversity in monocultures is playing some role in increasing pest problems (van Emden & Williams 1974, Way 1973, Murdoch 1975, Gibson & Jones 1977, McVicar & MacKenzie 1977, Potts 1977). At the same time it appears that the lack of genetic integration is also contributing to instability in agro-ecosystems. More extensive investigations are needed to evaluate the relative importance of each factor in contributing to pest population outbreaks in crop monocultures.

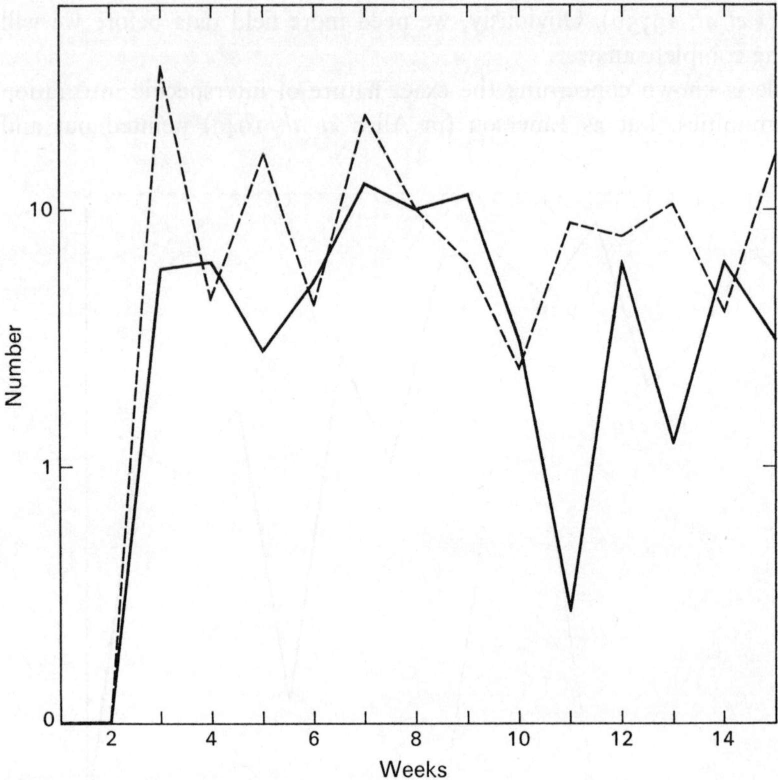

Weeks

Figure 4.   The log number of Lepidoptera per unit plant area in the mixed (– – – – –)
and monoculture (————) plantings. (After Pimentel 1961b.)

Genetic diversity

Based on numerous examples it is clear that many parasites have the genetic
variability to evolve and overcome 'single-factor' resistance in their host.
Hence, although parasites associated with hosts in natural situations appear
to be genetically stable, in agricultural ecosystems when the parasite is
stressed by a single factor, the parasite can often evolve and overcome host
resistance and cause serious damage to crops. For example, parasitic stem
rust and crown rust have been found to overcome genetic resistance bred into
their oat host. Since 1940, oat varieties have been changed in the corn belt
region every four to five years to counter the changes in the races of stem
rust and crown rust (Stevens & Scott 1950, van der Plank 1968).

Recently the Southern corn leaf blight parasite overcame resistance in
corn (Thurston 1973). The use of Texas sources of cytoplasmically inherited
male sterility (TMS) narrowed the resistance character in about 85% of the

corn grown in the United States to almost genetic homogeneity (Moore 1970, Roane 1973). Then in 1970, favourable environmental conditions resulted in selecting race T of *Helminthosporium maydis* (Southern corn leaf blight) which is virulent on all plants with TMS cytoplasm. The result was an epidemic causing devastating losses in the genetically homogeneous corn host (Nelson *et al.* 1970).

Evidence suggests that in some situations the Hessian fly (*Mayetiola destructor*) has adapted and overcome some introduced resistant wheat varieties (Suneson & Noble 1950, Allan *et al.* 1959, Hatchett & Gallun 1968, 1970, Gallun & Reitz 1971). Wheat resistance to the Hessian fly is inherited as a single dominant character in the wheat while the ability of the insect to survive on resistant wheats is inherited as a recessive character (Hatchett & Gallun 1970). This illustrates that when a parasite evolved and overcame the resistance in a host plant, the resistance in the host was often due to a single factor.

Genetic diversity however, within a given crop prevents a pest from overcoming natural plant resistance (Wolfe 1968). For example, when a wheat variety Eureka, resistant to wheat stem rust races was grown in progressively larger acreages, the incidence of the rust races attacking Eureka also increased (Fig. 5). The prime reason for the increase was that given the greater distribution of the Eureka variety, rust races could now be easily transmitted from host to host in an environment with abundant hosts. It is worthy of note that when the abundance and distribution of Eureka wheat declined, the incidence of rust infections by the special rust race also declined (Fig. 5). This example of a pathogen–wheat host system clearly illustrates the benefits of genetic diversity in agricultural crops.

Pest outbreaks have occurred in 'green revolution' wheat and rice varieties and have been associated with planting of a single variety over wide regions (Frankel 1971, Ida Oka, personal communication). The need for genetic diversity of resistant characters in host-plants for both 'green revolution' and U.S. monocultures has been well documented by Pathak (1970), Adams *et al.* (1971), Smith (1971), and Day (1973).

In experiments with an animal and simulated plant model, genetic stability was attained in the animal–plant relationship when six resistant factors were present in the plant population (Pimentel & Bellotti 1976). In these experiments the animal evolved tolerance to each of the six characters when exposed individually to the characters. Evidence suggested that a combination of three interrelated elements are involved in co-evolution and genetic stability among herbivore–plant and other parasite–host associations. These are: (1) the number and diversity of genetic characters in the crop-plant host; (2) ample gene flow between colonies of herbivores associated with the plant-host, if colonies occur in the population system; and (3) appropriate selection coefficients on herbivore population associated with its host population.

Utilizing our knowledge of genetic diversity and ecological stability can aid agriculture in preventing parasite pests from overcoming resistance bred into crop plants.

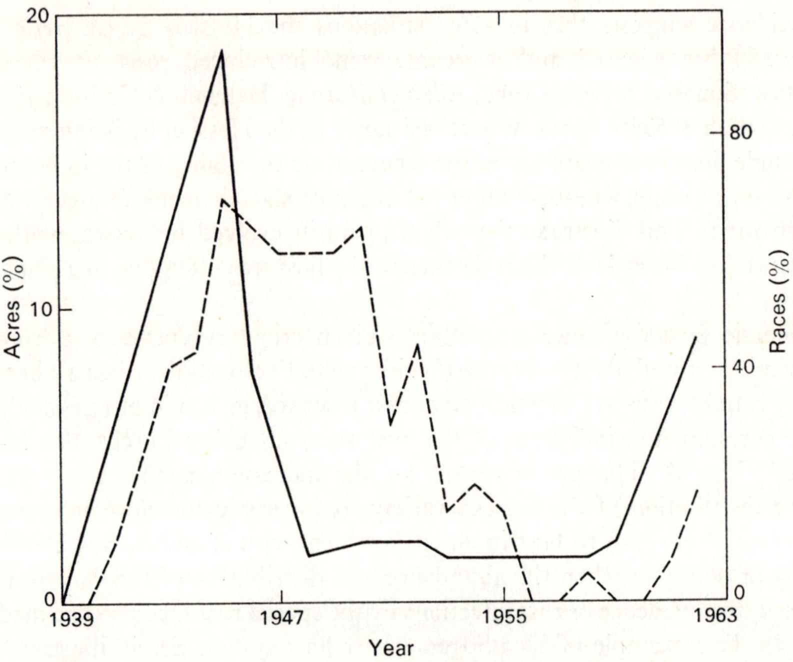

Figure 5.   Acres of Eureka wheat (————) grown in northern New South Wales as a percentage of total wheat acres and the percentage of wheat stem rust races which are able to attack Eureka (– – – – –). (Data of Watson & Luig 1963.)

Plant spacings

In cultivated fields, crop plant densities are carefully controlled to obtain the maximum population possible for optimal growth resulting in maximum economic yield. Seldom are the spacings of such plants similar to those in the wild. The new plant spacings result in an ecological situation that encourages pest outbreaks.

Numbers of herbivores may be limited if plants are either widely distributed from other individuals of the same species or exist in high density colonies (Pimentel 1961c). The inaccessibility of plants and other hosts in space and time may limit the numbers of feeding animals or parasites (Andrewartha & Birch 1954, Pimentel *et al.* 1963, 1965). These investigators pointed out that great spatial distances between individual hosts and a shortage of time relative to the exploiting animal's capacity for searching and finding its food-host are important factors in limiting exploiter populations. This proposition has been documented both experimentally and theoretically by several researchers (Andrewartha & Birch 1954, Huffaker 1958, Pimentel *et al.* 1963, 1965, Paine 1969, Levin & Paine 1974, 1975).

In addition, plant and animal hosts utilize other strategies to limit the impact of predatory animals feeding upon them. Plants and other hosts for example, existing in relatively dense stands may overwhelm the numbers of herbivores. Lebedev (1942) reported that planting dense stands of turnips reduced the numbers of flea beetles (*Phyllotreta* spp.) on the turnip plants. Also, Pimentel (1961c) investigated the impact of three different planting densities on animals feeding on cabbage, collards, broccoli, Brussels sprouts, and kale. Plants in the *dense* planting were spaced 15 by 15 cm from one another and totalled about 443,000 plants per hectare. Plants in the sparse planting were spaced 46 by 91 cm from one another (normal planting distance) and totalled about 26,000 plants per hectare. Plants in the dispersed planting were spaced 2·7 by 2·7 m from one another and totalled about 1,200 per hectare. Although there were more total herbivores in the dense planting than in either the sparse or dispersed hectare equivalent region, herbivores per plant surface area in both the sparse and dispersed plantings were more than five times more abundant than in the dense planting. Feeding pressure and impact of herbivores depends directly upon the relative abundance of herbivores per unit biomass (plant surface area). Thus, plant survival may depend upon a spatial pattern of dense design.

In summary, alteration of natural plant spatial patterns from either dense to sparse or sparse to dense may increase the relative abundance of exploiters feeding on the host population and, therefore, a serious pest problem may result from the new spatial pattern.

Continuous culture

Grown in the wild, annual and biennial plants often do not remain in the same location for many years but disperse and become established in new locations. This movement from one location to another helps some plants avoid attack from harmful insects and pathogens.

In agriculture, when crops and livestock are maintained in the same area year after year, pests associated with the crops or livestock tend to increase in number and this compares in severity. For example, if cole crops are cultured for several years in the same soil, club root (*Plasmodiophora brassicae*) organisms increase rapidly and can totally ruin production (Walker *et al.*

1958). Effective control of these pests is to plant non–cole type crops on the land for several years.

Also, if soybeans are grown on some land more than once every three to four years, brown stem rot (*Cephalosporium gregatum*) may be a serious problem (Cartter & Hartwig 1963). In the U.S. corn belt, corn and small grains make good rotation crops and reduce problems from the brown stem rot in soybeans.

Rotating corn with soybeans helps control the most serious insect pest of corn, corn rootworms (G. Musick & R. Treece, personal communication). Corn must be rotated every year to provide effective control of rootworms, and this rotation combines nicely with soybeans and small grains.

The practice of rotating pastures used by livestock can also reduce pests attacking the animals. For instance, this technique has been used effectively to reduce the number of dippings necessary to control cattle ticks (*Boophilum microplus*). British cattle in Australia with conventional dipping had to be treated about twenty times whereas with periodic 'pasture spelling' the cattle had to be dipped only seven times (Wharton *et al.* 1969).

A related ecological problem has developed as a result of the new multiple-cropping system in the tropics. Unfortunately, raising three crops of IRRI rice on the same land during one year has resulted in increased insect pest problems (M. Pathak, personal communication). With a continuous supply of food, insect pest populations also continue to increase, and the severity of the pest problem increases proportionally to the increase in insect populations.

Soil and host plant nutrients

All parasites and predators have specific nutrient requirements. Altering the nutrient level in the soil and then in the host plant influences the parasites and predators feeding on the plant. An improvement in nutrients often results in parasite and predator population numbers increasing and a decline in nutrients results in the reverse.

For example, Haseman (1946) reported that the grain aphid (*Macrosiphum granarium*) feeding on small grain plants with high nutrients (high nitrogen) produced an average thirty-three progeny per aphid whereas on plants with low levels of nitrogen, progeny production averaged only thirteen per aphid. Barker & Tauber (1951) found that the pea aphid (*M. pisi*) feeding on the garden pea with a high nutrient level (high nitrogen) produced an average of sixty-nine progeny per aphid whereas on low nitrogen, progeny production averaged forty-seven per aphid.

Aphids generally respond to increased nitrogen in host plants because aphids have difficulty obtaining sufficient nitrogen (protein) from plant sap. Many aphids have elaborate physiological mechanisms to aid them to suck large quantities of sap, filter the protein and other necessary nutrients from the sap, and discard the remainder (heavy in sugar) as honey dew.

Of course, differences exist in exactly how individual species respond to various nutrients and nutrient levels. For example, purple scale (*Lepidosaphes beckii*) on citrus increases with increased amounts of magnesium (Thompson 1940) and a similar response occurred with pea aphid on peas (Barker & Tauber 1951). However, although the pea aphid on peas increased with increasing phosphorus (Barker & Tauber 1951), the chinch bug (*Blissus leucopterus*) on sorghum decreased with increasing phosphorus (Dahms & Fenton 1940). Mites (*Tetranychus telarius*) on tomatoes were likewise found to be negatively correlated with phosphorus dosage (Rodriguez 1960). Also, contrary to aphids, greenhouse thrips (*Heliothrips haemorrhoidalis*) do poorly on spinach with high levels of nitrogen (Haseman 1950) and mites on tomatoes likewise do poorly with high nitrogen (Rodriguez 1960).

Dutch elm disease development has been found to be reduced in elms receiving a complete N–P–K fertilizer whereas excessive amounts of fertilizer resulted in more severe disease (Parker *et al.* 1947).

The nutritional level in host plant leaves may affect insect distribution on the plant host. For example, the cabbage aphid (*Brevicoryne brassicae*) primarily colonizes the heart and young leaves on crucifers whereas the green peach aphid (*Myzus persicae*) colonizes the ageing and senescing leaves (van Emden & Wearing 1965).

The biological control of pests can be influenced by the host plant on which the pest is present. A parasite (*Lypha dubia*) of the winter moth (*Operophtera brumata*) for instance, was found to have an average parasitism rate of only 2·4% in an oak area and an average rate of 15·1% in a mixed wood area (Cheng 1970). This change in parasitism rates was explained as differences in seasonal developmental times in the host plants.

Fertilizers may also alter the growth and development patterns of plants and make them more susceptible to pest attack. With cotton for example, maintaining high nitrogen levels late in the growing season encourages the continued growth of the cotton plant instead of maturing the buds and bolls. Continuing the growth of the cotton plant keeps it present in the field longer and furnishes food so the insect populations can continue to increase at the same time. Therefore, the longer the cotton plant grows, the more severe the insect problem becomes (L. Falcon, personal communication).

Planting times

Some plants in nature begin to grow early or late in the growing season and thereby manage to escape the attack of certain pests. Wild radishes have been observed to germinate early in the spring and make most of their growth before the cabbage maggot fly emerges and attacks the radishes. Damage to the radishes under these conditions is usually minimal (Pimentel, unpublished).

In other situations plants may escape attack by starting growth after the pest population has emerged and begun to die. This strategy has been employed by agriculturalists to reduce the attack on wheat by the Hessian fly. Delaying the seeding of wheat fields is effective in reducing the attack from the dangerous fall brood of the Hessian fly (PSAC 1965). The spring brood of the Hessian fly is less serious and delayed planting does not reduce the spring brood attack.

Growth stage may be an important factor in vulnerability to attack. Corn for example, is more susceptible to corn borer attack at certain stages of growth. Young corn (15·2 cm high) is relatively unattractive to corn borers compared with corn that has reached 45·7 cm high (Whitman 1975). If a farmer wants to reduce corn borer attack, corn should be planted to have its growth stage either very small or nearly full grown before the borer moths emerge.

Host plant associations

Some parasites and predators can attack and feed on several species of host plants. These parasites and predators can move from one host plant to another when one of the host plant populations declines in abundance for some reason.

In the United States for example, plant bugs (*Lygus* spp.) feed on alfalfa and cotton. When alfalfa is mowed for hay and eliminated as a food source for the bugs, they will move on to cotton in large numbers (Stern 1969). Thereafter, the bugs will cause damage to cotton if present in sufficient numbers. The plant bugs generally prefer alfalfa and therefore if the alfalfa is carefully managed by planned cutting, the *Lygus* bug population on cotton can be kept to a minimum. The successful strategy is to cut only a portion of the alfalfa at a time leaving sufficient alfalfa to attract the bugs and keep them away from the cotton. Another strategy is to plant narrow strips of alfalfa (6 m wide) for every 91–122 m of cotton in the cotton field. This not only attracts the plant bugs but may provide a source of natural enemies of such pests as cotton bollworms.

The leaf miner problem associated with spinach in California was in part solved by planting spinach considerably after the other crops that usually harboured the leaf miner (H. Lange, personal communication). Without a suitable host, the leaf miner population was reduced significantly before the spinach was planted. Another way to control such a pest is to reduce the growing of crops that act as an alternate host for the leaf miner.

Pesticides alter crop plant physiology

Pesticides are potent biocides and in some cases adversely affect the physiology of crop plants. Any change in the physiology of a crop plant may result either in making the plant more resistant or more susceptible to attack by its parasites and predators. Since crop plants which are not physiologically stressed are generally more effective in resisting parasite and predator attack, any chemical that alters normal physiology is likely to increase the susceptibility of the crop plant. This was demonstrated when calcium arsenate was used on cotton. McGarr (1942) reported that aphids on untreated cotton plants averaged 0·91 per 6 cm^2 of leaf area whereas aphids numbered 6·05 on plants treated with calcium arsenate at about 6·7 kg ha^{-1}.

Herbicides have also been found to increase insect pest and pathogen problems associated with corn. For example when corn plots were treated with a regular dosage of 0·55 kg 2,4-D ha^{-1}, aphid numbers on the corn averaged 1,679 whereas on the untreated they averaged only 618 (Oka & Pimentel 1976). Corn borer infestation averaged 28% in the 2,4-D treated corn population compared with only 16% in the untreated corn population.

In laboratory investigations of the impact of 2,4-D on the relative resistance of corn plants to pathogens, exposed corn plants were significantly more susceptible to corn smut disease (*Ustilago maydis*) (Oka & Pimentel 1974) and to southern corn leaf blight (*Helminthosporium maydis*) (Oka & Pimentel 1976).

Summary

The ecological basis for insect pest, pathogen, and weed problems is complex and is often the result of a combination of ecological factors. Some of these include monoculture; introduction of new crops and livestock into new biotic communities; introduction of pest species; movement of crops and livestock into different climatic regions; crop and livestock breeding; species diversity; genetic diversity; plant spacings; continuous crop culture; soil nutrients; planting times; and pesticide alteration of crop plant physiology.

Acknowledgments

This study was supported in part by grants from the National Science Foundation (Ecology) BMS 74-07900 and the Ford Foundation programme in the Ecology of Pest Management.

References

ADAMS M.W., ELINGBOE A.H. & ROSSMAN E.C. (1971) Biological uniformity and disease epidemics. *Bio. Sci.* **21**, 1067–70.

ALLAN E.R., HEYNE E.G., JONES E.T. & JOHNSON C.O. (1959) Genetic analysis of ten sources of Hessian fly resistance, their interrelationships and association with leaf rust reaction in wheat. *Bull. Kans. agric. Exp. Stn.* 104.

ALLEE W.C., EMERSON A.E., PARK O., PARK T. & SCHMIDT K.P. (1949) *Principles of Animal Ecology*. Saunders, Philadelphia.

ANDREWARTHA H.G. & BIRCH L.C. (1954) *Distribution and Abundance of Animals*. Univ. of Chicago Press, Chicago.

ANON. (1973) The screw worm strikes back. *Nature, Lond.* **242**, 493–4.

BARBEY A. (1931) Une invasion de la tordeuse du sapin dans les forêts d'Alsace. *Re. Eaux et Forêts* **69**, 132–6.

BARKER J.S. & TAUBER O.E. (1951) Fecundity of and plant injury by the pea aphid as influenced by nutritional changes in the garden pea. *J. econ. Ent.* **44**, 1010–12.

BONSMA J.C. (1944) Hereditary heartwater-resistant characters in cattle. *Farming in S. Africa* **19**, 71–96.

BROOKHAVEN SYMPOSIUM IN BIOLOGY (1969) Diversity and stability in ecological systems. Number 22.

BROWN L. & ECKHOLM E.P. (1974) *By Bread Alone*. Praeger Publishers, New York.

BUSH G.L. (1969) Sympatric host race formation and speciation in frugivorous flies of the genus *Rhagoletis* (Diptera, Tephritidae). *Evolution* **23**, 237–51.

CARTTER J.L. & HARTWIG E.E. (1963) The management of soybeans. In *The Soybean: Genetics, Breeding, Physiology, Nutrition, Management* (Ed. by A.G. Norman), pp. 162–226. Academic Press, New York and London.

CHENG L. (1970) Timing of attack by *Lypha dubia* Fall. (Dipt.: Tachnidae) on winter moth *Operophtera brumata* (L.) (Lep.: Geometridae) as a factor affecting parasite success. *J. Anim. Ecol.* **39**, 313–20.

CRAFTS A.S. (1975) *Modern Weed Control*. Univ. Calif. Press, Berkeley.

CRAMER H.H. (1967) Plant protection and world crop production. *Pflanzenschutz-nachrichten* **20**, 1–524.

CROMARTIE W.J. Jnr. (1975) The effect of stand size and vegetational background in the colonization of cruciferous plants by herbivorous insects. *J. appl. Ecol.* **12**, 517–33.

DAHMS R.G. (1948) Effect of different varieties and ages of sorghum on the biology of the chinch bug. *J. agric. Res.* **76**, 271–88.

DAHMS R.G. & FENTON F.A. (1940) The effect of fertilizers on chinch bug resistance in sorghums. *J. econ. Ent.* **33**, 688–92.

DAHMS R.G. & PAINTER R.H. (1940) Rate of reproduction of the pea aphid on different alfalfa plants. *J. econ. Ent.* **33**, 482–5.

DAY P.R. (1973) Genetic variability of crops. *A. Rev. Phytopathol.* **11**, 293–312.

DEBACH P.H. (Ed.) (1964) *Biological Control of Insect Pests and Weeds.* Reinhold, New York.

EDDY G.W. & BUSHLAND R.C. (1956) Screw worms that attack livestock. In *Yearbook of Agriculture.* U.S. Dept. Agr., Washington, D.C.

EIDMANN H. (1942) Der tropische Regenwald als Lebensraum. *Kolonialforstl.* 5, 91–147.

EIDMANN H. (1943) Zur Oekologie der Tierwald des afrikanischen Regenwaldes. *Beitr. Kolonialforsch.* 2, 24–45.

ELTON C.S. (1927) *Animal Ecology.* Sidgwick and Jackson, London.

ELTON C.S. (1958) *The Ecology of Invasions by Animals and Plants.* Methuen, London.

EMDEN H.F. VAN & WEARING C.H. (1965) The role of the aphid host plant in delaying economic damage levels in crops. *Ann. appl. Biol.* 56, 323–4.

EMDEN H.F. VAN & WILLIAMS G.F. (1974) Insect stability and diversity in agroecosystems. *A. Rev. Ent.* 19, 455–75.

FORBUSH E.H. & FERNALD C.H. (1896) *The Gypsy Moth.* Porthetria dispar (*Linn.*). Wright and Potter, Boston.

FOWLER D.L. & MAHAN J.N. (1975) The pesticide review. U.S. Dept. Agr, Agr. Stab. Cons. Ser., Washington, D.C.

FRANKEL O.H. (1971) Genetic dangers in the green revolution. *Wld. Agr.* 19, 9–13.

FRIEDRICHS K. (1928) Waldkatastrophen in biozönotischer Betrachtung. *Anz. Schädlingsk.* 4, 139–42.

GALLUN R.L. & REITZ L.P. (1971) Wheat cultivars resistant to races of Hessian fly. *Prod. Res. Rep.* #134, Agr. Res. Ser., U.S. Dept. Agr.

GIBSON I.A.S. & JONES T. (1977) Monoculture as the origin of major forest pests and diseases, especially in the tropics and southern hemisphere. In *Origins of Pest, Parasite, Disease and Weed Problems* (Ed. by J.M. Cherrett & G.R. Sagar), pp. 139–61. Blackwell Scientific Publications, Oxford.

GRAHAM S.A. (1915) *The biology and control of the white-pine weevil* (Pissodes strobi). Unpublished M.F. thesis, Cornell University, Ithaca, N.Y.

GRAHAM S.A. (1926). Biology and control of the white-pine weevil. *Pissodes strobi* Peck. *Bull. Cornell Univ. agric. Exp. Stn.* 449, 3–32.

HARRAR J.G. (1961) Socio-economic factors that limit needed food production and consumption. *Fed. Proc.* 20, 381–3.

HASEMAN L. (1946) Influence of soil mineral on insects. *J. econ. Ent.* 39, 8–11.

HASEMAN L. (1950) Controlling insects through their nutritional requirements. *J. econ. Ent.* 43, 399–401.

HATCHETT J.H. & GALLUN R.L. (1968) Frequency of Hessian fly, *Mayetiola destructor*, races in field populations. *Ann. ent. Soc. Amer.* 61, 1446–9.

HATCHETT J.H. & GALLUN R.L. (1970) Genetics of the ability of the Hessian fly, *Mayetiola destructor*, to survive on wheats having different genes for resistance. *Ann. ent. Soc. Amer.* 63, 1400–7.

HAWKES J.G. (1944) Potato collecting expeditions in Mexico and South America. *Imperial Bureau of Plant Breeding & Genetics.* School of Agriculture, Cambridge, England.

HUFFAKER C.B. (1958) Experimental studies in predation: (II) Dispersion factors and predator–prey oscillations. *Hilgardia* 27, 343–83.

HUFFAKER C.B. (Ed.) (1971) *Biological Control.* Plenum Press, New York.

HUFFAKER C.B. (1974) Some implications of plant–arthropod and higher-level, arthropod–arthropod food links. *Environ. Entomol.* 3, 1–9.

LAYCOCK G. (1966) *The Alien Animals.* Natural History Press, Garden City, New York.

LEBEDEV V.A. (1942) (On means for combatting garden insects of the genus *Phyllotreta* and on the effect of these on the growth and yield of plants.) *Zachsh. Rast.* **1**, 131–8. (In Russian; translated by E. Matthews.)

LEVIN S.A. & PAINE R.T. (1974) Disturbance, patch formation and community structure. *Proc. nat. Acad. Sci., Wash.* **71**, 2744–7.

LEVIN S.A. & PAINE R.T. (1975) The role of disturbance in models of community structure. In *Ecosystem Analysis and Prediction* (Ed. by S.A. Levin), pp. 56–67. SIAM, Philadelphia, Pa.

LUPTON F.G.H. (1977) The plant breeders' contribution to the origin and solution of pest and disease problems. In *Origins of Pest, Parasite, Disease and Weed Problems* (Ed. by J.M. Cherrett & G.R. Sagar), pp. 71–81. Blackwell Scientific Publications, Oxford.

MACALONEY H.J. (1930) The white-pine weevil. Its biology and control. *Bull. N.Y. St. Coll. For.* (Tech. Publ. 28) **3**, 1–87.

MANGELSDORF P.C. (1966) Genetic potentials for increasing yields of food crops and animals. In *Prospects of the World Food Supply. Symp. Proc. nat. Acad. Sci.*, Wash. 66–71.

MARCHAL P. (1908) The utilization of auxiliary entomophagous insects in the struggle against insects injurious to agriculture. *Pop. Sci. Mon.* **72**, 352–419.

MATTHYSSE J.G. (1959) An evaluation of mist blowing and sanitation in Dutch elm disease control programs. *Cornell Misc. Bull.* **30**, N.Y.S. Coll. Agric.

McGARR R.L. (1942) Relation of fertilizers to the development of the cotton aphid. *J. econ. Ent.* **35**, 482–3.

McVICAR A.H. & MACKENZIE K. (1977) Effects of different systems of monoculture on marine fish parasites. In *Origins of Pest, Parasite, Disease and Weed Problems* (Ed. by J.M. Cherrett & G.R. Sagar), pp. 163–82. Blackwell Scientific Publications, Oxford.

MOORE W.F. (1970) Origin and spread of southern leaf blight in 1970. *Plant Dis. Reptr.* **54**, 1104–8.

MURDOCH W.W. (1975) Diversity, complexity, stability and pest control. *J. appl. Ecol.* **12**, 795–807.

NAS (1968) Weed control. Principles of plant and animal pest control. Vol. 2. Washington, D.C.

NELSON R.R., AYERS J.E., COLE H. & PETERSEN D.H. (1970) Studies and observations on the past occurrence and geographical distribution of isolates of Race T of *Helminthosporum maydis*. *Plant Dis. Reptr.* **54**, 1123–6.

NORTON G.A. & CONWAY G.R. (1977). The economic and social context of crop pest, disease and weed problems. In *Origins of Pest, Parasite, Disease and Weed Problems* (Ed. by J.M. Cherrett & G.R. Sagar), pp. 205–226. Blackwell Scientific Publications, Oxford.

OKA I.N. & PIMENTEL D. (1974) Corn susceptibility to corn leaf aphids and common corn smut after herbicide treatment. *Environ. Entomol.* **3**, 911–5.

OKA I.N. & PIMENTEL D. (1976) Herbicide (2,4-D) increases insect and pathogen pests on corn. *Science.* **193**, 239–40.

PAINE R.T. (1969) The *Pisaster–Tegula* interaction: prey patches, predator food preference and intertidal community structure. *Ecology* **50**, 950–61.

PARKER K.G., TYLER L.G., WELCH D.S. & PAPE S. (1947) Nutrition of the trees and the development of Dutch elm disease. *Phytopathology* **37**, 215–24.

PATHAK M.D. (1970) Genetics of plants in pest management. *Conf. Principles Pest Management*, Raleigh, North Carolina.

PENFOUND W.T. & EARLE T.T. (1948) The biology of the water hyacinth. *Ecol. Monogr.* 18, 447–72.

PIERSON H.B. (1922) Control of the white-pine weevil by forest management. *Bull. Harv. Forest* 5, 1–42.

PIMENTEL D. (1961a) Animal population regulation by the genetic feedback mechanism. *Am. Nat.* 95, 65–79.

PIMENTEL D. (1961b) Species diversity and insect population outbreaks. *Ann. ent. Soc. Amer.* 54, 76–86.

PIMENTEL D. (1961c) The influence of plant spatial patterns on insect populations. *Ann. ent. Soc. Amer.* 54, 61–9.

PIMENTEL D. (1963) Introducing parasites and predators to control native pests. *Canad. Ent.* 95, 785–92.

PIMENTEL D. (1973) Genetics and ecology of population control. Addison-Wesley Modular Program, Addison-Wesley Publ. Co., Reading, Mass. 10, 1–39.

PIMENTEL D. (1974) Energy use in world food production. *Environ. Biol.* Cornell University, Ithaca, New York. Report 74-1.

PIMENTEL D. (1976) World food crisis: energy and pests. *Bull. ent. Soc. Am.* 22, 20–6.

PIMENTEL D. & BELLOTTI A.C. (1976) Parasite–host population systems and genetic stability. *Am.Nat.* 110, 877–88.

PIMENTEL D., DRITSCHILO W., KRUMMEL J. & KUTZMAN J. (1975a) Energy and land constraints in food-protein production. *Science* 190, 754–61.

PIMENTEL D., FEINBERG E.H., WOOD P.W. & HAYES J.T. (1965) Selection, spatial distribution and the coexistence of competing fly species. *Am. Nat.* 99, 97–109.

PIMENTEL D., LEVIN S.A. & SOANS A.B. (1975b) On the evolution of energy balance in exploiter–victim systems. *Ecology* 56, 381–90.

PIMENTEL D., NAGEL W.P. & MADDEN J.L. (1963) Space–time structure of the environment and the survival of parasite–host systems. *Am.Nat.* 97, 141–67.

PIMENTEL D. & SOANS A.B. (1971) Animal populations regulated to carrying capacity of plant host by genetic feedback. In *Dynamics of Numbers in Populations* (Proc. Adv. Study Inst.), Oosterbeek, Netherlands. pp. 313–26.

PIMENTEL D. & STONE F.A. (1968) Evolution and population ecology of parasite–host systems. *Canad. Ent.* 100, 655–62.

PODHORSKY J. (1933) Kann ein vollständig sich selbst überlassener Hochgebirgswald durch Überhandnehmen von Forstinsekten zu einer Gefahrenherd werden? *Allg. Forst- u. Jagdztg.* 109, 151–6.

POTTS G.R. (1977) Some effects of increasing the monoculture of cereals. In *Origins of Pest, Parasite, Disease and Weed Problems* (Ed. by J.M. Cherrett & G.R. Sagar), pp. 183–202. Blackwell Scientific Publications, Oxford.

PRADHAN S. (1971) Revolution in pest control. *World Sci. News* 8, 41–7.

PSAC (1965) Restoring the quality of our environment. Rep. of Environmental Pollution Panel, Pres. Sci. Adv. Comm., The White House.

PUTNAM W.L. (1963) The codling moth (*Carpocapsa pomenella* L.) (Lepidoptera: Tortricidae). A review with special reference to Ontario. *Ontario Entomol. Soc. Proc.* 93, 22–60.

ROANE C.W. (1973) Trends in breeding for disease resistance in crops. *Ann Rev. Phytopath.* 11, 463–86.

RODRIGUEZ J.G. (1960) Nutrition of host and reaction to pests. In *Biological and Chemical Control Plant and Animal Pests* (Ed. by L.P. Reitz), pp. 149–67. *Amer. Ass. Advan. Sci. Publ.* 61. Washington, D.C.

ROHRL D. (1928) Waldkatastrophen. *Forstwiss. ZentBl.* 73, 293–315.

ROOT R.B. (1973) Organization of a plant–arthropod association in simple and diverse habitats: the fauna of collards (*B. oleracea*). Ecol. Monogr. 43, 95–124.

ROOT R.B. (1975) Some consequences of ecosystem texture. In *Ecosystem analysis and prediction* (Ed. by S.A. Levin), pp. 83–97. Soc. Ind. Appl. Math., Philadelphia.

SCHENK C.A. (1924) Der Waldbau des Urwalds. *Allg. Forst.- u. Jagdzetg.* 100, 377–388.

SCHIMITSCHEK E. (1940) Die Massenvermehrung der *Ips sexdentatus* Börner im Gebiete der orientalischen Fichte. *Z. angew. Ent.* 26, 545–88.

SCHNEIDER F. (1939) Ein Vergleich von Urwald und Monokultur in Bezug auf ihre Gefährdung durch phytophage Insekten, auf Grund einiger Beobachtungen an der Ostkuste von Sumatra. *Schweiz. Z. Forstwes.* 90, 41–55, 82–9.

SIMMONDS F.J. & GREATHEAD D.J. (1977) Introductions and pest and weed problems. *Origins of Pest, Parasite, Disease and Weed Problems* (Ed. by J.M. Cherrett & G.R. Sagar), pp. 109–24. Blackwell Scientific Publications, Oxford.

SMITH H.H. (1971) Broadening the base of genetic variability in plants. *J. Hered.* 62, 265–76.

SMITH L.B. & HADLEY C.H. (1926) The Japanese beetle. Dept. Cir. U.S. Dept. Agric. 363, 1–66.

STEAD D.G. (1935) *The rabbit in Australia.* Winn, Sydney.

STERN V.M. (1969) Interplanting alfalfa in cotton to control lygus bugs and other insect pests. *Proc. Tall Timbers Conf. Ecol. Anim. Contr. Habit. Mgmt.* 1, 55–69.

STEVENS N.E. & SCOTT W.O. (1950) How long will present spring oat varieties last in the central corn belt? *Agron. J.* 42, 307–9.

SUNESON C.A. & NOBLE W.B. (1950) Further differentiation of genetic factors in wheat for resistance to the Hessian fly. *U.S.D.A. Tech. Bull.* 1004.

TAHVANAINEN J.O. & ROOT R.B. (1972) The influence of vegetational diversity on the population ecology of a specialized herbivore, *Phyllotreta cruciferae* (Coleoptera: Chrysomelidae). *Oecologia* 10, 321–46.

THOMPSON W.L. (1940) Pest control studies: effect of fertilizers on purple scale development. *Fla. Agr. Exp. Sta. Rep.*

THURSTON H.D. (1969) Tropical agriculture. A key to the world food crisis. *Bio. Sci.* 19, 29–34.

THURSTON H.D. (1973) Threatening plant disease. *A. Rev. Phytopathol.* 11, 27–52.

TRÄGÅRDH I. (1925) On some methods of research in forest entomology. *Trans. III. Internatl. Ent. Kongr.* Zurich 2, 577–92.

UN (1974) Assessment of the world food situation. United Nations, World Food Conference, Rome.

USDA (1965) Losses in Agriculture. U.S. Department of Agriculture. Agr. Handbook No. 291, Agr. Res. Serv. U.S. Govt. Print. Off.

USDA (1968) Extent of farm pesticide use on crops in 1966. Agr. Econ. Rep. No. 147, Econ. Res. Ser.

USDA (1970) Quantities of pesticides used by farmers in 1966. Agr. Econ. Rep. No. 179. Econ. Res. Serv.

USDA (1971) The pesticide review 1970. Agr. Stab. and Cons. Ser.

USDA (1974) The world food situation and prospects to 1985. Foreign Agric. Econ. Rep. No. 98.

USDA (1975) Farmers' use of pesticides in 1971 . . . extent of crop use. Econ. Res. Serv., Agric. Econ. Rep. No. 268.

VAN DER PLANK J.E. (1968) *Disease Resistance in Plants.* Academic Press, New York and London.

VON HASSEL G. (1925) Das Geheimniss des Urwaldes. *Anz. Schädlingsk.* 1, 42–3.

VOÛTE, A.D. (1935) *Cryptorhynchus gravis* F. und die Ursachen seiner Massenvermehrung in Java. *Archs. néerl. Zool.* 2, 112–43.

VOÛTE A.D. (1946) Regulations of the density of the insect-populations in virgin-forests and cultivated woods. *Archs. néerl. Zool.* 7, 435–70.

WALKER J.C., LARSON R.H. & TAYLOR A.L. (1958) Diseases of cabbage and related plants. USDA, Agr. Handbook No. 144.

WATSON I.A. & LUIG N.H. (1963) The classification of *Puccinia graminis* var. *Eritici* in relation to breeding resistant varieties. *Proc. Linn. Soc. N.S.W.* 88, 235–58.

WAY M.J. (1973) Objectives, methods and scope of integrated control. In *Insects: Studies in Population Management* (Ed. by P.W. Geier, L.R. Clark, D.J. Anderson & H.A. Nix), pp. 137–52. *Ecol. Soc. Aust.* (memoirs 1): Canberra.

WHARTON R.A., HARLEY K.L.S., WILKINSON P.R., UTECH K.B. & KELLEY B.M. (1969) A comparison of cattle tick control by pasture spelling, planned dipping and tick-resistant cattle. *Aust. J. agr. Res.* 20, 783–97.

WHITMAN R.J. (1975) *Natural control of the European corn borer.* Ostrinia nubilalis (*Hübner*), *in New York.* Unpublished Ph.D. thesis, Cornell Univ., Ithaca, N.Y.

WOLFE M.S. (1968) Physiological race changes in barley mildew 1964–67. *Pl. Path.* 17, 82–7.

Population dynamics

The relevance of population dynamic theory to pest status

T. R. E. SOUTHWOOD *Department of Zoology and Applied Entomology, Imperial College, London*

The etymology of pest implies numbers, but the modern concept of pest that has been described by Norton & Conway (1977) has other components and this should be emphasized lest the impression is gained that I believe pest status is entirely determined by population dynamics. The level of trouble or disbenefit due to a species that will give it the status of a 'pest' depends on (a) the numbers of the organism present and (b) the sensitivity or tolerance of the victim (crop, animal, person) to the pest's attack. This tolerance is composed of the socio-economic factors (i.e. market tolerance) reviewed by Norton & Conway (1977) and what may be called the biological tolerance of the victim. Genetic variation in pest and host both contribute importantly to differences in biological tolerance. In new host relationships the biological tolerance of the host is often low: Garnham (1977) points this out in relation to protozoal diseases. In instances where new insect-host plant relationships have been established the plants often seem particularly susceptible to damage. Examples are provided by the mite *Phyllocoptes destructor* on tomatoes in California (Bailey & Keifer 1943), the introduced scale insect, *Lepidosaphes newsteadi* on the endemic conifer *Juniperus bermudiana* in Bermuda (Bennett & Hughes 1959) and the endemic mirids on cocoa in West Africa and the Far East discussed later.

The spread of a species, the evolved change of host plant or a change of virulence, which is most frequent in pathogens, are features of its ecology and will indeed influence aspects of its population dynamics. Thus the biological tolerance of a host for its pest cannot be completely separated from population dynamics, but these inter-relations will not be considered in detail here.

The *r–K* continuum

Applied biologists have long been concerned with changes in numbers, an aspect emphasized by Andrewartha & Birch (1954). Other biologists have been impressed with the relative stability of animal numbers, more especially when viewed in the light of their prodigious powers of increase (Darwin 1872). We can now recognize that there is no conflict between these views; there is a whole continuum of types of population dynamics. The frit fly (*Oscinella frit*) is present in late sown oats in southern Britain in most years in numbers to give it pest status (Strickland 1960, Southwood & Jepson 1961); locust outbreaks occur at irregular intervals and last for a few seasons (Waloff 1966); the larch tortrix (*Zeiraphera griseana*) defoliates its host trees in the Engadine for two to three successive years in every eight to nine years (Baltensweiler 1968, 1971); the moth, *Dendrolimus pini*, reached very high levels in a German pine forest in 1888, but did not approach such numbers (about 100 times average density) again until 1937 (Schwerdtfeger 1941); the numbers of white storks (*Ciconia ciconia*) nesting at Oldenburg, Germany remained within the range 138–548 over the years 1928–1963 (Lack 1966); the numbers of wandering albatross (*Diomedea exulans*) nesting on Gough Island are believed to have remained within a narrow range, around 4000, for the last eighty years (Elliott 1971). This spectrum of change to stability is associated with increasing duration stability of the habitat (Southwood 1962 1976, Southwood *et al.* 1974) and may be conveniently equated with the *r–K* continuum (MacArthur & Wilson 1967, Pianka 1970). *r*-strategists are opportunists, selected for obtaining maximum food intake by the exploitation of their ephemeral habitats. *K*-strategists have been selected for harvesting food efficiently in a crowded environment, their populations remain near the carrying capacity of their habitats. *r*-strategists tend to be small, mobile and with a short generation time, *K*-strategists are larger, with less migratory tendency and with a long generation time (Connell 1975, Shapiro 1975, Southwood 1976).

The synoptic model

The associated spectra of bionomic strategies and of habitat characteristics conveniently described by the term *r–K* continuum are expressed in the form of the population dynamics of the species at different points of the continuum. A model may be constructed with three axes: population growth ($N_{t+1}-N_t/N_t$), population density and habitat duration stability. This 'synoptic model' provides a framework within which to consider the comparative population dynamics of different species (Southwood 1975, Southwood & Comins 1976) (Fig. 1).

Briefly one can recognize three regions. The *r*-strategists whose habitats are ephemeral and whose numbers are characterized by 'boom and bust'. Large-scale migration dominates this strategy and with it go massive losses unrelated to the numbers migrating; new populations are continually developing from small beginnings, from a handful of colonisers. At the other extreme the *K*-strategists maintain a steady population at or near the carrying capacity of the habitat; they are in equilibrium with their resources,

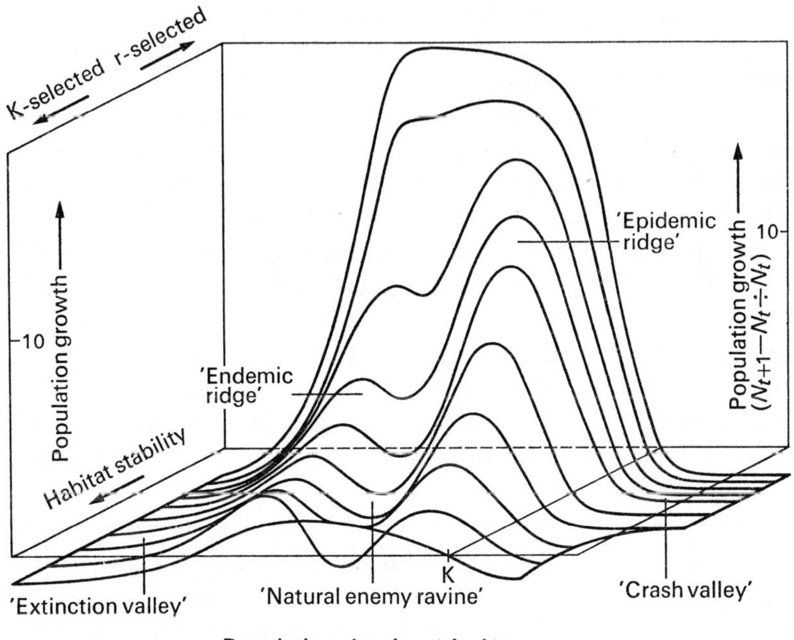

Figure 1.   The synoptic population model (after Southwood & Comins, 1976). K = carrying capacity (as in K-selected).

Note: equilibrium points only occur where an 'east facing slope' cuts a zero population growth contour, as only here does negative feedback operate.

whose renewal they do not adversely affect. Recruitment, migration and mortality rates are low, there is little exposure to new habitats and less opportunity for adaptation to changed environments. These organisms are specialized to their particular environments and if their numbers are reduced to low levels, unlike *r*-strategists, they may become extinct. Both the extreme *r* and *K*-strategists have one stable equilibrium point, the upper one at the population density of the carrying capacity.

The middle region is characterized by the existence of the 'natural enemy ravine'. Where this dips below the zero population growth contour there is a second equilibrium point and where it rises through the contour on the other side of the ravine is the escape or release point from natural enemies. Beyond this point, in the absence of density independent catastrophes, the population rises to the upper equilibrium point where intraspecific competition mechanisms, including disease, operate. Clark (1964) recognized these two levels in the population dynamics of the eucalyptus psyllid, *Cardiaspina albitextura*, he termed the lower one the endemic level and the upper one the epidemic level. The influence of different components of the predator's response on the form of the ravine and its effect on the dynamics of the prey population in deterministic and stochastic models is described by Southwood & Comins (1976).

At the *r*-end of the natural enemy ravine the prey and their predators 'play hide and seek': the highly mobile prey colonize new habitats (these, being of short duration, are continually arising) and the population may develop before specialist predators arrive. Polyphagous predators are therefore likely to be important in such situations, as they may be present already in the habitat. Thus at this end of the ravine it is the numerical response that is inadequate to prevent the population rising to epidemic levels. At the *K*-end the predators will have had plenty of time to respond numerically, but the prey will have evolved defence mechanisms so that it is the functional response that 'fails'. Because of the relatively high numerical response the lower equilibrium point is proportionally closer to the upper equilibrium than near the *r*-end.

These characters should not be thought of as absolute. The studies of Anderson (1970) and Berry (1970) on the house mouse (*Mus musculus*) show how a species may move, to a limited extent, along the continuum in different situations. As Way (1977) has pointed out, a species must be seen as part of the ecosystem in which it lives, changes in this (either by man or other agencies) will alter the population dynamics of the species. For example if the natural enemies are removed, then the natural enemy ravine is 'filled in' and the population will grow to the level where it is limited by resources.

Conway (1976) has shown how the concepts of the synoptic model with its *r*–*K* continuum can usefully be used to provide insights into the population dynamics of pests and suitable control strategies. I will explore this approach further.

r-Pests

These organisms are always so numerous at certain times in certain places that they may destroy their habitat. For example microorganisms such as smallpox virus, and anthrax (*Bacillus anthracis*) that are usually lethal to their hosts (Bradley 1977) can survive only because they infect their hosts randomly and at a rate that is less than the recruitment rate of the host (=renewal rate of the resource). At another trophic level many flies are *r*-organisms: house (*Musca*) or flesh (*Sarcophaga*) flies in decaying materials, mycetophilid gnats in fungal fruiting bodies and certain drosophilids in fleshy fruits, and with larger scale habitats, so are locusts and the army worm (*Spodoptera exempta*). Among plants true *r*-strategists are represented by short-lived ruderals, such as groundsel (*Senecio vulgaris*) which produces about 1,000 propagules with a generation time of often little more than five weeks (Salisbury 1961).

r-strategists often become pests because man's activities have increased the infection rate, by providing more contacts (greater proximity of sources) or more breeding sites (more invaders or inoculum) or both.

Uvarov (1957) shows how the mismanagement, especially overgrazing, of many arid regions has greatly increased the suitable breeding ground for locusts. The many artificial containers associated with urbanization greatly enhance the breeding sites for several mosquitoes, such as *Aedes aegypti*. The high cattle population of Australia has markedly increased the breeding sites of the bushfly (*Musca vetustissima*); an insect that recolonizes much of its range annually (Hughes 1970, Hughes & Walker 1970).

One of the most characteristic features of *r*-strategists is their high migratory tendency, essential for their movement from a 'dying habitat' to a new one. As mentioned above their populations therefore regularly build-up in any one habitat from a low to a high level; they are exogenous or immigrant pests for a particular crop. Because of the ephemeral nature of their local habitats it is often useful to consider their population on a regional scale (Mackauer & Way 1976). It must be stressed that on this scale an extremely high proportion of the total mortality will be density-independent associated with migration, although the level of migration itself may be density-related. When the regional population is considered it is often apparent that there is a low level, representing an ecological bottleneck, at certain seasons. This low level (trough), frequently occurs on a wild plant and as it is of no direct economic importance, has been neglected which is unfortunate because such bottlenecks present potential control opportunities. Several *r*-pests, that are towards the beginning of the natural enemy ravine on the model, show this phenomenon: *Aphis fabae* overwinters on spindle (Way & Banks 1968, Way & Cammell 1973) and the frit fly on wild grasses (Southwood & Jepson 1962), many heteroptera (e.g. *Dysdercus*, *Acanthomia*) in eastern and southern Africa pass the dry season on a limited number of wild hosts.

Intermediate pests

These organisms are in the area of the natural enemy ravine in the model. They range from species at the r-end, such as cereal aphids to species at the K-end such as the bullfinch and others discussed in the next section. The dominant feature of their population dynamics is that they are generally held at a level lower than the carrying capacity of their habitat by the action of natural enemies. The numerical response improves as one moves along the valley to the K-end: this is because the increasing duration stability of the habitat provides more time for this response and thus underlines the role of time as a resource (Watt 1971). Occasionally organisms will become epidemic in natural situations because environmental noise, such as climatic irregularities, will alter the relationships sufficiently for the population to pass the release or escape (Voûté 1946) point. The frequency with which this will occur will depend on the level of the environmental noise and the width of the ravine (Southwood & Comins 1976).

Thus in contrast to extreme r-strategists that are naturally pests virtually all the time, intermediate species will achieve pest status, in natural situations, only occasionally. The changes in population dynamics that occur as one moves forward across the landscape, along the natural enemy ravine, may be illustrated by reference to aphids. Temperate cereal aphids represent a condition close to the r-end of the spectrum: the studies of Potts (1977) (also reviewed by Southwood & Comins 1976) show the importance of the autochthonous general predators in preventing outbreaks, in holding populations at a very low level. If the aphids escape from this control their numbers will normally rise to high densities (Hagen & van den Bosch 1968). As in many situations in arable crops the specific natural enemies, coccinellids, syrphids, chrysopids, seem mainly effective when the population is already, in items of the synoptic model, in 'crash valley' (van Emden *et al.* 1969, Mackauer & Way 1976). A contrast is provided by the tropical aphid *Toxoptera aurantii* on cocoa: a habitat of greater duration stability and hence further from the r-end than cereal aphids. This species, which is present in often small colonies on the underside of cocoa leaves but as a rule 'causes little damage' (Entwistle 1972), was studied by Firempong & Kumar (1975). They found specific natural enemies, particularly coccinellids and syrphids, to be effective, frequently attacking colonies soon after establishment. However some aphids survive, particularly when a syrphid larva is the predator. In summary this comparison suggests the greater probability of an outbreak in an arable crop, but a higher endemic (non-outbreak) population density in a tree crop.

Intermediate species may become pests because:

(i) The disbenefits to man of even occasional outbreaks are significant. Many forest pests come into this category.

(ii) Changes in the environment, normally due to man, have reduced or eliminated the natural enemy ravine. One must note that this does not mean that natural enemies are absent, but that their functional or numerical responses to prey numbers have been reduced: theoretically moderate changes in one parameter (e.g. attack rate) may eliminate the ravine (Southwood & Comins 1976). The many instances of 'secondary' or 'upset' pests, arising as a result of the destruction of their natural enemies (usually by pesticides), are examples of this e.g. Massee 1958, Entwistle *et al.* 1959, Conway & Wood 1964). The discontinuity created in the ecosystem by crop plantings and associated agricultural operations may reduce the populations of predators and the role they play in holding their prey at low densities (Way 1976, Potts 1977). The practice in biological control of introducing predators and prey along with the crop, pioneered by Huffaker & Kennett (1956), serves to minimize this discontinuity and preserve the 'ravine'.

(iii) Man has introduced a species to a new area where it is free of its natural enemies. There are many examples of this and the topic is reviewed by Simmonds & Greathead (1977). Among plants, St. John's wort or Klamath weed (*Hypericum perforatum*) in N. America and the prickly pear (*Opuntia* spp.) in Australia and South Africa are classic examples (Holloway 1964, Huffaker 1967). Other introductions, where the organism is not a pest in its country of origin, include the gipsy moth (*Lymantria dispar*) to North America (Turner 1963), the wood wasp, *Sirex gigas* to Australasia (Morgan 1968) and the giant African snail, *Achatina fulica*, in Pacific Island (Elton 1958).

The restoration of the ravine by the introduction of natural enemies is the role of classical biological control, discussed by Simmonds & Greathead (1977). The challenge is to establish a wide enough ravine. Failures may arise from a number of causes:

(i) The numerical response of the predator (or parasitoid) is inadequate because it is not adapted to the climate of the new area. Examples are provided by the failure of *Aphytis maculicomis* alone to control the olive scale, *Parlatoria oleae* (Huffaker & Kennett 1966) and the attempts to control *Aleurocanthus woglumi* in Mexico by *Eretmocerus serius* (De Bach 1964).

(ii) The numerical response of the predator is inadequate because its numbers suffer at certain times from competition from other predators. The classic instance of this is the failure of the control of the coconut leaf mining beetle *Promecotheca* after the introduction of a predatory mite (Taylor 1937). Frazer & van den Bosch (1973) describe how the introduction of the highly efficient parasitoid (*Trioxys pallidus*) of the walnut aphid (*Chromaphis juglandicola*) led to the reduction in the resident population of predators;

Trioxys itself is now adversely affected by the ant *Iridomyrmex humilis* and renewed aphid outbreaks are probable.

(iii) An additional natural enemy or other factor is necessary to widen the ravine adequately to allow the reduction from the high equilibrium level to one sufficiently low for the species to be no longer regarded as a pest. Instances of this in Canada are the winter moth (*Operophtera brumata*) and the European spruce sawfly (*Diprion hercyniae*): in spite of the introduction of natural enemies, the outbreak of the latter was not reduced to a low level until a virus disease was introduced.

Sufficient population data have been published on the epidemic of *Diprion hercyniae* and its suppression (Balch 1939, Balch & Bird 1944, Bird & Elgee 1957) for it to be described quantitatively in terms of the synoptic model. The mortality due to disease and the real mortality due to parasitoids after 1940 were calculated from these papers and plotted against log larval density (Fig. 2a). The mortality curves have been fitted by eye and it will be seen that parasitoids act as a density dependent factor up to larval densities of only two per ft^2.* Prior to the introduction of the virus high mortalities do not seem to have been achieved until severe intraspecific competition for food commenced at densities of 150–200 larvae per ft^2 (Balch 1939). This information is described in terms of a section through the synoptic model in Fig. 2b. The zero width line (dotted) has been inserted taking Balch & Bird's (1944) calculation that 97·2% mortality would lead to a constant population and assuming an arbitrary 7·2% density independent mortality and that mortality is constant at all densities. The significant population points (equilibrium and release) (see Fig. 7 of Southwood 1975) can be inserted and it is noted that in the absence of disease the natural enemy ravine does not dip below the zero growth contour. The population will rise until larval densities of about 200 per ft^2 lead to defoliation and intraspecific competition, as is recorded by Balch (1939). However the introduction of the disease reduces the upper equilibrium point to a lower level (U') and by deepening the ravine, establishes a lower (natural enemy) stability point (S) and a release point (R). Environmental fluctuations will from time to time move the population out of the domain of stability of S, that is beyond R to U'. The densities (larvae/ft^2) determined from Fig. 2b may be compared with the actual field data:

Graphical model		*Field data*	
		average	range
$U' - 10$	Epidemic years[†]	5·5	1·7–11
$S\ - 0·95$	'Endemic years'	0·6	0·2–1·4

* = 0·19 per m^2 but SI units are not used as this would impede comparisons with the original papers. † Regarded as 'epidemic years' by Bird & Elgee (1957).

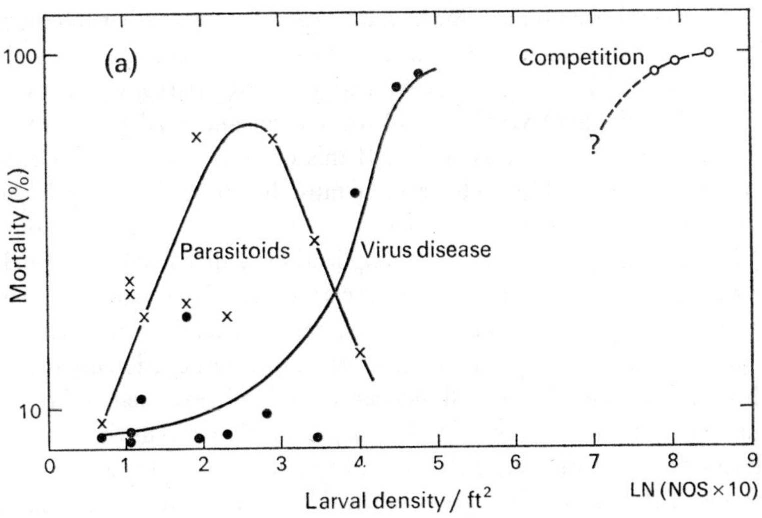

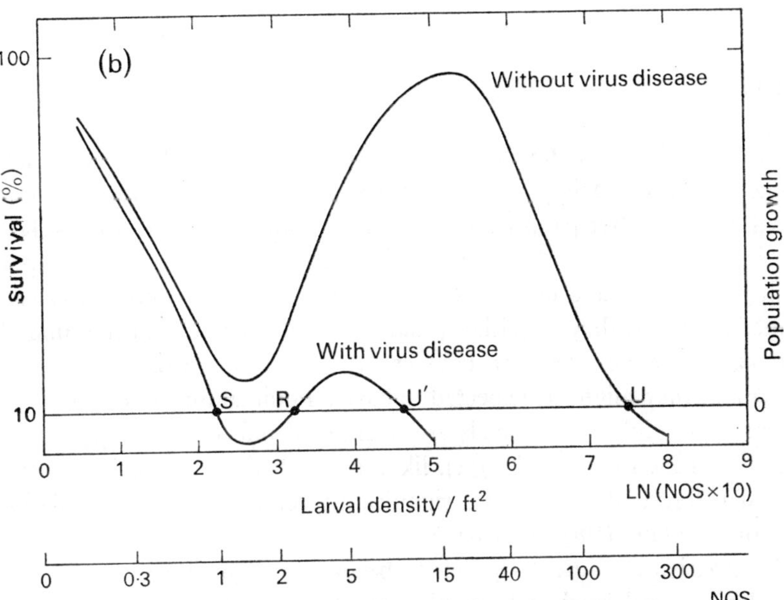

Figure 2. The description in terms of the synoptic model of the dynamics of the European spruce sawfly, *Diprion hercyniae*, in New Brunswick.

(a) Mortalities, at various larval densities, due to parasitoides, virus disease and intra-specific competition (data from Balch 1939, Balch & Bird 1944, Bird & Elgee 1957).

(b) The population growth density curves based on (a), for the population without and with the virus disease; U = equilibrium point (prior to disease), U' = upper equilibrium point (after disease), R = release point, S = lower (natural enemy) equilibrium point.

The value for the release point from the model (2·25), seems rather too high, the field data suggests somewhere between 1·4 to 1·7; however this depends on following Bird & Elgee's (1957) classification of 1953 (larval density 1·7) as an 'outbreak (epidemic) year'; it was the second and final year and so could have been below the release point. If this change is made the release point lies between 1·7 and 3·2. However it must be stressed that with the inherent variation in field data (the delayed density dependence of some relationships and, in this instance, the complication of diapause) the model, which is deterministic, cannot provide precise values. The model merely describes how since the advent of the disease, natural enemies will normally keep the population below 2·25 larva per sq ft^2, however occasionally it will exceed this release value, then it will increase to an 'outbreak' that will reach maximum densities in the region of 10 larvae per sq ft^2; without the disease, density levels of 200 larvae per sq ft^2 would be attained. One must also note that, in terms of the model, the advent of the disease has moved the form of the population dynamics towards the K-end, which is appropriate for a forest insect.

K-pests

In general, within their taxon, K-strategists are relatively large organisms: the size of the niche imposes restrictions and this is particularly true of internal parasites, but parasitic platyhelminths and nematodes can be large for their phyla. By definition their populations remain at or near their upper equilibrium density and do not fluctuate widely. Thus in contrast to r- and intermediate pests, they are either pests all the time or none of the time. As K-strategists are considered to have evolved to become closely adapted to their habitat they might be expected to have a minimal impact on their food sources: most adult wild animals are tolerant of infection by *Trypanosoma brucei* (Garnham 1971, Ford 1971), likewise the fruiting of tropical trees in Indonesian forests does not seem significantly reduced by the depredations of the orang-utan (*Pongo pigmaeus*).

Man's activities cause K-species to become pests in three ways:

(i) The normal level of harvesting by the K-strategist is economically intolerable to man.

(ii) Man disturbs the natural regulation of the K-strategist increasing its density.

(iii) The mixing of the flora and fauna occasioned by man leads to the establishment of new trophic relationships: the biological tolerance of the new host or host-plant is much less than in the original relationship.

Only the second category explicitly involves a change in population size, but the third may lead to this and in any particular situation more than one mechanism may be involved.

Conway (1976) cites the codling moth (*Cydia pomonella*) as a *K*-pest: it represents an excellent example of a *K*-pest's normal harvesting having no real impact on the species being harvested, but the level being intolerable to man (Clark *et al.* 1967). Elephants always have more impact on their habitat than the codlin moth does on its habitat, but in forest situations in both Africa and Indonesia the natural populations do not seem to cause significant degradation. However when man cultivates small areas in these regions 'raids' by elephants immediately place the latter in the pest category. The skylark (*Alauda arvensis*) is a third example of a relatively *K*-species becoming a pest because, economically, man cannot tolerate its feeding. In the spring in East Anglia skylarks often remove a small number of seedlings of sugar beet. Formerly there were excess seedlings and the stand was thinned by hoeing, but recent developments of drilling to a stand with specially bred sugar beet fruits containing a single fertile seed have made this bird, the poet's 'blithe spirit', a pest.

A striking instance of the damage caused by a *K*-species after man has disturbed its regulation is provided by the African elephant (*Loxodonta africana*) when its population is confined to game parks. This compression of the population, together with the shooting of the leader-matriarchs of the small herds, has led to the formation of large herds of several hundreds of individuals: these mill around on the periphery of the parks where their activities are causing serious destruction of the trees and other vegetation (Laws *et al.* 1975).

A slightly different scenario is provided by the bullfinch (*Pyrrhula pyrrhula*) which is now regarded as one of the major pests of fruit growing in southern Britain. Its large size for a finch and territorial, non-migratory behaviour are *K*-characteristics. For much of the year the adults are seed feeders, but in the spring the flowering buds of trees form a major part of their diet; where they are available the buds of certain fruit trees are favoured. An individual will take between 15 and 45 buds/minute and will feed for about five hours/day (Newton 1964). Thus a single bullfinch may take some thousands of buds per day and a relatively low population may cause a marked reduction in the subsequent yield of fruit, a characteristic of a *K*-pest. It seems that shortage of food resources in the spring often limits the size of the bullfinch population (Newton 1964, Lack 1966) and thus we may suppose that the development and planting by man of fruit trees with large buds has increased the survival and hence population level of the bullfinch in fruit growing and garden areas. It has been recognized as a pest for over 300 years, but the upsurge in its importance in the last two decades may reflect a further increase in population size due to reduced predation, especially by man (shooting and egg-collecting), but also by raptores. Another reason for its increased importance in modern crops may be relative: because the amount of damage sustained from other causes has been greatly reduced in the last thirty years.

In tropical forests ants frequently play a dominant role in the invertebrate ecosystem: modification of their distribution may allow the increase of other insects, such as the coreids *Amblypelta* and *Pseudotheraptus* on coconuts, which then cause extensive damage to their hosts (O'Connor 1950, Way 1953a and b).

Undoubtedly the introduction of plants and animals to new areas is the commonest cause of K-pests. The cocoa tree has been introduced into West Africa and the Far East and in both areas among the indigenous insects that have become pests are certain large-sized mirids: *Sahlbergella singularis*, *Distantiella theobroma*, *Pseudodoniella* spp. and *Platyngomiroides apiformis* (Leston 1970, Conway 1971). In West Africa the two former cause considerable damage at relatively low densities (2–4 per tree) to the young shoots of cocoa; their native hosts are various trees, *Cola* spp., *Adansonia* and other malvales, and these suffer less than cocoa (Entwistle 1972). The transfer to cocoa in West Africa seems to have occurred around 1900 and is undoubtedly a reflection of the general relationship between the number of insect species attached to a tree and its abundance (Southwood 1961, Strong 1974): it is noteworthy that there is now some evidence that there may be separate races (Entwistle 1972). *Platyngomiroides apiformis* was first found on cocoa in Sabah in 1962 feeding on pods, later young shoots (chupons) were attacked (lesions extending 'up to a foot' from the feeding site) and in 1964 'severe and extensive damage to the trunk and branches' was seen (Conway 1971). The sensitivity of the new host, to the extent that trees may be killed by relatively small numbers, is a feature of K-pests arising by this mechanism and it is noteworthy that in Brazil, where the relationship with the mirid *Monalonion* presumably has a long evolutionary history, the major damage is to the pods: unless this is at a very high level it is unlikely to influence the food supply of subsequent generations—a feature of K-strategists.

Turning to a completely different part of the trophic web many zoonoses may be regarded as examples of K-pests. These are discussed by Garnham (1971, 1977); for the purpose of this argument it is necessary only to point out the good adaptation of the parasite to its natural host whose numbers it seldom depletes (the K-strategy), and the extreme sensitivity of the new ('unnatural' or adventitious) host, in which the parasite may develop new races (c.f. cocoa mirids) and where it behaves essentially as an r-strategist.

Extremely interesting examples of two K-pests are provided by the African trypanosomes and their vectors, tsetse flies (*Glossina* spp.), Conway (1976) has already pointed out the K-strategy characters of tsetse flies and it is noteworthy that they are considered to have evolved in association with that *K*-group *par excellence*, the large reptiles: on the latter's extinction (apart from crocodiles) tsetse transferred to Suidae, which occupied similar habitats and were abundant in the Eocene and Oligocene, or to other *K*-strategists—elephants and rhinoceros (Ford 1971). The association with antelopes and other mammals is believed to have come later; buffalo may have been the first, again because of habitat similarities. The Salivaria group of trypanosomes probably arose in leeches, were passed to the Suidae and from them infected *Glossina*. The high degree of adjustment to infection of pigs and buffaloes (parasitaemia is infrequent), supports this view (Ford 1971). Although some other game animals are more sensitive, especially when young, for most of those regularly bitten by tsetse trypanosomiasis is not lethal. Into this situation man and his domestic animals intruded, firstly the negroid races, subsequently the Arabs and Europeans. In a fascinating account Ford (1971) shows how the present status of trypanosomiasis in Africa may be related to the disturbances occasioned by the arrival of Europeans, their associated diseases (rinderpest, smallpox) and political actions. Man and his domestic animals, being the most recently acquired hosts, are the most sensitive to infection by the trypanosomes and in these hosts new races have developed (Garnham 1971).

Elm disease (*Ceratocystis ulmi*) has generally behaved like an *r*-strategist killing most of the trees it attacks: its origin is uncertain, but we do know that as it is carried around by man more virulent strains develop. Both the fungus and its beetle vector have been carried to the North American continent, exposed to different species of elm and now a new aggressive strain introduced to Europe (Brasier & Gibbs 1973). Perhaps the suckering habit of elm allowed such a lethal parasite to survive or perhaps it was originally more of a '*K*-strategist', killing limbs on old trees over a long period of time. Whichever the case, the present pandemic may enable us to witness an ill-adapted, over-virulent race of a parasite exterminating its host and itself, the race resulting from the disturbances to evolved relationships caused by man's movement of flora and fauna around the world.

When consideration has been made for the different biological circumstances of the cocoa mirids, the trypanosomes and elm disease fungus the parallels in the increased sensitivity, the new races and the exploitation (*r*-strategy) after disturbance are striking.

Pest control strategy and the *r*–*K* continuum

Applied biologists have recognized that habitat characters are important indicators for pest management strategies (Southwood & Way 1970, Fig. 4) and Conway (1976) has shown that as the *r–K* continuum is related to habitat characteristics it is relevant to decisions on the choice of control strategy, control being used in the sense of reducing the impact of the pest to a level when it no longer justifies pest status. Normally this is done by reducing the population, but it may be achieved by breeding for more tolerance (less sensivity) in the victim.

r-PESTS

The levels of these populations are always fluctuating, therefore any appeal to the 'stability of natural ecosystems' is foredoomed to failure. With arable crop pests in this category the annual re-invasion is the key population pathway (Southwood 1975) and a powerful control strategy is to maximize the difficulties in this pathway. This is analysed by Way (1977): the degree of isolation may be increased by greatly enlarging the scale of the crop monoculture and so minimizing the 'edge effect' in the crop, and changes can be made in the overwintering, aestivating or alternate hosts' sites. The manipulation of planting dates to minimize the synchronization of host plant and pest migration time is another cultural control technique. Attention should also be given to the ecological bottlenecks in the regional population dynamics of such species, these may be manipulated to make them 'narrower'. Direct reduction, the removal of alternate hosts, is one of the oldest techniques of pest and disease control. More subtly the pest could be 'overcrowded' in its refuge so bringing intraspecific mechanisms into play and natural enemies might be encouraged in these sites, as Way (1976) suggests for the aphid *Myzus persicae*. If pest numbers cross the economic injury threshold (Norton & Conway 1977) pesticides will have to be used to protect the crop: the pest population will of course 'bounce-back', this is the nature of *r*-strategists. The crop may be harvested before this happens. A more serious long-term problem is the potential development of resistance to the pesticide. The high reproductive rates of *r*-species will favour this but, as Comins (1976) shows, provided much of the population remains untreated the high level of migration will retard the development of resistance. From economic and biological viewpoints pesticide application must be related to need, this highlights the importance of methods of forecasting and monitoring the pest population (Smith 1970). Sometimes forecasts can be based on information about the pest in its overwintering site (refuge or foci of infection) as with the Russian work on the 'functional-morphological condition' of the bug *Eurygaster integriceps* (Fedotov & Bocharova 1955) or the studies on *Aphis fabae* in southern Britain (Way & Cammell 1973). More frequently, variations in the success of hibernation, migration and colonization require monitoring the invading population (Taylor 1974) or that in the field (Matthews & Tunstall 1968).

INTERMEDIATE PESTS

The essence of the dynamics of these species is that there is a natural enemy ravine: the frequency with which the pest is able to cross this will depend on many factors, particularly its width (which up to a certain point is proportional to the duration stability of the habitat), the degree of climatic instability (i.e. the environmental noise) and the organism's own intrinsic rate of increase. It is assumed that the lower equilibrium point (see Fig. 2b) is sufficiently low that the organism is no longer a pest. This could be invalidated by the increasing emphasis on cosmetic control (Norton & Conway 1977), and when this happens the application of integrated pest control will be difficult. Host resistance to attack inevitably lowers the intrinsic rate of increase and so strengthens the ravine.

When the pest is not endemic then the introduction of appropriate natural enemies will establish the ravine. Most successful examples of biological control fall into this category and occur in habitats characteristic of intermediate to K-species, rather than r-pests. This is shown in Table 1

Table 1.   Cases of biological control of insects by imported natural enemies grouped according to habitat characters (on r–K continuum); based on the analysis of Table 12 of De Bach (1964).

Habitat type	All cases		Cases graded 'Complete Control'	
	No.	% total	No.	% total
r-end				
Cereals and forage crops (wheat, lucerne, etc.)	6	4	0	0
Vegetables (tomatoes, asparagus, taro, etc.)	21	14	0	0
Sugar cane, cotton, pasture	19	12	2	8
Trees (citrus, coconut, oak, olive, grape, etc.)	107	70	23	92
K-end				

which is biased against tree crops by the exclusion of the data in De Bach's table 13 which lists another 64 instances of biological control (42 of them 'complete') for four pests of trees (*Aleurocanthus*, *Enosoma*, *Icerya* and *Pseudaulacaspis*) that have been controlled in many different countries.

When intermediate pests are native species it is important that management strategies keep the natural enemy ravine intact and widen it if possible: this is the essence of integrated control.

K-PESTS

The extreme *K*-strategist is a specialist and will be sensitive to the disturbance of its habitat. Therefore *K*-pests may be reduced by cultural procedures that change the form of the habitat, for example altering the amount of shade or, for many parasites, reducing the numbers or longevity of another organism (the vector). The majority of *K*-strategists fail to make the transition from natural to agroecosystems. The low recruitment rate of these organisms and, frequently, their complex reproductive tactics make them susceptible to methods of reproductive control, e.g. sterile male techniques.

These indications for control strategies are summarized diagrammatically in Fig. 3.

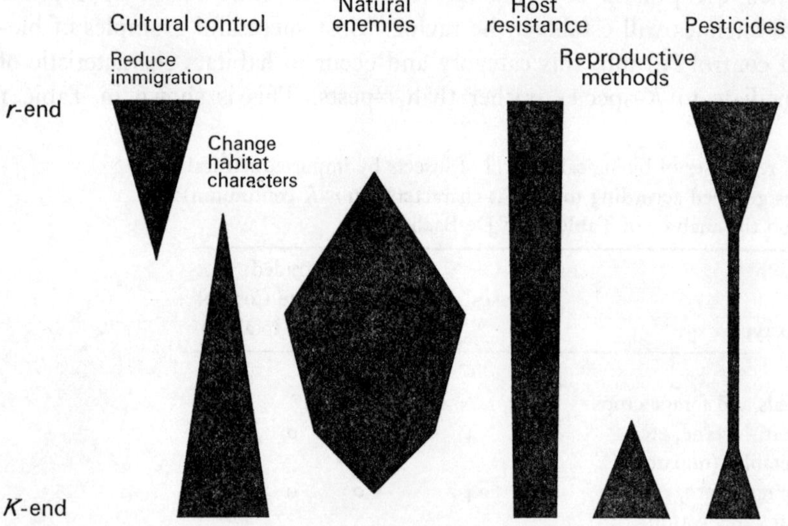

Figure 3. Diagrammatic representation of the relative importance of various pest control strategies in relation to the *r-K* continuum.

Conclusion

Population dynamics theory has now reached the stage where certain generalizations are beginning to appear. The synoptic model is one way of expressing these. It is based on equations for natality, intraspecific competition and natural enemy action which are varied along an axis related to the duration of the habitat and to various bionomic characters of the organism (size, etc.) conveniently described as the *r–K* continuum. It allows the diversity of nature to be expressed whilst maintaining the uniformity of the underlying equations. Pest organisms from anthrax to elephants can be viewed against this framework. It gives useful insights into their origins and the probable success of potential control strategies (Fig. 3).

References

ANDERSON P.K. (1970) Ecological structure and gene flow in small mammals. *Symp. zool. Soc. Lond.* No. 26, 299–325.

ANDREWARTHA H.G. & BIRCH L.C. (1954) *The Distribution and Abundance of Animals.* University of Chicago Press, Chicago.

BAILEY S.F. & KEIFER H.H. (1943) The tomato russet mite, *Phyllocoptes destructor* Keifer: its present status. *J. econ. Ent.* 36, 706–12.

BALCH R.E. (1939) The outbreak of the European spruce sawfly in Canada and some important features of its bionomics. *J. econ. Ent.* 32, 412–18.

BALCH R.E. & BIRD F.T. (1944) A disease of the European spruce sawfly, *Gilpinia hercyniae* (Htg.) and its place in natural control. *Sci. Agric.* 25, 65–80.

BALTENSWEILER W. (1968) The cyclic population dynamics of the grey larch tortrix *Zeiraphera griseana* Hübner (=*Semasia diniana* Guenée) (Lepidoptera: Torticidae). In *Insect Abundance* (Ed. by T.R.E. Southwood), *R. ent. Soc. Lond. Symp.* 4, 88–97, Blackwell Scientific Publications, Oxford.

BALTENSWEILER W. (1971) The relevance of changes in the composition of larch bud moth populations for the dynamics of its numbers. *Proc. Adv. Study Inst. Dynamics Numbers Popul.* (Oosterbeek 1970), 208–19.

BENNETT F.D. & HUGHES I.W. (1959) Biological control of insect pests in Bermuda. *Bull. ent. Res.* 50, 423–36.

BERRY R.J. (1970) Covert and overt variation, as exemplified by British mouse populations. *Symp. zool. Soc. Lond.* 26, 3–26.

BIRD F.T. & ELGEE D.E. (1957) A virus disease and introduced parasites as factors controlling the European spruce sawfly, *Diprion hercyniae* (Htg.) in central New Brunswick. *Canad. Ent.* 89, 371–8.

BRADLEY D.J. (1977) Human pest and disease problems; contrasts between developing and developed countries. In *Origins of Pest, Parasite, Disease and Weed Problems* (Ed. by J.M. Cherrett & G.R. Sagar), pp. 329–45. Blackwell Scientific Publications, Oxford.

BRASIER C.M. & GIBBS J.N. (1973) Origin of the Dutch elm disease epidemic in Britain. *Nature, Lond.* 242, 607–8.

CLARK L.R. (1964) The population dynamics of *Cardiaspina albitextura* (Psyllidae). *Aust. J. Zool.* 12, 362–80.

CLARK L.P., GEIER P.W., HUGHES R.D. & MORRIS R.F. (1967) *The Ecology of Insect Populations in Theory and Practice.* Methuen, London.

COMINS H.N. (1977) The management of pesticide resistance. *J. theor. Biol.* (in press).

CONNELL J.H. (1975) Some mechanisms producing structure in natural communities: a model and evidence from field experiments. In *Ecology and Evolution of Communities* (Ed. by M.L. Cody & J.M. Diamond), pp. 460–90. Harvard, Camb., Mass.

Conway G.R. (1971) *Pests of Cocoa in Sabah and their Control.* Kementerian Pertonian dan Perikanan, Sabah.

Conway G.R. (1976) Man versus Pests. In *Theoretical Ecology: Principles and Applications* (Ed. by R.M. May), pp. 257–8. Blackwell Scientific Publications, Oxford.

Conway G.R. & Wood B.J. (1964) Pesticide chemicals—help or hindrance in Malaysian agriculture? *Malay. Nat. J.* 18, 111–19.

Darwin C. (1872) *Origin of Species by Means of Natural Selection,* 6th edition (1945 reprint, Watts, London).

DeBach P. (1964) Successes, trends, and future possibilities. In *Biological Control of Insect Pests and Weeds* (Ed. by P. DeBach), pp. 673–713. Chapman & Hall, London.

Elliott H. (1971). Wandering albatross. In *Birds of the World* (Ed. by J. Gooders), pp. 29–31, vol. 1. I.P.C., London.

Elton C.S. (1958) *The Ecology of Invasions by Animals and Plants.* Methuen, London.

Emden H.F. van, Eastop V.F., Hughes P.D. & Way M.J. (1969) The ecology of *Myzus persicae. A. Rev. Ent.* 14, 197–270.

Entwistle P.F. (1972) *Pests of Cocoa.* Longmans, London.

Entwistle P.F., Johnson C.G. & Dunn E. (1959) New pests of cocoa (*Theobroma cacao* L.) in Ghana following applications of insecticides. *Nature, Lond.* 184, 2040.

Fedotov D.M. & Bocharova O.M. (1955) [Dependence of morpho functional conditions of *Eurygaster integriceps* Put. on environment. (In Russian.)] *Vrednaya cherepachka* 3 (Ed. by D.M. Fedetov), pp. 7–67.

Firempong S. & Kumar R. (1975) Natural enemies of *Toxoptera aurantii* (Boy.) (Homoptera: Aphididae) on cocoa in Ghana. *Biol. J. Linn. Soc.* 7, 261–92.

Ford J. (1971) *The Role of the Trypanosomiases in African Ecology.* Oxford University Press.

Frazer B.D. & van den Bosch R. (1973) Biological control of the walnut aphid in California: the interrelationship of the aphid and its parasite. *Environ. Entomol.* 2, 561–8.

Garnham P.C.C. (1971) *Progress in Parasitology.* Athlone Press, London.

Garnham P.C.C. (1977) The origins, evolutionary significance and dispersal of haemosporidian parasites of domestic birds. In *Origins of Pest, Parasite, Disease and Weed Problems* (Ed. by J.M. Cherrett & G.R. Sagar), pp. 57–70. Blackwell Scientific Publications, Oxford.

Hagen K.S. & van den Bosch R. (1968) Impact of pathogens, parasites and predators on aphids. *A. Rev. Ent.* 13, 325–84.

Holloway J.K. (1964) Projects in biological control of weeds. In *Biological Control of Insect Pests and Weeds* (Ed. by P. DeBach), pp. 650–70. Chapman & Hall, London.

Huffaker C.B. (1967) A comparison of the status of biological control of St. Johnswort in California and Australia. *Mushi* 39, 51–73.

Huffaker C.B. & Kennett C.E. (1956) Experimental studies on predation: predation and cyclamen-mite populations on strawberries in California. *Hilgardia* 26, 191–222.

Huffaker C.B. & Kennett C.E. (1966) Studies of two parasites of olive scale, *Parlatoria oleae* (Colvée). IV. Biological control of *Parlatoria oleae* (Colvée) through the compensatory action of two introduced parasites. *Hilgardia* 37, 283–335.

Hughes R.D. (1970) The seasonal distribution of bushfly (*Musca vetustissima* Walker) in south-east Australia. *J. Anim. Ecol.* 39, 691–706.

Hughes R.D. & Walker J. (1970) The role of food in the population dynamics of the Australian bushfly. In *Animal Populations in Relation to their Food Resources* (Ed. by A. Watson), pp. 255–67. Blackwell Scientific Publications, Oxford.

Lack D. (1966) *Population Studies of Birds.* Clarendon Press, Oxford.

LAWS R.M., PARKER I.S.C. & JOHNSTONE R.C.B. (1975). *Elephants and their Habitats.* Oxford University Press.

LESTON D. (1970) Entomology of the cocoa farm. *A. Rev. Ent.* 15, 273–94.

MACARTHUR R.H. & WILSON E.O. (1967) *The Theory of Island Biogeography.* Princeton University Press, Princeton, N.J.

MACKAUER M. & WAY M.J. (1976) *Myzus persicae* Sulz. an aphid of world importance. In *Studies in Biological Control* (Ed. by V.L. Delucchi), pp. 51–119. Cambridge University Press.

MASSEE A.M. (1958) The effect on the balance of arthropod populations in orchards arising from the unrestricted use of chemicals. *Int. Cong. Ent.* X (3), 163–8.

MATTHEWS G.A. & TUNSTALL J.P. (1968) Scouting for pests and the timing of spray applications. *Cott. Grow. Rev.* 45, 115–27.

MORGAN F.D. (1968) Bionomics of Siricidae. *A. Rev. Ent.* 13, 239–56.

NEWTON I. (1964) Bud-eating by Bullfinches in relation to the natural food supply. *J. appl. Ecol.* 1, 265–79.

NORTON G.A. & CONWAY G.R. (1977) The economic and social context of pest, disease and weed problems. In *Origins of Pest, Parasite, Disease and Weed Problems* (Ed. by J.M. Cherrett & G.R. Sagar), pp. 205–26. Blackwell Scientific Publications, Oxford.

O'CONNOR B.A. (1950) Premature nutfall of coconuts in the British Solomon Islands Protectorate. *Agric. J. Fiji* 21, 1–22.

PIANKA E.R. (1970) On *r*- and *K*-selection. *Am. Nat.* 104, 592–7.

POTTS G.R. (1977) Some effects of increasing the monoculture of cereals. In *Origins of Pest, Parasite, Disease and Weed Problems* (Ed. by J.M. Cherrett & G.R. Sagar), pp. 183–202. Blackwell Scientific Publications, Oxford.

SALISBURY E.J. (1961) *Weeds and Aliens.* Collins, London.

SCHWERDTFEGER F. (1941) Über die Ursachen des Massenwechsels der Insekten. *Z. angew. Ent.* 28, 254–303.

SHAPIRO A.M. (1975) The temporal component of butterfly species diversity. In *Ecology and Evolution of Communities* (Ed. by M.L. Cody & J.M. Diamond), pp. 181–95. Harvard, Camb., Mass.

SIMMONDS F.J. & GREATHEAD D.J. (1977) Introductions and pest and weed problems. In *Origins of Pest, Parasite, Disease and Weed Problems* (Ed. by J.M. Cherrett & G.R. Sagar), pp. 109–24. Blackwell Scientific Publications, Oxford.

SMITH R.F. (1970) Pesticides: Their use and limitations in pest management. In *Concepts of Pest Management* (Ed. by R.L. Rabb & F.E. Guthrie), pp. 103–18. N.C. State University, Raleigh.

SOUTHWOOD T.R.E. (1961). The evolution of the insect–host tree relationship—a new approach. *Int. Cong. Ent.* XI(1), 651–5.

SOUTHWOOD T.R.E. (1962) Migration of terrestrial arthropods in relation to habitat. *Biol. Rev.* 37, 171–214.

SOUTHWOOD T.R.E. (1975) The dynamics of insect populations. In *Insects, Science and Society* (Ed. by D. Pimentel), pp. 151–99. Academic Press, New York and London.

SOUTHWOOD T.R.E. (1976) Bionomic Strategies and Population Parameters. In *Theoretical Ecology: Principles and Applications* (Ed. by R.M. May), Chapter 3 Blackwell Scientific Publications, Oxford (in press).

SOUTHWOOD T.R.E. & COMINS H.N. (1976) A synoptic population model. *J. Anim. Ecol.* 45, 949–65.

SOUTHWOOD T.R.E. & JEPSON W.F. (1961) The frit fly—a denizen of grassland and a pest of oats. *Ann. appl. Biol.* **49**, 556.

SOUTHWOOD T.R.E. & JEPSON W.F. (1962) The productivity of grasslands in England for *Oscinella frit.* (L.) (Chloropidae) and other stem-boring Diptera. *Bull. ent. Res.* **53**, 395–407.

SOUTHWOOD T.R.E., MAY R.M., HASSELL M.P. & CONWAY G.R. (1974) Ecological strategies and population parameters. *Am. Nat.* **108**, 791–804.

SOUTHWOOD T.R.E. & WAY M.J. (1970) Ecological background to pest management. In *Concepts of Pest Management* (Ed. by R.L. Rabb & F.E. Guthrie), pp. 6–28. N.C. State University, Raleigh.

STRICKLAND A.H. (1960) Ecological problems in crop pest control. In *Biological Problems Arising from the Control of Pests and Diseases* (Ed. by R.K.S. Wood), Symp. Inst. Biol. **9**, 1–13.

STRONG P.R. (1974) Rapid asymptotic species accumulation in phytophagous insect communities: The pests of cacao. *Science* **185**, 1064–6.

TAYLOR L.R. (1974) Monitoring change in the distribution and abundance of insects. *Rep. Rothamsted exp. Stn.* 1973 (2), 202–39.

TAYLOR T.H.C. (1937) *The Biological Control of an Insect in Fiji.* Imp. Inst. Ent., London.

TURNER N. (1963) The Gipsy moth problem. *Bull. Conn. agric. Exp. Stn.* **655**, 1–36.

UVAROV B.P. (1957) The aridity factor in the ecology of locusts and grasshoppers of the Old World. In *Human and Animal Ecology. Reviews of Research.* Paris, Unesco, pp. 164–98. (Arid zone research VIII.)

VOÜTÉ, A.D. (1946) Regulation of the density of the insect-populations in virgin-forests and cultivated woods. *Archs. néerl. Zool.* **7**, 435–70.

WALOFF Z. (1966) The upsurges and recessions of the desert locust plague: an historical survey. *Anti-Locust Mem.* **8**, 1–11.

WATT K.E.F. (1971) Dynamics of populations: A synthesis. *Proc. Adv. Study Inst. Dynamics Numbers Popul. (Oosterbeek,* 1970), 568–80.

WAY M.J. (1953a) Studies on *Theraptus* sp. (Coreidae); the cause of the gumming disease of coconuts in East Africa. *Bull. ent. Res.* **44**, 657–67.

WAY M.J. (1953b) The relationship between certain ant species with particular reference to biological control of the coreid *Theraptus* sp. *Bull. ent. Res.* **44**, 669–91.

WAY M.J. (1976) Concluding remarks. In *Studies in Biological Control* (Ed. by V.L. Delucchi), Int. Biol. Prog. **9**, 229–36. Cambridge University Press.

WAY M.J. (1977) Pest and disease status in mixed stands vs. monocultures; the relevance of ecosystem stability. In *Origins of Pest, Parasite, Disease and Weed Problems* (Ed. by J.M. Cherrett & G.R. Sagar), pp. 127–38. Blackwell Scientific Publications, Oxford.

WAY M.J. & BANKS C.J. (1968) Population studies on the active stages of the black bean aphid, *Aphis fabae* Scop., on its winter host. *Euonymus europaeus* L. *Ann. appl. Biol.* **62**, 177–97.

WAY M.J. & CAMMELL M.E. (1973) The problem of pest and disease forecasting—possibilities and limitations as exemplified by work on the bean aphid, *Aphis fabae. Proc. 7 Br. Insecticide Fungicide Conf.* 933–54.

Evolution and distribution

The origins, evolutionary significance and dispersal of haemosporidian parasites of domestic birds

P. C. C. GARNHAM *Department of Zoology and Applied Entomology, Imperial College, London*

Any ecological study of parasites must inevitably include their hosts, and often the life cycle takes place, as in the haemosporidian parasites, both in invertebrate and vertebrate hosts. Moreover the parasitological phenomena may involve, in addition, reservoir hosts. Thus the ecology of at least three and sometimes four primary factors are concerned; the disease is the result of their inter-action. In man, epidemics and pandemics of infections follow, and their course is much influenced by the unnatural circumstances of human life. A less complicated picture is presented by diseases of domestic animals, and I have taken as examples, haemosporidian infections of birds. Before discussing the latter, I shall briefly relate the history of some common human infections in order to place in perspective the avian parasites which are the main subject of my paper.

Origins and dispersal of typical human diseases

These examples are taken from bacterial, viral and protozoal infections.

The spread of pests over great distances has long been recognized, especially those like plague which have such terrifying results. Few scientific observations were made on the nature of the events until the concept of contagion was formed, and it was only in the last twenty-five years that the significance of an animal reservoir was appreciated. The original source is always a wild animal and the later condition in man is usually a zoonosis, i.e. an infection in man acquired from an animal.

Plague was about the only disease whose pandemic course was studied at an early date and it is interesting that the old Arabian authors (see Dols 1974) were able to plot accurately the progress of the disease from its nidus in the steppes of central Asia along the overland trade route (as followed by

Marco Polo in the thirteenth century), to the Crimea and thence to Byzantium, and southwards to Aleppo in Syria, Palestine, Cairo and Alexandria. They described also its spread to India and westwards into Europe. This trail of horror kept much to the same route, throughout the centuries almost up to our own day, and even now we cannot be sure that the route is permanently blocked.

There has been much talk in recent years about travellers from the tropics bringing back malaria to their homes in temperate countries. And quite rightly, because the parochial physician may fail to recognize in time that the disease from which his patient is suffering, is not influenza but malignant malaria. Fortunately, the disease will not spread because of the changed environment in which people live today. But in the tropics, an infected person may be the source of new outbreaks in a country which has painfully and at much cost eradicated the disease; such re-introductions have taken place in Sri Lanka, Mauritius, Guyana and Venezuela.

If the pest is taken to a far-off region, like quartan malaria or leishmaniasis which were probably transported from the Mediterranean to the Americas, a change, probably of genetic nature, is liable to arise. In the new locality, new vertebrate reservoirs and insect vectors become involved, within a different macro-environment. In quartan malaria, it is probable that the parasite (*Plasmodium malariae*) eventually became established (and slightly altered into the so-called *P. brasilianum*) in the local monkeys, thus producing a zoonosis in reverse (see Table 1). In leishmaniasis, *Lutzomyia* spp., instead of *Phlebotomus* spp. transmit the parasite (*Leishmania infantum* as it is called in the Mediterranean) through the wild fox reservoir and the parasite in the New World is named *Leishmania chagasi*.

Another instance is provided by yellow fever, a virus disease of African origin. The infection is present in the monkeys of the tropical rain forest and is transmitted by acrodendrophilic mosquitoes. The monkeys suffer only slightly from yellow fever. Man intrudes into the cycle and human epidemics follow. Some centuries ago, probably in the course of the slave trade, the infection was carried across the Atlantic in mosquito-infested ships from West Africa to the Americas. The disease spread both among the human population and eventually to the local monkeys and once again a zoonosis in reverse was set up. A totally different kind of mosquito transmitted the disease to which the New World monkeys proved to be highly susceptible. It was a bitter experience to travel in these forests after an epizootic had decimated the simian population.

An unexplained feature in the ecology is the failure of yellow fever to spread to Asia where *Aedes aegypti* and other vectors are present in high density. In recent years the strict treatment of aircraft with insecticide and insistence on the immunization of travellers may have hindered the spread

Table 1.   Parasites evolving in new hosts. (Modified from Baker 1974.)

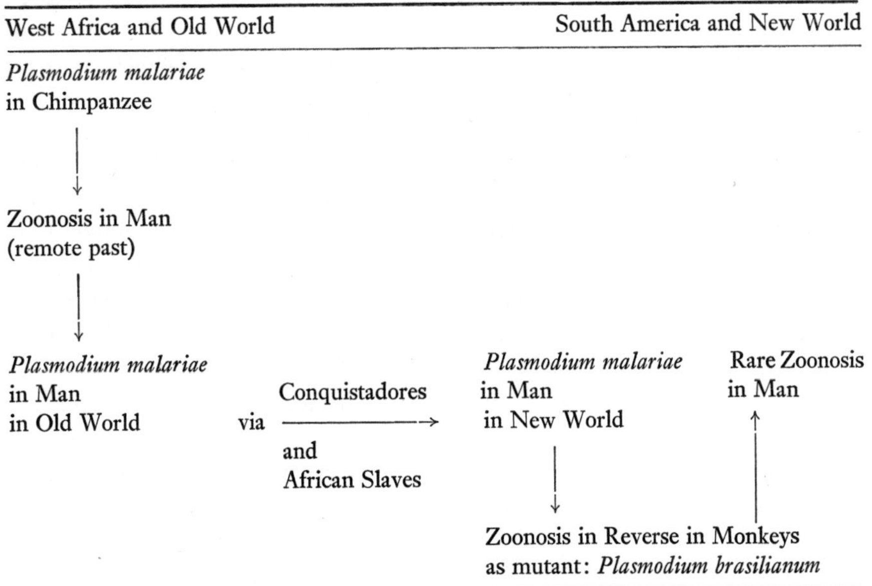

of the disease into India and further east, but in the past the virus could have been taken by dhow traffic across the Indian Ocean—but this never happened.

I discussed in the Heath Clark Lectures the evolutionary significance of these and other migrations (Garnham 1971).

Origin, evolutionary significance and dispersal of avian infections

I shall discuss five important or interesting protozoal infections of birds, as follows:

Plasmodium gallinaceum Brumpt, of chickens, wild reservoir and mosquitoes.

Plasmodium juxtanucleare Versiani & Gomes, of chickens, wild reservoir and mosquito.

Plasmodium durae Herman, of turkeys, wild reservoir and mosquito.

Leucocytozoon simondi Mathis & Leger, of ducks and geese, wild reservoir and simuliids.

Leucocytozoon sp. of parakeets in aviaries, wild reservoir and (?) simuliids.

PLASMODIUM GALLINACEUM

This malarial parasite causes an acute and often fatal disease in exotic breeds of hens (*Gallus gallus*) imported into limited areas in S.E. Asia, extending eastwards from Sri Lanka and southern India (Madras) to western Malaysia, Indo-China and the Philippines. The reservoir hosts are different species of jungle-fowl—*G. sonnerattii* (Madras), *G. lafayettei* (Sri Lanka) and *G. spadiceus* (Malaysia). The infections in the domestic fowl are termed 'veterinary zoonoses' (Garnham 1969). In the feral host, the infection is mild; in the local chickens, there is little mortality, but in exotic breeds, the infection fulminates and many birds die of cerebral malaria. There is a progressive diminution of natural resistance to the infection, from the jungle fowl through the local chickens and finally to imported breeds. A parallel situation exists in holoendemic human malaria in Africa and elsewhere in the tropics where there is much natural immunity in the native population, but newcomers die of cerebral malaria. Once established in a flock of chickens, *P. gallinaceum* spreads amongst them without the intervention of the sylvatic reservoir.

The fourth component of the biocenose is the mosquito, of which many species are capable of acting as definitive hosts. The natural vector in Sri Lanka is *Mansonia crassipes*. At least forty species belonging to six genera (Garnham 1966), are more or less suitable hosts for this malarial parasite, or can be adapted to it (Corradetti *et al.* 1966). Several species, e.g. *Aedes aegypti*, become 100% infected.

Much experimental work has been done on *P. gallinaceum*, and under different conditions its morphology is seen to change, the capacity to produce malaria pigment is lost, and certain phases (exoerythrocytic and sexual) become suppressed. It is doubtful if such substrains represent true mutations, as the new characters disappear when the parasite is no longer exposed to the abnormal conditions. An unusual feature, present in some strains, is the possession of elongate as well as spherical gametocytes, and this rather important morphological feature of the sexual stage may indicate that speciation is taking place (Fernando & Dissanaike 1975).

P. gallinaceum thus originates in the jungle fowl and becomes dispersed in the domestic hen in foci in tropical parts of Asia. The limited distribution of the parasite is puzzling. The limitation coincides with the distribution of the reservoir host, which is confined to S.E. Asia, and it may be that in the absence of the jungle fowl, *P. gallinaceum* is unable to become established. On the other hand, epizootics spread rapidly in the domestic flocks, and the parasite is one of the easiest to transmit in the laboratory, and by the ubiquitous domestic mosquito, *Aedes aegypti*. Chickens are transported by train, boat or road on the widest scale, and the infection might be expected to have accompanied them, but it has not done so, even into North India, or along the western lands bordering the Indian Ocean.

PLASMODIUM JUXTANUCLEARE

The other malarial parasite of chickens is at the opposite extreme to *P. gallinaceum* in having an almost cosmopolitan distribution in the tropics. The infection was first discovered in Brazil in 1940 and soon afterwards in Mexico and Uruguay; later, chickens infected with this parasite were found in various places in the Old World—Sri Lanka, eastern and western Malaysia, Philippines, and Japan. There are doubtful records also from tropical Africa.

The general ecology of this parasite is however similar to that of *P. gallinaceum* (Bennett *et al.* 1966). Again the reservoir host is the jungle fowl —at least in some parts of the Old World, e.g. Sri Lanka and western Malaysia, but not in Japan nor the Americas where the gallinaceous bird is absent, and where no other reservoirs have been found. In Taiwan, Manwell (1966) found *P. juxtanucleare* commonly in the Bamboo Partridge, but conditions must be unsuitable for a veterinary zoonosis for the chickens never contract the infection. The apparent absence of a reservoir in the wild birds in the Americas suggests that the true home of *P. juxtanucleare* is the Old World and that it was transported in domestic chickens to the New World, presumably not by the Conquistadores a few centuries ago from southern Europe where the parasite is absent in the local chickens, but in 'rafts' across the Pacific in pre-Columbian times.

An essential difference between the two species of parasite must be mentioned: *P. gallinaceum* gives rise to acute infection, *P. juxtanucleare* to chronic infections. Thus the latter is better able to survive long journeys in the vertebrate host. On the other hand, *P. juxtanucleare* has a much more restricted range of mosquito hosts (*Culex sitiens*, *C. gelidus*, *C. pipiens pallens*, *Mansonia crassipes*).

The wide dispersal of *P. juxtanucleare* has exposed the parasite to differing environments; it is not surprising that the biological features and morphology of the parasite vary, and that, by isolation, these changes become stabilized. Thus the *juxtanucleare*-like species in Japan was given the new name *P. japonicum*, by Ishiguro (1957) on account of its relative non-pathogenicity and its failure to infect turkeys; New World strains are more lethal to chickens than those occurring in Asia. At present the use of a single name is recommended.

PLASMODIUM DURAE

Small malarial parasites of turkeys were first described from Kenya in 1941, where they seemed to be restricted to a few foci in the highlands. The turkey is an exotic bird for Africa, and it seemed fairly certain that the parasite must have been derived from some indigenous gallinaceous bird.

In 1967, Southgate (unpublished) had to investigate an epizootic in Kenya which was decimating the President's flocks of turkeys on one of his farms at Kiboko (between Nairobi and Voi). He found *P. durae* in their blood and showed that the yellow-necked francolin also harboured the parasite. Here at last seemed to be the long sought reservoir from which the exotic turkey had apparently acquired the infection.

Discoveries in the last few years may invalidate this conclusion. Telford and Forrester (1975) examined the blood of wild turkeys in their natural home in Florida and found a parasite which closely resembles *P. durae*. The wild turkey is only found in the southeastern United States and Mexico and it is necessary to speculate on how its malarial parasite could have reached Africa.

Strains of the wild turkey had long been domesticated by the Aztecs in Mexico (Schorger 1966) and soon after the arrival of the Conquistadores in the New World turkeys were sent back to Spain (as early as 1498). They quickly became distributed throughout Europe. At the beginning of the seventeenth century, the Portuguese took them to India, and about the same time to West Africa and even earlier to St. Helena.

Various factors facilitate the transmission of *P. durae*: the parasite gives rise to very chronic infections in turkeys, it is capable of surviving in the blood and would endure long sea voyages; it develops easily in *Culex pipiens fatigans* (de Jong 1971), a common mosquito both in southern Portugal and Africa where opportunities for transmission must be numerous. Thus, *P. durae* (in the turkey) could easily have been transported from the New World, have maintained itself on the southern coasts of the Iberian peninsula and established new foci when finally taken to Africa. Turkeys were bred extensively in missions and on farms.

There is no evidence of malaria in the turkeys of southern Europe, and *P. durae* itself was first described from East Africa. A few vague reports of turkey malaria arose from West Africa and in 1961, Dr Arnon Gunders and I examined blood films of about a hundred turkeys on the lower Niger River. We found very scanty parasites in a single specimen. Finally Barnes (1974) in northern Nigeria was more successful: he isolated strains of a *durae*-like parasite in turkeys, francolins and peacocks, and studied their behaviour for several years.

The question now arises, are these various forms of turkey malaria of a single origin or are the African parasites, examples of a 'veterinary zoonosis' derived from wild birds (e.g. francolins)?

First we must note that turkeys are susceptible to other species of avian malaria parasites, including *P. fallax, P. lophurae, P. pinottii, P. juxta-*

nucleare and *P. gallinaceum*, while another species (*P. griffithsi* Garnham 1966) occurs naturally in this bird in Burma. This rather wide range of susceptibility suggests the possibility that turkeys may acquire infections easily from an outside source. But *P. durae* bears no resemblance to any of these species.

There are two possible origins of the francolin parasites: they may be truly indigenous in the francolin, or this bird may have acquired the infection from the introduced turkeys as a 'zoonosis in reverse'. A *P. durae*-like parasite has been reported in francolins in South Africa and in Ethiopia (Ashford *et al.* 1970); but the turkeys are uninfected.

The most critical evidence of identity is provided by the morphology and biological behaviour of the individual strains. This is not the place in which to discuss the problem in detail, but it is worth while emphasizing that we should come to a full stop here in our ecological analysis if the essential taxonomic details were lacking. But ecology is as important as taxonomy because changes in the environment can provoke profound speciation, and this is what may be happening in turkey malaria. (The strains are compared in Table 2.)

Table 2. Malaria parasites in turkeys.

	Plasmodium durae		Francolin parasite		*P. hermani*	*P. griffithsi*
	Kenya	Nigeria	Kenya	Nigeria	Florida	Burma
Type of erythrocyte	Mature	Mature	Mature	Mature	Immature	Mature
Normal merozoite number	8	4	10	4	8	20
Size of schizont	Small	Small	Smaller	Smaller	Small	Large
Gametocyte	Elongate	Elongate	Shorter	Shorter	Irregular elongate	Spherical
Sporogony in *Culex*	Present	?	Present	?	Present	?

The conservative school of taxonomists, including Dr. Carlton Herman and Professor Marshall Laird regard the Floridan, Nigerian and Kenyan parasites as strains of *P. durae*; others including myself, think that major biological differences and minor morphological ones are sufficient to warrant giving separate specific rank to the Floridan parasite. This course was followed by Telford & Forrester (1975) who named the latter *P. hermani* in honour of Dr. Carlton Herman; he disagreed with this nomenclature! No decision has yet been reached about the West African strains although the strain in the turkey is probably *P. durae*.

The origin of the African parasites in the turkey, francolin and peacock is still a matter of speculation. A line of descent from imported American turkeys (with several 'bars sinister'!) is an attractive hypothesis. The original mosquito host in Florida is now known to be *C. nigripalpis*; has been replaced in Africa by *C. antennatus* and *C. pipiens* where the francolin acts as a feral reservoir in place of the wild turkey. These changes over the last three centuries, involving innumerable sexual processes may well have led to mutations and speciation. The solution to the problem will have to await further experimentation and possibly the plotting of the isoenzyme patterns of the different 'strains'.

LEUCOCYTOZOON SIMONDI

This 'malaria' parasite is common in ducks and geese in many parts of the world, including Canada and the United States. Both domestic and wild birds of the Anatidae are infected. The incidence of any malaria parasite is determined by the prevalence of its invertebrate host (in *Leucocytozoon*, a species of *Simulium*) and the ambient temperature, for below a certain temperature, the parasite cannot develop in the insect which itself depends on rainfall and temperature for breeding. But apart from these external factors, which start to operate in the summer, the parasite (*L. simondi*) exhibits a characteristic chronobiology. Relapses occur in the spring, as the result either of hormone changes in the bird preparatory to egg production, or of stress induced by migration to the breeding grounds. In March 1967, I travelled to Toronto to catch at the predicted moment relapses of *Leucocytozoon* in mallards. We examined the blood of the birds daily and on March 31st detected a sudden invasion of the blood by the tiny merozoites, rupturing out of the re-awakened exoerythrocytic stages of the parasite (Desser, Fallis & Garnham 1968).

L. simondi is responsible for devastating epizootics in geese unused to the infection. Domestic geese were introduced into the subarctic region of Canada in an attempt to augment the protein resources of the inhabitants. *Simulium* spp., which had fed on local birds infected with the parasite, bit the geese and caused such a heavy mortality that the project was abandoned (Laird & Bennett 1970).

LEUCOCYTOZOON SP. IN PARAKEETS

One of the most interesting problems in parasitology that I have encountered in recent years was the outbreak of a fatal disease in parakeets living in outside aviaries in scattered parts of southern England and as far north as Huntingdon and Northampton. The most severe epizootics have occurred yearly since 1971 in the large aviaries at Leeds Castle in Kent (Garnham 1974). The disease is strictly seasonal, being confined to the months of July, August and September. The causative organism was at first thought to be *Sarcocystis*, as macroscopic lesions were found in the muscle of the heart and oesophagus, but later Walker & Garnham (1972) showed that these lesions were due to megaloschizonts of a species of *Leucocytozoon* accompanied by extensive haemorrhage in the surrounding tissue. However, the blood cells never became invaded by the millions of merozoites produced by the tissue forms, and we postulated that the infections were aberrant in the parakeet, an abnormal host in which the more primitive (exoerythrocytic) stages of the parasite were able to develop profusely in the tissues of this host, but the progeny were unable to enter the blood corpuscles. Such an anomaly is commonly observed experimentally in other abnormal hosts; for instance the *liver* of a chimpanzee may be extensively invaded by exoerythrocytic stages of *Plasmodium falciparum* (a human malaria parasite), but its *erythrocytes* remain immune.

The parakeets affected by this disease had been imported largely from Australia where the parasitic condition is unknown, and the origin of the infection in England is thought to be a reservoir in some of the local wild birds. Dr. J. B. Baker and I showed that at Leeds Castle, *Leucocytozoon* sp. was present in the blue tit (*Parus coeruleus*) and the blackbird (*Turdus merula*); we assumed (but failed to verify) that *Simulium* spp. or *Culicoides* spp. bit the local infected birds, and these insects later bit the parakeets which subsequently developed the disease.

Parakeets are costly and it was decided that prophylactic measures must be instituted before our researches on this condition were completed. Since 1973, the drug Whitsyn S (pyrimethamine and sulphaquinoxaline) has been added to the food from June to September, and practically no cases have occurred in the last two years. Other exotic birds are kept under wild conditions at Leeds Castle and in 1975, a black swan was found dead with its organs showing the typical appearances of leucocytozoonosis; again the blood remained free of the infection.

Discussion

This paper has dealt with the reservoir hosts of avian haemosporidia and the eventual involvement of domesticated birds. The 'K' pest (Southwood 1977) becomes transformed into an 'r' pest as follows: in the reservoir, conditions are constant, the rate of reproduction is controlled by premunition and the capacity for mutation is probably minimal. Once domestic birds become involved, there is a rapid rate of increase of the parasite in the absence of immunity, a high mortality, but more opportunities for mutation. The transport of edible birds to all corners of the earth (Hyams 1972) brings the parasites into contact with a wide variety of new vertebrate and invertebrate hosts and speciation is a likely sequel.

The haemosporidian infections in birds which I have described and other examples in animals in general, present interesting problems in zoogeography, complicated by the existence of the four primary factors operating in haemo-parasites. The first factor—the parasite itself—is completely dependent upon the other three, and zoogeography does not directly concern it. The three remaining factors—the vertebrate host, the reservoir vertebrate host and the invertebrate vector are all influenced by zoogeography. As Müller (1974) states, isolation leads progressively from races of a polytypic species to sub-species and eventually to the birth of a new species. The evolution takes place in the three hosts and eventually operates on the parasite. Some of these changes have probably occurred in recent times, as the result of human intervention, whereby animals and their parasites are transported across oceans. At the other extreme, continental drift millions of years ago, may have left the animals stranded in isolation or equally, the separation may have taken place before one or other of the components of the biocenose had evolved. I have not attempted, however, to follow the pathway through Gondwanaland which may have affected the evolution and distribution of malaria in galliform birds.

The haemosporidian parasites are thought to have originated either from coccidia escaping from the intestine of a vertebrate animal into its blood-stream or from natural parasites of the invertebrate host. In the absence of fossil evidence, it is unlikely that this question can be answered, but probably the former hypothesis fits the facts best. My paper is devoted to later events in which speciation of the established haemosporidian infection is involved.

With some exceptions—particularly in species of *Leucocytozoon*—the so-called natural hosts seem to be little affected by the malaria parasite, but information about pathogenicity of parasites in wild populations is as sketchy as it was in 1968 when, at a symposium on diseases in free-living wild animals (McDiarmid 1969) I stressed how little we knew about the subject. Unfortunately studies on infections in wild birds in captivity are inevitably accompanied by stress, which invalidates any conclusions.

However, some light is thrown on the subject by the reaction of hens to *P. gallinaceum*. The effect of the parasite on the natural host—the jungle-fowl—is unknown, but when the infection reaches the local chickens, few die and it is only when imported breeds contract the disease that a heavy mortality follows. The phenomenon is seen even more strikingly when

Leucocytozoon simondi gets into a flock of ducklings which may be healthy in the morning, ill in the afternoon and dead in 24 hours (Levine 1975); similarly with the 'lightning haemorrhagic disease' of turkeys in Rangoon, which is caused by *P. griffithsi* acquired from an unknown feral source.

These dangerous situations are likely to arise under conditions of instability, including (1) the introduction of new breeds of birds to a place where the disease is endemic (as in *P. gallinaceum* enzootics or amongst Caucasians into a region of holoendemic *P. falciparum* in Africa), (2) the introduction of totally exotic birds (e.g. parakeets) to a place where the parasite (*Leucocytozoon* sp.) is enzootic in the local birds and a fatal, aberrant malady is the result, and (3) the introduction of new parasites in birds into a place where hosts exist and the environment is suitable for transmission (e.g. *P. juxtanucleare* to the Americas).

Some attempts have been made to cope with these dangers, and some countries have controlled the importation of the parasites, e.g. *P. gallinaceum*, into the United States. Laird & Hoogstraal (1975) were responsible for a unanimous Resolution of the 3rd International Congress of Parasitology in Munich, stressing the serious disease hazard posed by the dispersal of exotic birds, leading to catastrophic mortalities in poultry stocks. It recommended control of the importation of avian hosts of known parasites into areas where they are not already enzootic.

Although I have emphasized the much greater susceptibility of the newcomer to the parasites in an endemic region and thus the greater pathogenicity to the domestic as compared to the feral host, this is not always the case, and I have pointed out elsewhere (Garnham 1971) that a disease is often less virulent in man when it is acquired directly from a wild animal, i.e. as a zoonosis, than after inter-human transmission; similarly many parasites take several passages in the new host to flourish well—*Toxoplasma gondii* is a good example and certain avian malaria parasites react in the same way.

Such transfers from one host to another probably encourage speciation of the parasite, as I have already mentioned in regard to human and avian species in nature. In the new environment, the pressure of natural selection is felt, but it is often difficult to know if the original populations were mixed and an unidentified constituent selected or if a real mutation had occurred.

Experimental work on bird malaria nevertheless indicates that speciation takes place in these circumstances. Corradetti (1973) has observed the phenomenon in the laboratory, whereby *P. vaughani* in the type host (*Turdus migratorius*) from America was compared with the parasite in *Turdus merula* from Italy and in *Leithrix luteus* from Asia. The last two had been given different names owing to minor changes in morphology and biological behaviour, but probably all are examples of a *P. vaughani* complex in process of speciation. Corradetti *et al.* (1966) selected by repeated passage a strain of *P. gallinaceum* in *Anopheles stephensi* to which it is normally practically insusceptible; they concluded that such a mechanism may occur in nature under appropriate conditions and lead to speciation.

A good example is provided by *P. subpraecox* of the owl (*Athene noctua*), a member of the cosmopolitan *P. relictum* complex, usually parasitic in passerine birds, but which has become adapted to a wide range of others. *P. subpraecox* is found in owls in America, Europe, Egypt and Asia. I studied a strain in Cairo, and like Giovannola (1939) found that the pigment was black, the merozoites limited to 12 or fewer, and the sporozoites were 11 μm in length. These characters are all different from those exhibited by *P. relictum*. Yet if the owl parasite is inoculated into a passerine bird (canary) it immediately changes its morphology so that it now resembles *P. relictum* (Corradetti 1942)—but the owl is insusceptible to the latter parasite. Both Corradetti and I feel that those strains of *Plasmodium* which become isolated in nature in certain vertebrate species, should be given subspecific rank. Thus *P. subpraecox* becomes *P. relictum subpraecox*. The isolation is not confined to the vertebrate host, but also to the mosquito host—owls are nocturnal birds and are bitten by different mosquitoes from those which bite the diurnal passerines.

Experiments involving the transfer of a species of *Plasmodium* from one bird to another are beset with anomalies. For instance, Herman (see McDiarmid 1969) reported that a mynah strain of *P. circumflexum* was established in canaries in Sri Lanka and successfully inoculated into geese in the U.S.A.; but when the infected blood of the latter was re-inoculated into canaries, it failed to infect them.

In nature, the potentialities for the spread of avian malaria must have been innumerable over the millennia; migration takes the birds into different environments where they are bitten by different mosquitoes which in their turn bite birds of other species. In most instances, transmission probably fails to take place; in a few, the parasite survives in the new host and gradually becomes established. Different iso-enzymes are probably produced with speciation, and although these have not yet been detected in avian species of *Plasmodium*, Carter (1973) has established the existence of a number of species and subspecies in rodent malaria on this basis.

References

ASHFORD R.W., PALMER T.T., ASH J.S. & BRAY R.S. (1976) Blood parasites of Ethiopian birds. I. General Survey. *J. Wildlife Diseases* **12**, 409–26.
BAKER J.R. (1974) In *Actual. Protozool.* **1**, 191–2. Univ. Clermont.

BARNES H.J. (1974) *Plasmodium* sp. infecting turkeys in northern Nigeria. *Vet. Rec.*, Sept. 7th, pp. 218–19.

BENNETT G.F., WARREN M. & CHEONG W.H. (1966). Biology of the Malaysian strain of *Plasmodium juxtanucleare* Versiani & Gomes, 1941. *J. Parasit.* 52, 565–9, 547–59.

CARTER R. (1973) Enzyme variation in *Plasmodium berghei* and *Plasmodium vinckei*. *Parasitology* 66, 297–307.

CORRADETTI A. (1942) Dimostrazione del esistenza del *Plasmodium praecox* nelle civette e rapporti di questo plasmodio con la specie denominata *P. subpraecox*. *Riv. Parassit.* 6, 177–85.

CORRADETTI A. (1973) Speciation problems in bird malaria parasites. *Proc. 9th Int. Congr. trop. Med. Malaria* 1, 241-2.

CORRADETTI A., DI DELUPIS G.L.D. & PALMIERI C. (1966) Selezione di un ceppo di *Plasmodium gallinaceum* adatto a svilupparsi in *Anopheles stephensi*. *Parassitologia* 8, 183–91.

DESSER S.S., FALLIS A.M. & GARNHAM P.C.C. (1968) Relapses in ducks chronically infected with *Leucocytozoon simondi* and *Parahaemoproteus nettionis*. *Can. J. Zool.* 45, 1061–5.

DOLS M. (1974) Ibn Al-Ward's *Risalah Al-Naba' 'An Al-Waba'*. A translation of a major source for the history of the black death in the Middle East. *Near Eastern Numismatics, Iconography, Epigraphy and History*, pp. 443–5. American University, Beirut.

FERNANDO M.A. & DISSANAIKE A.S. (1975) Studies on *Plasmodium gallinaceum* and *Plasmodium juxtanucleare* from the Malayan jungle fowl *Gallus gallus spadiceus*. *Southeast Asian J. trop. Med. publ. Hlth.* 6, 25–32.

GARNHAM P.C.C. (1966) *Malaria parasites and other haemosporidia*. Blackwell Scientific Publications, Oxford.

GARNHAM P.C.C. (1969) Malaria parasites as medical and veterinary zoonoses. *Bull. Soc. Path. exot.* 62, 325–32.

GARNHAM P.C.C. (1971) *Progress in Parasitology*. Athlone Press, London.

GARNHAM P.C.C. (1974) *Unusual hosts for parasites under natural and experimental conditions*. Actual. Protozool. (4th Congrès International de Protozoologie) Université de Clermont 1, 191–202.

GIOVANNOLA A. (1939) I plasmodi aviari. *Riv. Parassit.* 3, 221–66.

HYAMS E. (1972) *Animals in the Service of Man: 10,000 Years of domestication*. Dent, London

ISHIGURO H. (1957) *Plasmodium japonicum*, a new species of malaria parasite pathogenic for the domestic fowl. *Bull. Fac. Agric. Yamaguchi Univ.* 8, 723–33.

DE JONG A.C. (1971) Some haemosporidian parasites of parakeets and francolins. Unpublished Ph.D. Thesis, University of London.

LAIRD M. & BENNETT G.F. (1970) The subartic epizootiology of *Leucocytozoon simondi*. *J. Parasit.* 56, 198.

LAIRD M. & HOOGSTRAAL H. (1975) Disease hazards associated with bird importations. *Environm. Conserv.* 2, 2.

LEVINE N.D. (1975) *Protozoal Parasites of Domestic Animals and Man*, 2nd edn. Burgess, Minneapolis.

McDIARMID A. (1969) *Diseases in free-living animals*. Symp. Zool. Soc. Lond., 24. Academic Press.

MANWELL R.D. (1966) *Plasmodium japonicum*, *P. juxtanucleare* and *P. nucleophilum* in the Far East. *J. Protozool.* 13, 8–11.

MÜLLER P. (1974) *Aspects of zoogeography*. Junk, The Hague.

SCHORGER A.W. (1966) *The Wild Turkey. Its History and Domestication.* Norman, University of Oklahoma Press.

SOUTHWOOD T.R.E. (1977) The relevance of population dynamic theory to pest status. In *Origins of Pest, Parasite, Disease and Weed Problems* (Ed. by J.M. Cherrett & G.R. Sagar), pp. 35–54. Blackwell Scientific Publications, Oxford.

TELFORD S.R. & FORRESTER D.J. (1975) *Plasmodium (Huffia) hermani* sp. n. from wild turkeys (*Meleagris gallopavo*) in Florida. *J. Protozool.* **22**, 324–8.

WALKER D. & GARNHAM P.C.C. (1972) Aberrant Leucocytozoon infection in parakeets. *Vet. Rec.* **91**, 70–2.

The plant breeders' contribution to the origin and solution of pest and disease problems

F. G. H. LUPTON *Plant Breeding Institute, Trumpington, Cambridge*

Plant pests and diseases are rarely obvious in a natural environment, and it was only when systems of monoculture were introduced by man that a situation was created in which they could thrive and spread to cause epidemic losses. In considering the role of the plant breeder as a contributor to the origin and solution of pest and disease problems, it is therefore necessary to remember that the problems which the breeder is attempting to control were caused by man in the first place. Furthermore the breeder was probably brought in originally in order to control them.

It should also be remembered that if it is effective, the introduction of a resistant variety is the cheapest way in which to control a disease at least from the grower's point of view, as someone else has paid for the work. The only cost he is likely to meet will be in the form of royalty or other increased charges for the seed of the resistant variety.

Examples of successful breeding

One of the earliest examples of fully effective control of a plant disease by breeding was the introduction of resistance to powdery mildew of vines (*Plasmopara viticola*) towards the end of the last century. In this case resistance was obtained by hybridizing European stocks with American varieties. The example is of particular interest because it occurred before the rediscovery of Mendel's laws, so that the genetic principles concerning the inheritance of resistance were not understood.

Another frequently quoted example of successful disease control is that of wart disease of potatoes (*Synchytrium endobioticum*), when it threatened to destroy the potato crop at the beginning of the present century. Resistance was in this case achieved by selection of unaffected tubers from potato crops; it was later shown to be determined by a simply inherited genetic factor (Black 1935) which has subsequently been widely used in potato breeding.

The breeding of cotton varieties resistant to jassids (Knight 1954) illustrates the effective use of plant breeding in the control of insect pests. Resistance in this case is determined by the presence of a dense covering of hairs on the under side of the leaves, and is controlled by two major genes, H_1 and H_2, both of which occur in a number of primitive cotton varieties. Both these genes are linked to undesirable agronomic features in the cotton plant, but Knight was able to break these linkages by backcrossing, and to introduce a form of resistance which has subsequently maintained its effectiveness.

Resistance to insect pests may also be achieved by the introduction through breeding of a factor unattractive to the pest. Such resistance may be only partial, as in the case of breeding for resistance to frit fly (*Oscinella frit*) in oats. Here Bingham & Lupton (1958) found that egg laying preferences, shown when the ovuliferous female was offered a choice of varieties sown in small plots, were not maintained under field conditions using larger plots. Effective control of susceptibility to attack by Colorado beetle (*Leptinotarsa decemlineata*) in potatoes has, however, been obtained in crosses using the wild species *Solanum chacoense* (Ross 1966). Resistance in this case is caused by factors giving a bitter taste to the foliage, though its value is limited as it often appears to be associated with a taint in the flavour of the tubers.

Other examples of the introduction of effective and apparently permanent resistance to pests and diseases by breeding could be quoted, though it should be noted that most of these involve the introduction of a physical barrier to disease entry or spread. Resistance to loose smut (*Ustilago nuda*) in barley, for example, is shown by varieties which are pollinated apogamously (Pederson 1960). This is because infection is caused by spores which land on the stigmas at flowering time, but which cannot penetrate to an apogamous flower. Similarly resistance to smudge of onions (*Colletotrichum circinans*) is associated with varieties with pigmented outer scales containing catechol and protocatechuic acid (Link & Walker 1933).

Breakdown of resistance

Attempts to breed for disease resistance have not always been so successful. This has sometimes been because breeders have been unable to find a suitable source of resistance to use in their programmes. No satisfactory resistance is known, for example, to take-all (*Ophiobolus graminis*) in cereals, or to wireworms (*Agriotes* spp.) in potatoes. Of greater concern, however, has been the breakdown of varietal resistance due to the appearance of new strains of the pathogen. This phenomenon was first observed by Stakman & Piemesel (1917) who reported a new strain of black rust (*Puccinia graminis*) capable of attacking previously resistant wheat varieties. The subsequent history of wheat breeding in North America has largely been concerned with the exploitation of successive sources of resistance each capable of over-coming newly arising virulent races of the pathogen, until the introduction of the variety Selkirk in 1954 following the epiphytotic of race 15B. The instability of a situation in which the rapid spread of a previously unknown or unimportant race of pathogen can have such drastic effects on one of the world's major food crops needs no emphasis. It should, however, be noted that the resistance of Selkirk has continued to be effective since the variety was introduced, and that race 15B was known for some years before it caused serious losses, so that work on breeding for resistance had already reached an advanced stage by 1954. Curiously the gene concerned, Sr_6, has proved less effective when used in Australia and elsewhere.

The history of breeding for pest and disease resistance is chequered with reports of the introduction of new genes for resistance to counter newly arising sources of virulence. Apart from the cereal rusts, to which greater reference will be made later, examples include potato blight (*Phytophthora infestans*), flax rust (*Melampsora lini*), club root of crucifers (*Plasmodiophora brassicae*), apple scab (*Venturia inaequalis*) and even wart disease of potatoes (*Synchytrium endobioticum*), though work on this disease was earlier quoted as an example of successful breeding, as the spread of the new races, which can only be carried on infected tubers or soil, has been prevented by strict plant hygiene regulations. Similar considerations apply to plant losses caused by animal pests such as potato root eelworm (*Heterodera rostochiensis*), raspberry aphid (*Amphorophora rubi*) and hessian fly of cereals (*Phytophaga destructor*) and to virus diseases such as tobacco mosaic virus. In this case it is possible to associate the spread of a new strain of virus directly with the introduction of new varieties. Tomato varieties with monogenic resistance to tobacco mosaic virus were introduced commercially in Britain in 1966, but a new strain of virus capable of attacking them appeared during the next three years (Pelham, Fletcher & Hawkins 1970). Pelham and his colleagues collected samples of infected leaves from growers during these years, noting whether they had previously grown resistant varieties. On analysis, they found that the proportion of the resistance breaking strain increased greatly when the new variety had been grown in previous years. Furthermore, when growers had reverted to older varieties, the proportion of the resistance breaking strain fell rapidly, showing that it has poor competitive value on these varieties.

Circumstantially, these examples suggest that the plant breeder is responsible for many of the disease problems which he seeks to solve. This is not to say that he has been ineffective, or that the disease problems are any worse than they would have been if he had not attempted to control them. Indeed, he has been most successful in dealing with many of the problems with which he has been faced, and even if his success with others has been of a more temporary nature, they have been associated with considerable gains in crop productivity. It was for example estimated by Craigie (1944) that the use of resistant wheats had increased the yield of crops in Manitoba and Saskatchewan by 1·15 million tonnes annually.

There are major problems in breeding for resistance to diseases in which new races of the pathogen tend to arise as each new source of resistance is introduced. Such sources of resistance are, however, attractive to the breeder because they are determined by simply inherited major genes which are relatively simple to introduce in a breeding programme, especially when tests can be carried out on seedling plants on which little cultural effort has been expended. They are also in many cases associated with hypersensitive resistance, in which the invading fungus kills the infected portion of the host tissue, and is then itself killed because, as an obligate parasite, it can only survive on living tissue.

The first of these problems is that the number of known major genes available to the breeder is limited, and a time is rapidly approaching when new sources of resistance will not be available as new races of fungus evolve. An equally serious problem is that the introduction of major genes for disease resistance has been associated with a loss of the minor background genes previously present in most crop varieties. In his review of the problems of breeding for disease resistance, van der Plank (1963) points out that before the introduction of the first varieties bred specifically for their disease resistance, Marquis and other wheat varieties in general cultivation in North America showed a reasonable level of resistance to rust. This general field resistance, which he referred to as 'horizontal' in contrast to 'vertical' resistance against specific genes, had he suggested been lost in the process of selection for specific resistance genes. He illustrated his point by comparing work on wheat breeding with that on maize, where rust caused by *Puccinia sorghi* has never been the cause of serious economic loss, though it is normally of widespread occurrence. van der Plank suggests that this is because resistance breeding in maize has been based throughout on field tolerance rather than on specific major genes. In the same way the rapid spread of *P. polysora* through the African maize crop, which caused widespread concern in the late 1950's, died away rapidly to insignificance as farmers selected and grew the most field tolerant varieties available (Cammack 1961).

Disease problems arising from plant breeding programmes

The examples considered so far have been concerned with problems encountered in programmes where the introduction of disease resistance was a primary objective. There are, however, examples where disease susceptibility has been inadvertently introduced into programmes where pathological problems were originally considered of less importance. A well-known example is concerned with the introduction of susceptibility to leaf blight in hybrid maize breeding.

This disease may be caused by either of two races, T and O, of the fungus *Bipolaris* (*Helminthosporium*) *maydis*. Resistance to race O, which is relatively mild in its effect, is based on nuclear genes and is qualitative in expression and polygenic in inheritance. Race T, which differs from race O in a single gene for toxin production and pathogenicity, was first noted in the U.S.A. in the late 1960's and spread rapidly in 1970. It is of special interest because resistance is inherited cytoplasmically, symptoms being caused by a phytotoxin which is highly detrimental to the cytoplasm of susceptible plants and is specific in pathogenicity to plants with the *Cms* T cytoplasm widely used in hybrid maize production (Hooker 1972). It has also been found that cell free mitochondrial preparations from plants with *Cms* T cytoplasm are inactivated when treated with race T pathotoxin, while the mitochondria of resistant plant are unaffected (Miller & Koeppe 1971, Petersen, Flavell & Barratt 1975). Resistance to race T is both cytoplasmic and nuclear, the nuclear genes probably being the same as those which determine resistance to race O. It is therefore still possible to use *Cms* T cytoplasm in appropriate crosses.

The rapid spread of race T was of great economic importance in the American corn belt because the *Cms* T cytoplasm, which also determines male sterility, had been almost exclusively used in hybrid maize production there. Total losses estimated at one billion dollars were caused during the years 1971 to 1973 and provide an unfortunate example in which the plant breeder, by building his programme on too narrow a basis, was the cause of a major disease problem.

Another example of a disease problem arising as a result of a breeding programme relates to the resistance of hybrid varieties of pearl millet (*Pennisetum typhoides*) to downy mildew, caused by *Sclerospora graminicola*. The male sterile variety used in the production of these varieties in India ('Tiftol 23a) was known to be susceptible to downy mildew, but experience had shown this susceptibility to be recessive to factors carried by the restorer varieties (S. C. Pokhriyal, personal communication). The first hybrid variety was released in 1965 and was followed by three others in 1967 and 1968, all of which were completely resistant when released. But a 10–12%

attack was reported from occasional crops in 1969, and the disease caused almost complete destruction of the crop in the Delhi region in 1970. By 1971 it had spread throughout northern India, and by 1972 through the country. As a result the national pearl millet yield which had increased from 3·66 to 8·00 million tonnes as a result of the introduction of the new hybrid varieties had fallen back to 3·65 million tonnes by 1973.

The susceptibility of Tiftol 23a appears to be genetically determined, so that resistance can be introduced by crossing with resistant lines. Resistance has also been induced by irradiation and a new line, MS 5071a, is being used in the development of new hybrids. As in the case of maize, there is, however, an urgent need for greater diversity in sources of male sterility.

A new approach to breeding for resistance

Although the breeder has been successful in limiting losses due to pests and diseases in many crops, the rapid adaptability of the pathogen in response to the introduction of new resistant varieties has made it necessary to adopt a new approach in breeding for disease resistance. The necessity for such an approach first became apparent in breeding potatoes for resistance to blight (*Phytophthora infestans*). This disease first appeared in Europe in the 1840's, where it spread rapidly, causing widespread famine in Ireland and other areas where the potato had become the mainstay of the national diet. These losses were greatly reduced in the next few years as a result of the use of fungicidal sprays combined with the selection of varieties showing field tolerance to blight. But breeders and growers were not satisfied with this form of resistance, and an attempt was therefore made to introduce single genes giving a very high level of hypersensitive resistance, derived from the wild species *Solanum demissum*. This resistance was shown by Black *et al.* (1953) to be determined by four genes, designated R_1 to R_4, occurring singly or in groups. The gene R_1 was the first to be exploited, and the varieties into which it was introduced showed a level of resistance previously unknown, when grown in breeders' nurseries. But when these varieties were multiplied for commercial use, occasional pustules soon appeared on them, the infection from which rapidly spread and investigation revealed a new race of blight. A similar sequence followed the introduction of each new gene for resistance, and it has now been accepted that such new races will inevitably appear as a result of sexual recombination or asexual mutation of the pathogen. The use of R-genes has therefore been discontinued, and breeders have in recent years selected for horizontal resistance in the field, though interactions between varieties and isolates of the pathogen still occur (Caten 1974).

The current situation can perhaps be best illustrated by reference to the leaf diseases of cereals with special reference to yellow rust of wheat and powdery mildew in barley, the most important diseases of these two crops in Britain. In both cases, breeders had until recently concentrated their efforts on the use of major genes determining hypersensitive resistance, but experience with the breakdown of resistance in such varieties as Heine VII, Rothwell Perdix, Maris Beacon and Maris Templar amongst the wheats and Maris Badger, Sultan and Impala in the barleys has shown the limitation of this procedure and led breeders to seek other methods of disease control.

Selection of plants showing hypersensitive resistance had normally been based on tests carried out on seedling plants in the glasshouse, since it was found that such resistance was maintained throughout growth. The presence of minor genes modifying disease reaction cannot however be detected in plants exhibiting hypersensitive reactions. A first step in selecting for more durable resistance was therefore to eliminate the hypersensitive genes by discarding all plants showing resistance as seedlings. Susceptible plants were then selected for low disease incidence in the field. Such resistance was supposedly non-race-specific and polygenic in inheritance. But the procedure was less successful than had been anticipated because it was found that the adult plant resistance so selected was sometimes both race-specific and simply inherited. The situation was highlighted by the variety Joss Cambier, which, when originally released, exhibited an apparently desirable combination of seedling susceptibility and field tolerance. After a few years, however, a new and specific field race appeared, capable of causing serious damage to this variety, and indicating the fallibility of the selection procedure.

What, then, should the breeder do? A first approach has been to analyse the mechanism of rust infection and epidemiology, in an attempt to identify points at which the build up of infection can be limited. Russell *et al.* (1975) have shown that the number of uredospores deposited per unit area of leaf surface varies with variety, as does also the percentage germination of spores, and the proportion of germinated spores which successfully penetrate the stomata to form substomatal vesicles. There are also differences in the time between penetration and the production of the next generation of spores and in the number of spores produced per unit of leaf area. Selection for characters of this nature is clearly tedious, and impracticable at least in the early stages of a breeding programme, but the demonstration of geno-typic variation in these characters, which are unlikely to be race-specific in their expression, indicates the possibility of obtaining varieties with a high expression of durable resistance.

Alternatively, the breeder can observe the incidence of rust on varieties which have been exposed to infection for a long time. Lupton & Johnson (1970) reported such an analysis from which it was found that a number of older varieties notably Little Joss, released in 1912, and Squarehead's Master, selected from an old land variety, showed a good level of resistance, although they had been exposed to infection for fifty years or more, during many of which they had been widely grown. Crosses involving these varieties are now being extensively handled, though selection within them is difficult as they are very tall and have other undesirable agronomic characters.

Although it is generally considered dangerous to use simply inherited hypersensitive resistance alone in a breeding programme, such resistance may be valuable if combined with non-specific resistance. Direct selection of such lines is normally impossible, because the non-specific will be masked by the specific resistance. Such lines may, however, be obtained either by back crossing genes for specific resistance into a genotype showing good non-specific resistance, or by selecting hypersensitive sibs from advanced families in crosses segregating for this character in a non-race-specific background. An alternative approach was suggested by Johnson & Law (1975), who tested the rust reactions of 20 monosomic lines in the highly resistant variety Bersee. One of these lines, monosomic for the chromosome $5B^S$-$7B^S$ was much more susceptible than the euploid, and the nullisomic for this chromosome was more susceptible than the monosomic. Further studies showed that chromosome $5B^S$-$7B^S$ carried at least two genes for rust resistance, one on each arm, and that this resistance was also carried by the corresponding chromosome on related varieties. The cytological identification of a chromosome determining rust resistance is of considerable importance as it opens the possibility of transferring such resistance into varieties whose genetic backgrounds already confer high levels of rust resistance. Furthermore the transfer can be achieved using cytological techniques, thus avoiding the need for testing for one form of resistance in the presence of another.

Use of multiline varieties

As was pointed out in the introduction, plant pests and diseases are rarely of importance in a natural environment, because they cannot readily build up sufficient inoculum potential to cause serious harm. This suggests the possibility that cultivation of varietal mixtures might offer another means of limiting disease loss. The use of such a mixture was first attempted by Rosen (1949) who grew a mixed population of oats, resistant to races of crown rust (*Puccinia coronata*) and Victoria blight (*Helminthosporium victoriae*), and was formalized by Jensen (1952) who suggested that a multiline variety should consist of pure lines chosen for uniformity of appearance and other agronomic characters, but that each should contribute additional desirable genetic factors without detracting from the phenotypic uniformity of the composite. The theme was again taken up by Borlaug (1958) who suggested that wheat breeders should introduce genetic factors for each race separately into a single acceptable variety, by back crossing, and that a mixture of such varieties should be used as a composite. Although this principle was never applied in his Mexican programme, it led to the introduction of the Columbian yellow rust resistant composites Miromar 63 and Miromar 65 (Browning & Frey 1969). Frey, Browning & Simons (1975) demonstrate very effective control of brown rust in oats using a composite of nine isolines carrying varying levels of resistance to three races of *Puccinia coronata*. It has been suggested that the use of a multiline where each component carries a single gene might result in the synthesis by the parasite of complex races, virulent over a wide range of resistance genes. Frey, Browning & Simons (1975) have, however, found no evidence of this taking place.

The use of composite varieties raises a number of administrative problems in Britain, as the marketing of a mixture would contravene regulations under the Plant Variety and Seeds Act (1964). But there could be considerable advantages of such varieties in controlling powdery mildew in barley and possibly yellow rust in wheat, even if it was necessary to multiply the components separately and mix them shortly before drilling. Experiments are therefore being conducted on the epidemiology of varietal mixtures and back crossing programmes conducted to develop suitable isogenic lines.

Conclusion

The pest and disease problems with which the modern farmer or horticulturist has to contend arise almost entirely from the systems of monoculture on which modern society depends. The plant breeder has done much to alleviate these problems, although it must be agreed that he has on occasions created others, especially when his work has been conducted on too narrow a genetic basis. It must also be agreed that his scope for future action has been to some extent undermined by his policy of successive exploitation of single resistance genes to which the pathogens have in turn become virulent.

I suggest therefore that the breeder should select for polygenetically inherited disease resistance, accepting that it may be necessary to tolerate a low level of disease attack. At the same time he should not insist on genetic uniformity in his crop, but should investigate the possibilities of multilines or varietal mixtures as a means of controlling disease spread.

References

BINGHAM J. & LUPTON F.G.H. (1958) Breeding spring oats for resistance to frit-fly attack. *Ann. appl. Biol.* **46**, 493–7.

BLACK W. (1935) Studies on the inheritance of resistance to wart disease (*Synchytrium endobioticum*). *J. genet.* **30**, 127–46.

BLACK W., MASTENBROEK C., MILLS W.R. & PETERSEN L.C. (1953) A proposal for an international nomenclature of races of *Phytophthora infestans* and of genes controlling immunity in *Solanum demissum* derivatives. *Euphytica* **2**, 173–9.

BORLAUG N.E. (1958) The use of multilineal or composite varieties to control airborne epidemic diseases of self-pollinated crop plants. *Proc. 1st Int. Wheat Genet. Symp., Winnipeg, Canada*, 12–27.

BROWNING J.A. & FREY K.J. (1969) Multiline cultivars. *A. Rev. Phytopath.* **7**, 355–82.

CAMMACK R.H. (1961) *Puccinia polysora*. A review of some factors affecting the epiphytotic in West Africa. *Report of the 6th Commonwealth Mycological Conference, Kew.* 134–8.

CATEN C.E. (1974) Inter-racial variation in *Phytophthora infestans* and adaptation to field resistance for potato blight. *Ann. appl. Biol.* **77**, 259–70.

CRAIGIE J.H. (1944) Increase in production and value of the wheat crop in Manitoba and Eastern Saskatchewan as a result of the introduction of rust resistant wheat varieties. *Scientific Agriculture* **25**, 51–64.

FREY K.J., BROWNING J.A. & SIMONS M.D. (1975) Multiline cultivars of autogamous crop plants. *S.A.B.R.A.O. Journal I*, 113–23.

HOOKER A.L. (1972) Proceedings 6th meeting EUCARPIA Maize and Sorghum Section, Weihenstephan 1971, pp. 333–67.

JENSEN N.F. (1952) Intra-varietal diversification in oat breeding. *Agron. J.* **44**, 30–4.

JOHNSON R. & LAW C.N. (1975) Genetic control of durable resistance to yellow rust (*Puccinia striiformis*) in the wheat cultivar Hybride de Bersee. *Ann. appl. Biol.* **81**, 385–91.

KNIGHT R.L. (1954) Cotton breeding in the Sudan. Part II, Egyptian cotton. *Emp. J. exp. Agr.* **22**, 81–92.

LINK K.P. & WALKER J.C. (1933) The isolation of catechol from pigmented onions. *J. biol. Chem.* **100**, 379–84.

LUPTON F.G.H. & JOHNSON R. (1970) Breeding for mature-plant resistance to yellow rust in wheat. *Ann. appl. Biol.* **66**, 137–43.

MILLER R.J. & KOEPPE D.E. (1971) Southern corn leaf blight: susceptible and resistant mitochondria. *Science* **173**, 67–9.

PEDERSEN P.N. (1960) Methods of testing the pseudo-resistance of barley to infection by loose smut, *Usilago nuda* (Jens.) Rostr. *Acta agric. Scand.* **10**, 312–32.

PELHAM J., FLETCHER J.T. & HAWKINS J.H. (1970) The establishment of new strains of tobacco mosaic virus resulting from the use of resistant varieties of tomato. *Ann. appl. Biol.* **65**, 293–7.

PETERSEN P.A., FLAVELL R.B. & BARRATT D.H.P. (1975) Altered mitochondrial membrane activities associated with cytoplasmically-inherited disease sensitivity in maize. *Theor. appl. Genet.* **45**, 309–14.

ROSEN H.R. (1949) Oat parentage and procedures for combining resistance to crown rust, including Race 45 and Helminthosporium blight. *Phytopathology* **39**, 20.

ROSS H. (1966) The use of wild *Solanum* species in German potato breeding of the past and today. *Amer. Potato J.* **43**, 63.

Russell G.E., Andrews C.R. & Bishop C.D. (1975) Germination of *Erysiphe graminis* f.sp. *hordei* conidia on barley leaves. *Ann. appl. Biol.* **81,** 161–9.

Stakman E.C. & Piemesel R.J. (1917) A new strain of *Puccinia graminis*. *Phytopathology* **7,** 73.

van der Plank J.E. (1963) *Plant Diseases: Epidemics and Control.* Academic Press, New York & London.

The dynamics of plant diseases

R. C. SHATTOCK *School of Plant Biology, University College of North Wales, Bangor, Gwynedd*

Studies of the dynamics of plant diseases are concerned with changes in the size and structure of populations of plant pathogens and the factors affecting these changes. Plant diseases are caused by microorganisms and the amount of disease is frequently used as a measure of the size of populations because there are either too many organisms to count (bacteria and viruses) or the individuals cannot be distinguished (fungi) (Zadoks 1972).

In this paper some of the factors affecting changes in the size and structure of pathogen populations in natural and agricultural ecosystems are discussed but particular attention is given to *Phytophthora infestans* the late blight fungus of potato; some of the techniques that have been used to analyse the quantitative and qualitative fluctuations are described. Throughout, the word *virulent* is used to mean ability to produce disease on a resistant host (Day 1960, 1974); the word *aggressiveness* is used to describe the intensity of infection. Thus for disease to occur, a pathogen must be virulent but virulence may be associated with strong or weak aggressiveness (van der Plank 1975).

Population size

An approach to the quantitative aspects of the dynamics of epidemics of diseases including the factors determining how disease multiplies and spreads through populations of susceptible plants has been made by van der Plank (1960, 1963, 1965, 1975) whose unique method is a landmark in plant pathology. van der Plank has analysed the data on epidemiology of plant pathogens on a mathematical basis; he has given particular attention to the infection rates of different pathogens and the relationship between the amount of inoculum and the amount of disease it produces. He showed that

83

many of the features of plant disease epidemics such as foliar rust diseases of cereals and late blight of potato with S-shape growth curves (Fig. 1) can be described with some slight modifications by the mathematics of continuous compound interest. The amount of disease at any one time (x) can be calculated from the equation $x = x_0\, e^{rt}$, where $x_0 =$ the amount of disease initially present (inoculum), $r =$ the rate at which infection spreads, $t =$ the time during which infection has occurred.

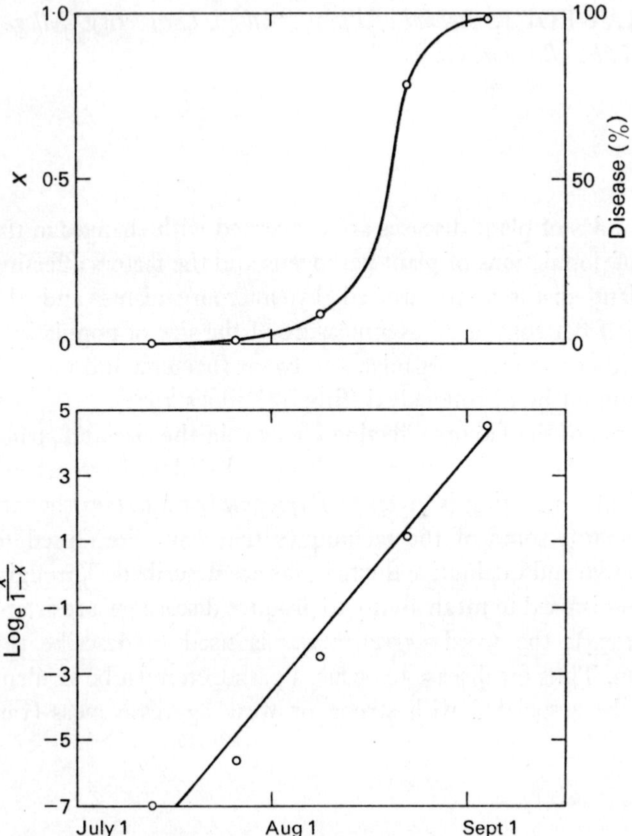

Figure 1.   Progress of an epidemic of potato blight (van der Plank 1965).

van der Plank (1963, 1965) has shown that increase of disease with time, the apparent infection rate r, is defined as $dx/dt = rx(1-x)$. The rate dx/dt is related both to the proportion of tissue that is infected (x) and to the proportion of healthy tissue available for infection ($1-x$). Fig. 1 shows that the progress of late blight disease in a potato crop during an epidemic is sigmoid. If $\log_e x/(1-x)$ is used instead of x as the measure of the amount of disease, then the rate of increase in disease, r, is the regression coefficient with time in days; in this case $r = 0.22$ per day (Fig. 1). As van der Plank (1965) has shown, this is no more than an estimate of the average value over the whole period. However, there is no reason why r should stay even roughly constant over a period of observations and it is therefore best estimated over short periods of time. van der Plank (1965) calculated that for late blight of potato and rust diseases of cereals r is commonly between 0.1 and 0.7 per day producing a doubling of the amount of disease in 7 and 1 days respectively in the early stages of an epidemic. In contrast, disease in perennial crops spreads more slowly. van der Plank calculated that in chestnut blight (*Endothia parasitica*) in the United States of America in any particular area during the early stages of attack $r = 1.9$ per year.

van der Plank's models adequately describe the rate of increase of certain diseases and are particularly accurate early in epidemics when there is a logarithmic increase in disease. Later, the reduction in healthy tissue available for infection affects the accuracy and usefulness of r but nonetheless the apparent infection rate remains a sensitive index and easy to estimate because it relates to the proportion of tissue that is infected. Many diseases, however, are discontinuous because the rate of disease progress can be affected by any component of the disease triangle—host, pathogen and environment—producing for example, intermittent infection or delay in sporulation. Newly infected tissue takes a period of time to become infectious and this is described as the latent period (p). Thereafter the diseased tissue, e.g. lesion or canker, remains infectious only for a period (i). van der Plank (1963, 1965, 1975) describes simple models incorporating p and i and then defines the basic infection rate (R).

Control of polycyclic diseases which increase exponentially with time can be achieved by reducing either x_0 (the initial inoculum), r (the rate at which plants are infected) or t (the time during which infection takes place). In practice either x_0 is reduced by cultural methods, e.g. reducing over-wintering sources of the pathogen or using major genes for resistance; or attempts are made to reduce the rate of infection r by employing polygenic (minor gene) resistance and using fungicides.

Other diseases are monocyclic and the rate of increase in disease resembles the increase of money at simple interest. Examples include *Ustilago* spp. producing loose smut of small grain cereals and root pathogens such as *Fusarium oxysporum* f. sp. *vasinfectum*, wilt of cotton. Here plants are infected from inoculum in the soil or carried on or in the planting material and the pathogen characteristically does not spread between plants during the growing season. Serious problems may arise, however, with monocyclic diseases if susceptible crops are repeatedly planted in the same place year after year allowing soil borne pathogens to increase dramatically.

Some root pathogens normally considered monocyclic can become poly-cyclic through mismanagement or the introduction of new techniques. Overwatering can lead to severe damage by *Pythium* spp. in seedling and nursery stock, whilst propagules of *Phytophthora* spp. and other pathogens are readily distributed amongst plants by irrigation water (Rotem & Palti 1969). Similarly *Pythium* spp. can cause total crop failure in cucurbits growing in thin film hydroponic culture systems (Cooper 1975).

Much use has been made of computers in the past decade to analyse disease epidemics and alternative models that have adapted, amended, expanded or replaced those of van der Plank (1963, 1965) are discussed by Kranz (1974b). Computers have also been used to simulate epidemics, forecast disease outbreaks and estimate crop losses (Butt & Royle 1974, Kranz 1974a & b, van der Plank 1975, Waggoner 1974, Zadoks 1971). Both disease forecasting and estimation of crop loss have been analysed by a multiple regression (MRA) technique (Butt & Royle, 1974). Schrodter & Ullrich (1965) used MRA to forecast outbreaks of late blight (*P. infestans*) in certain areas of West Germany and were able to ascribe approximately 56% of the variation of epidemic progress of the disease to two meteoro-logical variables, temperature and rainfall. James *et al.* (1972) used (MRA) to predict the effect of *P. infestans* on the yield of potato tubers in Canada. Some forecasting rules are entirely empirical and derived from simple weather observations, e.g. for potato late blight (Beaumont 1947, Smith 1956), for apple scab *Venturia inaequalis* (Mills & La Plante 1954).

In many ways the population dynamics of plant pathogens are only just beginning to be understood and the development of quantitative methods (Zadoks 1972), and more recently modelling (Kranz 1974a), are valuable tools to assist interpretation of the dynamics of plant diseases. However, studies of the population biology of microorganisms responsible for plant diseases have lagged behind those of pests and parasites associated with human and animal diseases and of plants and animals generally. Indeed, Zadoks (1972) states, 'phytopathology in general and . . . epidemiology in particular are service rendering sciences. An epidemic must be stopped; the question whether it will be understood ranks second'. Typically, control chemicals have usually been applied long before their mode of action was elucidated; the same is true of many of the resistance factors incorporated into crop plants since little is known of the genetics of many hosts and pathogens and their interactions. An understanding of the population biology of plant pathogens is necessary *per se* but absolutely essential if control of epidemics is to be achieved with restricted application of control chemicals by choice or necessity if resistant or tolerant pathogen strains should occur and if genetic sources of resistance are to be best employed.

Population structure

Genetical variation

Flor (1955) studied the genetics of both members of the host–pathogen interaction between flax (*Linum usitatissimum*) and flax rust (*Melampsora lini*) and proposed that for each resistance gene in flax there is a corresponding gene in the rust fungus for virulence. This is the gene-for-gene concept (Flor 1955, 1956, Person, Samborski & Röhringer 1962). The hypothesis has since been extended either by direct genetical evidence, e.g. covered smut (*Ustilago hordei*) of barley (Sidhu & Person 1971, 1972), or indirectly from epidemiological data, e.g. late blight (*P. infestans*) of potato (Toxopeus 1956, Person 1959) and downy mildew (*Bremia lactucae*) of lettuce (Crute & Johnson 1976), to many other host–parasite combinations including other fungi, both obligate and non-obligate biotrophs, bacteria, viruses, nematodes, insects and higher plant parasites (Flor 1971, Day 1974). Host resistance and avirulence in the pathogen are generally dominant whilst virulence is a recessive character. Loegering & Powers (1962) showed that four possible interactions occur between a pair of corresponding genes governing resistance in the host and virulence in the pathogen. This is called the quadratic check and for biotrophic fungi such as the rust fungi and potato

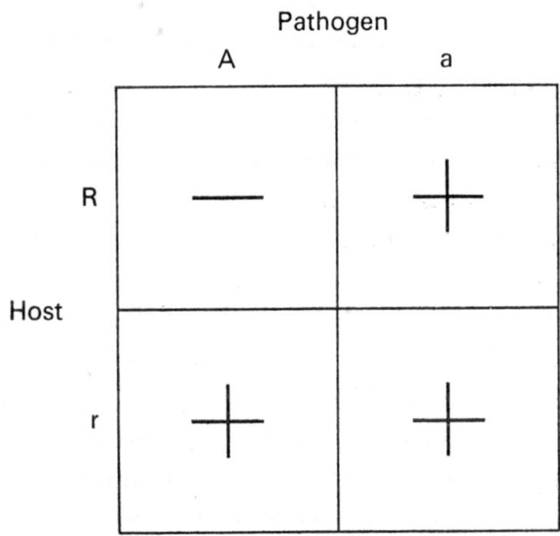

Figure 2. The quadratic check. Resistance R is dominant to susceptibility r; avirulence A is dominant to virulence a. + infection, − no infection.

D

late blight means that three of the four interactions will result in susceptibility (Fig. 2). This is because resistance is only expressed when the resistance gene is matched by the corresponding gene for avirulence in the
pathogen. Röhringer *et al.* (1974) and Howes, Samborski & Röhringer
(1974) have shown that host recognition of an avirulent strain of wheat
stem rust (*Puccinia graminis tritici*) is due to a specific ribose nucleic acid
(RNA) which is a product of the gene for avirulence. It has been suggested
that deletion of such genes for avirulence, and thus loss of synthetic ability,
may be the result of mutations in the rust fungi since according to the
gene-for-gene concept these mutations produce virulent races. In disease
caused by biotrophs, resistance appears to be an induced response. In other
diseases, especially those in which a selective pathotoxin is produced by
virulent strains of a pathogen, e.g. *Helminthosporium victoriae*, it is proposed
that products of genes for virulence are essential for susceptible reactions.
This is induced susceptibility (Wheeler 1975).

Population studies of *P. infestans*

Although *P. infestans* can easily be cultured *in vitro*, ecologically it behaves
like the rust and the downy and powdery mildew fungi as an obligate
biotroph. Races of *P. infestans* are named by their capacity to infect a range
of host genotypes (differentials) each possessing a different single major
R-gene for resistance to late blight (Black *et al.* 1953). Since eleven
such R-genes have now been identified the theoretical number of possible
different races is 2^{11}. These R-genes occur naturally in *Solanum demissum*,
a wild Mexican species and have since been bred into *S. tuberosum* (Malcolmson & Black 1966) and resistant plants respond to infection with a hypersensitive reaction. Resistance is characterized by disorganization and
death of cells at the infection site. Some commercial cultivars possess
specific monogenic R-genes, e.g. Orion (R1)*, Maris Peer (R1R2) and
Pentland Dell (R1R2R3). These cultivars also possess different levels of
general polygenic resistance which is assumed to be conditioned by many
minor genes and therefore can be viewed as the counterpart of aggressiveness
in the pathogen. General polygenic resistance probably acts by limiting the
rate of infection, growth and sporulation of the fungus. Cultivars of potato
lacking R-gene resistance and with a low level of polygenic resistance include
King Edward and Dunbar Standard whereas Majestic and Golden Wonder
exhibit a high level. Polygenic resistance is affected by variation in environmental conditions and thus cultivars of potato exhibiting a range of this
type of resistance from low to very high may all be severely attacked in

* R-genes.

seasons conducive to rapid spread of late blight. Resistance conferred by R-genes is not normally affected by environmental conditions and cultivars with R-genes, e.g. Pentland Dell, can only be attacked by races of the fungus with the corresponding virulence characters 1, 2 and 3.

R-gene resistance, however, has not been effective in controlling late blight throughout the potato growing countries of the world. For example, in the United Kingdom the cultivar Pentland Dell (R1R2R3) was released in 1961 and planted by many growers. Since it carried three major genes for resistance to late blight, routine spraying with protectant fungicides stopped. After two or three seasons this cultivar suffered attacks from *P. infestans* (Malcolmson 1969) and it has been necessary to re-introduce routine fungicide application on this cultivar throughout the growing season. Malcolmson (1969) collected and identified samples of *P. infestans* from infected crops of Pentland Dell and other cultivars throughout the United Kingdom during 1966, 1967 and 1968. She identified 70 different races, 23 of which occurred on Pentland Dell. A more recent survey of late blight races in N. Wales from 1970 to 1973 (Janssen 1973, Shattock 1974) also revealed a large number of different phenotypes in the populations of *P. infestans* many of which were widely virulent. These widely virulent races were identified by their capacity to attack many of the differential host genotypes containing the R-genes 5–11 although these resistance genes have never been introduced into commercial cultivars. The evidence from these two surveys suggests that the presence of virulence characters in populations of *P. infestans* is independent of the presence of the complementary R-genes in the host population. A similar situation occurs in wheat stem rust populations in New Zealand and Kenya (Martens 1975) and oat stem rust populations in Canada, and it must be assumed either that the virulence genes are selectively neutral, selectively advantageous as a result of their pleiotropic action or selected for other attributes which determine aggressiveness.

The evidence from the *P. infestans* potato investigations contrasts with data from other host–pathogen combinations such as wheat and stem rust (*Puccinia graminis tritici*) in Australia where the accumulation of genes in wheat for resistance to stem rust has been paralleled by the evolution by directional selection of complex pathogen races with complementary virulence (Luig & Watson 1970). Other examples of increased virulence following the introduction of major genes for resistance include barley powdery mildew (*Erysiphe graminis* f.sp. *hordei*) (Moseman 1966), tomato leaf mould (*Cladosporium fulvum*) (Kerr & Baily 1966), and lettuce downy mildew (*Bremia lactucae*) (Crute & Johnson 1976, Crute & Davis 1976).

Although race surveys allow identification of phenotypes present in pathogen populations the most important information is the frequency of individual virulence characters (Wolfe *et al.* 1973, Day 1974, van der Plank 1975). Wolfe & Schwarzbach (1975) have shown that for a haploid pathogen such as *E. graminis* where phenotype and genotype frequencies are the same, the expected frequency of virulence gene combinations can be determined simply as the product of the individual frequencies of the virulence genes, if the pathogen populations sampled are large and reassortment of virulence characters is random and independent. For pathogens of higher ploidy, e.g. the dikaryotic rust fungi and diploid *Phytophthora* spp. including *P. infestans* (Sansome & Brasier 1973), some further assumptions must be made notably that virulence alleles are either completely dominant or completely recessive to the non-virulence alleles and that the pathogen populations are in or close to Hardy–Weinberg equilibria (Wolfe *et al.* 1976). Any departure from the expected frequency is a measure of the non-independence of the virulence factors. The frequency of particular races can also be estimated by multiplying the product of the individual virulence frequencies by the product of the frequencies of the remaining known characters for avirulence. Table 1 shows the relative frequencies of four virulence characters in three race surveys of the pathogens *P. infestans* and

Table 1. (a) Relative frequencies of four common virulence factors occurring in pathogen populations in potato late blight (Malcolmson 1969 and Shattock 1974) and oat stem rust (Roelfs & Rothman 1974).

| Potato late blight | | Potato late blight | | Oat stem rust | |
Virulence	f	Virulence	f	Virulence	f
V_4	0·39	V_3	0·78	V_2	0·62
V_7	0·99	V_4	0·78	V_4	0·65
V_{10}	0·77	V_{10}	0·56	V_8	0·82
V_{11}	0·21	V_{11}	0·57	V_9	0·21
No. of isolates 146		54		447	

f = frequency in the population.

(b) Observed and estimated frequencies of all possible combinations of pairs of the virulences given above.

| Potato late blight | | | Potato late blight | | | Oat stem rust | | |
	obs.	est.		obs.	est.		obs.	est.
V_4V_7	0·38	0·39	V_3V_4	0·67	0·61	V_2V_4	0·61	0·40
V_4V_{10}	0·34	0·30	V_3V_{10}	0·46	0·44	V_2V_8	0·51	0·51
V_4V_{11}	0·08	0·08	V_3V_{11}	0·50	0·45	V_2V_9	0·15	0·13
V_7V_{10}	0·77	0·76	V_4V_{10}	0·46	0·44	V_4V_8	0·50	0·53
V_7V_{11}	0·20	0·21	V_4V_{11}	0·44	0·45	V_4V_9	0·19	0·14
$V_{10}V_{11}$	0·14	0·16	$V_{10}V_{11}$	0·41	0·32	V_8V_9	0·04	0·17

Puccinia graminis avenae (oat stem rust) and all possible combinations of the pairs of virulence.

Virulence analysis allows the best understanding of pathogen variation where observed and expected values differ significantly as in the case in the combination V8V9 in oat stem rust (Table 1), and may allow plant breeders to exploit this deficiency by selection of host lines with corresponding resistance genes (Wolfe *et al.* 1976). Different methods of virulence analysis in populations of pathogens are described by Wolfe & Schwarzbach (1975). One method, the use of mobile nurseries, has been used by Wolfe & Minchin (1976) to assess the variation in field populations of *Erysiphe graminis* f.sp. *hordei*, powdery mildew of barley.

Several of the individual virulence characters in the potato blight populations (Table 1) occur at high frequencies and consequently result in widely virulent races occurring in these pathogen populations (Janssen 1973, Malcolmson 1969, Shattock 1974). These results do not support the hypothesis of stabilizing selection (Fig. 3) proposed by van der Plank (1968),

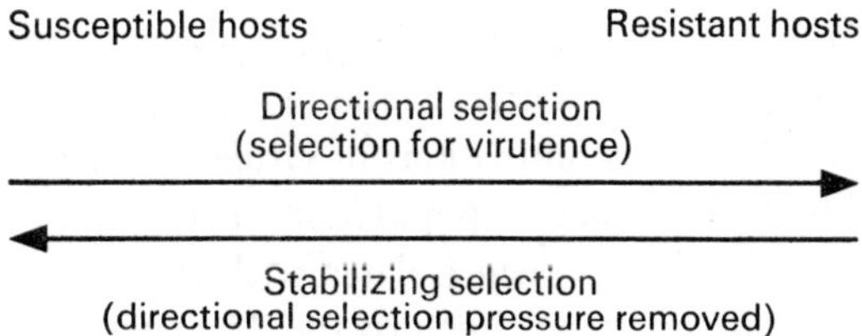

Figure 3. The opposite processes of directional selection in favour of more virulent races in the population of the pathogen when it is presented with resistant hosts, and of stabilizing selection with the loss of unnecessary virulence when the resistant hosts are removed (van der Plank 1968).

based on field data available at the time, of the competitive ability of isolates. According to this concept an inverse relationship exists between certain unnecessary genes for virulence and the fitness of the pathogenic strain to survive in populations. In order to explain exceptions to this relationship van der Plank proposed two types of resistance genes—strong and weak. Stabilizing selection only acts on virulence genes that are complementary to strong resistance genes. In potato the R-genes, 1, 2 and 3, are described by van der Plank (1968) as strong but the R4 gene as weak. If correct then the R-genes 5 11 would also be weak and this situation would account for the high frequency of highly virulent races in the populations of the late blight fungus. However, the universal occurrence of stabilizing selection has generally been disputed by data from many different fungal pathogens (Watson 1970, Nelson 1972, Brown 1975). Since unnecessary virulence does not appear to be disadvantageous it may be that certain virulence characters such as 7, 10 and 11 in the late blight populations are being carried in geno-types strongly selected for aggressiveness, competitive ability or survival ability.

The failure of cultivars with R-genes to remain free of blight has dis-couraged plant breeders from using single major gene resistance against *P. infestans* and the emphasis has shifted towards general polygenic resistance. van der Plank (1963, 1968) called this latter type of resistance horizontal resistance because no differential interaction occurred between the various isolates of different pathogens and their respective hosts. In the potato–late blight interaction van der Plank (1971) concluded from field data and the experimental data of Paxman (1963) that horizontal resistance was stable. However, Jeffrey *et al.* (1962), Upshall (1969) and Caten (1974) have pre-sented evidence to show that adaptation by *P. infestans* to cultivars carrying general polygenic resistance may occur. Similar evidence is presented by Johnson & Taylor (1972) who identified isolates of *Puccinia striiformis* (yellow rust of wheat) specifically adapted to the cultivar Joss Cambier but otherwise indistinguishable from previous isolates of the pathogen with similar virulence and by Habgood (1976) for isolates of *Rhyncosporium secalis* (leaf blotch of barley) to the cultivar Maris Otter. Some of the deviations between observed and expected values in combinations of virulence charac-ters revealed by virulence analysis of pathogen populations may indicate the occurrence of differential interaction between isolates with these viru-lence characters and general polygenic resistance in cultivars. The specificity of these pathogen strains differentially adapted to particular cultivars suggests that the gene-for-gene concept may extend to genes controlling polygenic resistance and pathogen aggressiveness (Person, Samborski & Röhringer 1962). However, little is known about the genetics of either system in most hosts or pathogens, but, one recent study of polygenic inheritance of aggressiveness in covered smut (*Ustilago hordei*) of barley (Emara & Sidhu 1974) revealed that the polygenes determining aggressive-ness act as modifiers of a particular virulence gene on a susceptible host.

The precise distinction made by van der Plank (1968, 1975) between vertical and horizontal resistance and the type of gene involved has been questioned by other workers. McIntosh, Luig & Baker (1967) and Knott (1968) showed that general polygenic resistance exhibited by certain wheat cultivars to stem rust (*P. graminis tritici*) can be accounted for by the joint action of several major genes for resistance. If these genes operate at different times and in different ways the phenotypic response of the host will appear to be quantitative rather than qualitative. Indeed, Wolfe (1972a), in dis-cussing the genetics of barley powdery mildew suggests that 'the complexity of the interaction between host and pathogen suggests that there may be no fundamental difference between resistance controlled by major or minor genes'.

To understand the population biology of plant pathogens it is desirable to understand the biology of the individuals which make up that population. At present, however, the physiology and genetics of many species are imperfectly understood. For example, there is still controversy on whether fungi of the class Oomycetes, which contains *P. infestans* and the downy mildew fungi, are haploid or diploid in their somatic phase. Also it is still unknown whether the determinants of virulence are nuclear or extra-chromosomal. Furthermore *P. infestans* is asexual throughout much of its geographical range including the United Kingdom and variation must arise by asexual mechanisms such as mutation, heterokaryosis and mitotic recombination. Recently virulent mutant strains of *P. infestans* resistant to, and dependent upon antibiotics have been isolated and may prove to be useful in genetical and physiological studies (Shattock & Shaw 1975, 1976). To study variation in virulence, gene flow and differential adaptation to particular cultivars lacking major gene resistance in populations of *P. infestans* use can be made of mobile walk-in polyethylene tunnels (Shattock 1976a). Several crops of potatoes can be grown each season using these structures and they also allow late blight to develop and continue to infect crops for a long period of time whilst containing disease spatially and pro-tecting the infected crops from outside sources of late blight.

In an epidemiological situation it is less important to consider how novel phenotypes arise than it is to understand the factors affecting the relative ability of strains of pathogens to survive in populations. Brown (1975) described some of these factors in populations of the same species of fungal pathogens. Firstly he pointed out that whilst plant breeding activities and widespread adoption of novel host genotypes have accelerated the evolution and selection of new strains of pathogens not all the changes in strain composition of pathogen populations are the result of changes in the genotype of the host population. Secondly the intraspecific competition and the effect of stabilizing selection (*sensu* van der Plank 1968) will affect the survival of different strains within populations. Intra-specific variation in the efficiency of infection and reproductive capacity is also likely to be important. Some of this variation may be due to environmental ecotypes which have different temperature optima, e.g. *P. graminis tritici* (Katsuya & Green 1967) and *P. infestans* (Martin 1949). Temperature may also affect the expression of cer-tain genes for resistance and therefore the survival ability of certain patho-genic isolates. For example, the Sr6 gene for resistance to stem rust in wheat is thermolabile and is ineffective above 24°C. Finally, laboratory experiments have shown that inter- and intra-specific interaction at infection sites can induce either cross-protection or increased susceptibility. The importance of these different factors, in particular the importance of stabilizing selection in different host–pathogen systems (van der Plank 1968), remains to be evaluated. Knowledge of these factors may lead to methods of directing and managing pathogen evolution by manipulation of host genotypes in space and time and thereby introduce stability into the host–parasite interaction in agricultural ecosystems (Knott 1972, Nelson 1972). One approach lies in adapting epidemic simulation models such as EPIDEM (Waggoner & Horsfall 1969) and EPIMAY (Waggoner, Horsfall & Lukens 1972) or developing future models to predict the relative survival ability of different strains in pathogen populations (Ogle, Taylor & Brown 1973, Waggoner 1974).

Disease in natural and agricultural ecosystems

Disease occurs in natural plant communities but seldom reaches epidemic proportions. This is because a dynamic equilibrium exists between the host and its parasites, their co-evolution producing parasites that are not so aggressive that hosts are destroyed and equally the host does not become so resistant that the parasite is destroyed.

The greatest stability between hosts and their parasites is likely to exist in the centres of origins of plants where host and parasite have been together for the longest period of time. Consequently the primary and secondary gene centres of cultivated plants are the best places to find genuine resistance to common diseases and insect pests (Leppik 1970).

The introduction of exotic pathogens into both natural and agricultural plant communities has resulted in devastating disease epidemics. This is because the host and pathogen have been disassociated for a long time. For example, the cultivated potato *S. tuberosum* was domesticated in the Andes, whereas the gene centre for the late blight fungus is Mexico and Guatemala on the weedy species *S. demissum* (Niederhauser & Cobb 1959). When the fungus was introduced into the north-eastern states of America and western Europe in the mid-19th century (Bourke 1964) the cultivars at that time were extremely blight susceptible and resulted in mass famine as witnessed especially in Ireland (Large 1940, Woodham-Smith 1963). Similarly the gene centre of the genus *Vitis* is in North America and all the species are highly resistant to downy (*Plasmopara viticola*) and powdery mildew (*Uncinula necator*) and also the root pest *Phylloxera devastratix*, but the cultivated grape and its progenitor are Eurasian in origin and introduction of these parasites has resulted in the past in catastrophic epidemics in European vineyards (Leppik 1970).

Leppik (1970) also describes the epidemics in natural perennial tree species in North America from imported fungal pathogens, e.g. white pine blister rust (*Cronartium ribicola*), chestnut blight (*Endothia parasitica*) and Dutch elm disease (*Ceratocystis ulmi*). A highly aggressive form of the latter disease has recently been introduced into the United Kingdom on infected elm logs from Canada and is largely responsible for the large-scale epidemic in England since 1970 (Gibbs, Heybroek & Holmes 1972, Brasier & Gibbs 1973, Gibbs & Brasier 1973, Brasier & Gibbs 1975, 1976). This particular disease highlights how strains of a pathogen of greater aggressiveness and virulence can evolve in secondary gene centres and produce major disease epidemics alone or after recombination with local endemic strains of the pathogen. Although the fungus *C. ulmi* can spread from tree to tree by root contact, the beetle vectors *Scolytus scolytus* and *S. multistriatus* are also agents of dispersal. An understanding of the population biology of vector borne pathogens including many plant viruses must be a function of the population biology of the vector (Thresh 1975). As well as changes in the biology of Dutch elm fungus (*C. ulmi*) affecting the present status of Dutch elm disease in the United Kingdom the importance of abiotic factors should be taken into account, e.g. mild winters affecting the numbers of beetles larvae and adults surviving the winter and the recent falling watertable putting extra water stress on infected trees. Since pathogens have tended to become selective among their hosts, and hosts have developed resistance against their special parasites, it must be assumed both the elm and *C. ulmi* in the case of Dutch elm disease will once again reach an evolutionary balance. At which point the present outbreak of Dutch elm disease is in the grand cycle of disease (Fig. 4) is impossible to judge and the presence of a vector complicates the matter.

In agricultural ecosystems, the balance established in nature rarely persists. The previous examples illustrate that when quarantines fail and exotic pathogens are introduced then epidemics can ensue. But more

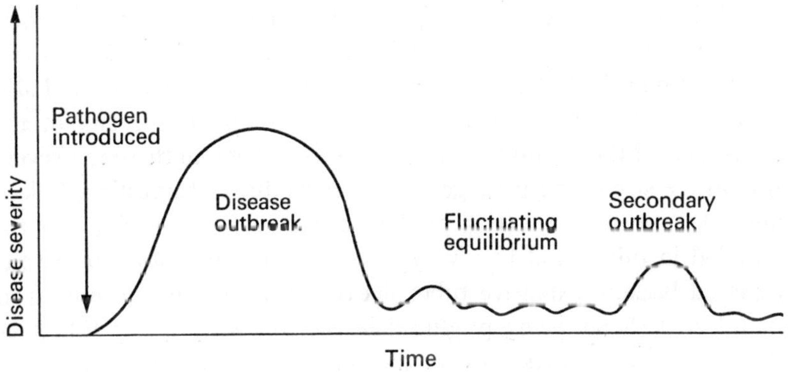

Figure 4.   The grand cycle of disease (Baker & Cook 1974).

importantly the attempts by plant breeders to limit (stabilize) disease by introducing resistance factors often obtained from gene centres of cultivated plant species (Leppik 1970) often fail as a long-term control of endemic diseases. Of course, if success in the short term is the aim of plant breeders then breeding for resistance has been successful but this implies unending genetic resources of plants. This unfortunately, is unlikely to be the case. Indeed Harlan (1972) argues strongly that 'the activities of modern plant breeders since 1900 have led to genetic "erosion" by assembling landrace populations and selecting pure lines from them. Far worse is the genetic "wipeout" that has occurred in the post-modern period with Mexican wheats, for example, washing over Asia with astonishing speed replacing major centres of diversity of wheat and the subsequent loss of landrace populations in Asia.' This is not so serious if the germ plasm of indigenous landraces in centres of origin has been collected, catalogued and maintained in gene banks, but Harlan (1972) maintains that the collections are 'grossly inadequate for the burden they will have to bear'.

Stability between host and parasite is possible in agricultural ecosystems since it has already existed in the cultivated landrace populations which are the direct ancestors of modern crop plants. These landrace populations of crop plants that became important over a large area were transported far from their centres of origin and established secondary centres of diversification. Variation is often very high in the latter (Leppik 1970) but nonetheless the plant populations became adaptively integrated and reached a genetic balance with local races of parasites.

In present day cropping systems stability between hosts and their parasites may develop if a certain amount of disease is acceptable. For example, the summer outbreaks of late blight in potato arise from primary foci of disease usually in the form of infected shoots arising from diseased tubers remaining by chance in the ground over the winter period (ground-keepers), in discard (cull) piles or in seed stocks. During blight epidemics there is likely to be directional selection for highly aggressive isolates, characterized by several factors such as faster leaf penetration, higher growth rate of mycelium in the leaf and rate at which sporangia are formed and also larger numbers of sporangia per unit area. However, at the end of the growing seasons of the potato crop in the United Kingdom there is a severe reduction in the size of the pathogen populations due to harvesting, lack of non-infected tissue and frost (Fig. 5). The fungus can only survive over the winter period in tubers and highly aggressive isolates may not survive the winter period because extensive tuber decay will either kill the fungus or prevent infected shoots being produced in the following spring (Shattock 1976b). Thus selection pressures on the fungus appear to be different at different stages of its life cycle and the end result should be that only moderately aggressive strains of the fungus predominate over a long time. It is possible that some of the highly aggressive strains developing during seasonal epidemics of late blight are those strains adapted to the general polygenic resistance of particular cultivars. Reduction in directional selection for highly aggressive strains of *P. infestans* over the summer period should

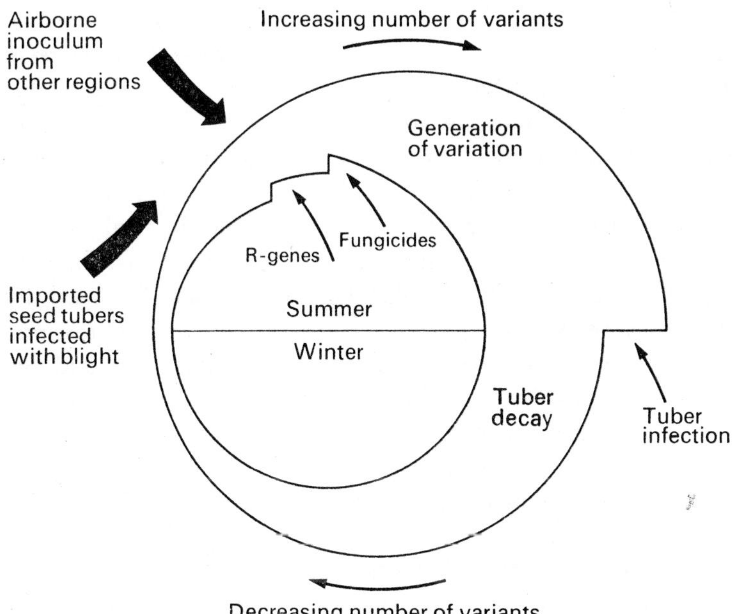

Figure 5. The possible factors affecting variation and selection at different stages in the life history of *Phytophthora infestans* in the United Kingdom.

also be possible if the genetic base of polygenic resistance in the host population is sufficiently wide and this could be achieved by making sure a sufficient number of cultivars with different polygenic resistance are available and planted each year. If this assumption is correct then selection is operating against excessive aggressiveness. For this selection to be effective on the basis of the above hypothesis it is essential that a particular crop should have a wide genetic base for polygenic resistance over the whole range of cultivars. However, the genetic base of many of our crop plants has narrowed considerably over the past 75 years (Harlan 1972, Horsfall *et al.* 1972, Day 1973). Nowhere is this better illustrated than in the United Kingdom winter wheat crop. The cultivar Maris Huntsman, because of its very high yield potential, occupied between 40 and 50% of the total winter wheat acreage in the United Kingdom in 1974–75 and it is also widely grown in France. Furthermore, even amongst the other thirteen winter wheat cultivars recommended by the National Institute of Agricultural Botany,

Cambridge for 1975 (Anon. 1975) nine have the cultivar Cappelle Desprez as a parent (Macer 1975).

The activities of the plant breeders have affected the distribution and frequency of virulent and aggressive strains of pathogens particularly fungal airborne foliar pathogens such as late blight (*P. infestans*) of potato, powdery mildew (*E. graminis*) of cereals and yellow rust (*Puccinia striiformis*) of wheat in the United Kingdom. In particular the use of major gene resistance and lack of genetic variability in the host populations has forced the evolution of plant pathogens by directional selection for greater virulence and probably also greater aggressiveness (Johnson 1961, Wolfe 1972b, Macer 1975). In the absence of stabilizing selection curbing the shifts to increased virulence in *P. infestans* and various species of *Puccinia* on cereals (Nelson 1972, Macer 1975) subsequent breeding programmes must be directed towards polygenic resistance and increased diversity of the genetic base in crop plants. One approach to diversification has been to use multiline cultivars which are heterogeneous as far as resistance factors are concerned (Browning & Frey 1969). However, the resistance factors are major genes and the attempts to produce stability may be unrewarding since simultaneous exposure of many resistance genes may lead to the development of super-virulent strains. In the absence of stabilizing selection (*sensu* van der Plank 1968), these latter strains will be no less competitive than less virulent strains especially if the background general polygenic resistance of the blends of the different genotypes of the multiline is the same.

With the exception of a few studies such as that of Leonard (1969) on stem rust (*P. graminis avenae*) in mixed plantings of oats and the work of Sumner & Littrell (1974) on southern leaf blight of maize (*Helminthosporium maydis*) in mixed populations of susceptible and resistant maize, there is generally a lack of experimental data on the dynamics of disease in heterogeneous host populations including not only agricultural multilines but also natural plant communities. Levels of disease are generally less in multilines than in monocultures and consequently higher yields are achieved in the presence of pathogens. Both Leonard (1969) and Sumner & Littrell (1974) concluded that it is the resistant plants alone in the plant mixtures that reduce the rates of increase in disease by reducing the number of effective dispersals. However, studies by Burdon & Chilvers (1975a & b, 1976) with the soil borne damping-off fungus *Pythium irregulare* in small mixed populations of susceptible garden cress *Lepidium sativum* and a resistant ryegrass *Lolium rigidum* demonstrated that the rates of both the multiplication and advance of the disease were determined mainly by the net density of susceptible seedlings and were little affected by the resistant plants present. Burdon & Chilvers (1976) comment that, although their results are based on controlled experiments with a soil borne fungus, 'the reduction in epidemic rates in multilines will be shown to be mainly a function of susceptible plant density'. Elsewhere they point out that reducing the numbers of primary disease foci of a soil fungus by lowering the number of host plants may not be acceptable agriculturally but in a natural plant community this kind of effect could be more important (Burdon & Chilvers 1975b).

Thus, it seems probable that both density-dependent factors as well as genetic feedback contribute to the equilibrium of hosts and pathogen in natural plant communities and the subsequent low levels of disease.

A better understanding of plant pathogen populations in agricultural cropping systems may come from studying the population biology of plant pathogens in natural plant communities. Several investigations into the biology of plant pathogens, particularly virulence and aggressiveness, have been conducted in centres of origin of cultivated plants, e.g. *P. infestans* on solanaceous hosts in Mexico (Niederhauser, Cervantes & Servin 1954, Niederhauser & Cobb 1959) and powdery mildew on graminaceous hosts including cereals in Israel (Eshed & Wahl 1970). But even fewer studies have been undertaken on specialized parasites in natural plant communities probably because only in the most important crop plants is the genetic behaviour of the host known and without a knowledge of the genetics of resistance the virulence characteristics cannot be identified in the parasite. Where studies have been conducted they have been directed towards investigating the effect of host-specific parasites on plant competition between co-dominant species (Chilvers & Brittain 1972, Burdon & Chilvers 1974) or the relationship between a host's recombination system and the parasite pressures it is likely to experience (Levin 1975). In the latter study Levin (1975) points out that pest and pathogen pressures are usually less severe in colonizing or fugitive species (r-selected species) which continually escape to new areas, whereas species constituting climax communities (K-selected species) must rely on a genetic strategy (open recombination systems) to resist parasite pressure by major genes for resistance. In returning to the gene-for-gene concept it has yet to be proved that the concept applies to host–parasite interactions in natural plant communities; it is possibly an artifact of breeding for resistance where major genes for resistance have been stripped of associated protective polygenic resistance (Day 1974). However, Day (1974) points out that the pioneer studies of Goldschmidt (1928) on virulence of anther smut (*Ustilago violacea*) to various *Melandrium* (*Lynchnis*) spp. showed that it was under simple genetic control.

An interesting study would be to look at pathogen specialization in plant species that are obligate apomicts. For example, *Taraxacum officinale* is one such case and several species of rust fungi and one species of powdery mildew are common pathogens in the United Kingdom. Certain varieties (morphs) of *T. officinale* appear to be resistant to either the common rust fungi, *Puccinia hieracii* and *P. variabilis* (Wilson & Henderson 1966) or powdery mildew (*Sphaerotheca fuliginea*) (Blumer 1967) or both.

It is interesting to speculate on the types of selection operating on these host–pathogen interactions because of the effectively closed recombination system in the host. It is unlikely that because the host has effectively reached an 'evolutionary dead-end' that the complementary genetic systems controlling virulence and aggressiveness in the pathogens have also stopped. Indeed the special asexual recombination systems in fungi, such as heterokaryosis and parasexuality, potentially allow for enormous variation to be generated during population explosions of fungal plant pathogens particularly airborne foliar pathogens, e.g. late blight of potato (Shattock 1976a). In this respect fungi differ from other organisms, e.g. aphids reproducing parthenogenetically during population explosions or apomictic and inbreeding r-selected plant species rapidly colonizing new habitats where it is supposed that new recombinants are prevented from becoming established and thus preventing the temporary forces of selection throwing the population off course. Because variation can be generated in these rust and powdery mildew fungi both sexually and probably asexually it is probable that frequency dependent selection is maintaining the equilibrium between *T. officinale* and its specialized pathogens by changes in the frequencies of the different host morphs in time and space. Thus, virulent and aggressive races of the pathogens will increase in frequency in the pathogen population by directional selection and attack susceptible host morphs which will subsequently fall to a low level in the host population. As the frequency of the host resistance falls so the value of the related factors controlling pathogenicity in the parasite will fall to a low level. This parallels the maintenance of balanced polymorphisms between host and pathogen as described for outbreeding species by Person (1966). Besides genetic feedback, the host density may effect some control over the frequency of the fungi and assist in maintaining the equilibrium between host and pathogen. This interplay of numbers of host and parasite is usually described as density-dependent negative feedback or ecological feedback.

Of course, any investigation of this sort faces many difficulties such as measuring changes in frequencies of the different host genotypes within a population and recognizing resistance and virulence factors in the host and pathogens respectively. This latter problem could be solved by initially investigating host–pathogen interactions in the centre of origin which in the case of *Taraxacum* species is the Himalayas (Richards 1973) and where outbreeding still occurs. Alternatively the phenotype–phenotype analysis of Wolfe *et al.* (1976) could be utilized because the genetic bases of the phenotypic observations are of no initial concern. The analysis can therefore be conducted at the host variety level rather than at the genotypic level.

As stated previously disease epidemics in natural plant communities are generally the result of attack by an exotic parasite. Unplanned introductions of parasites such as the highly aggressive strain of *Ceratocystis ulmi* and *Puccinia lagenophora*, a rust fungus on *Senecio* species (Wilson & Henderson 1966) first noticed in the United Kingdom in 1961, provide situations for study of population dynamics of pathogens in natural ecosystems. Even better systems for study may occur when pathogens are introduced on purpose as possible biological control agents (Hasan 1974). One such case is the rust fungus, *Puccinia chondrillina*, which has been introduced into Australia for the control of skeleton weed, *Chondrilla juncea*, a triploid obligate apomictic species, thus resembling and indeed related to *T. officinale* (Hasan 1972). Although extensive tests have been carried out to check that *P. chondrillina* does not attack plant species related or unrelated to *C. juncea* (Hasan 1972) it will be interesting to see how much variation is thrown up during the multiplication of the fungus which under Australian conditions is likely to be through recycling of the uredospore stage and, therefore, asexual.

In future, disease epidemics are likely to occur in both natural and crop plant communities from exotic pathogens introduced by chance or because of poor quarantine methods. Epidemics will also occur, particularly those caused by foliar airborne fungal pathogens, if breeders continue to use monogenic resistance and narrow the genetic base of cultivated plant species still further. Of course, not all breeding programmes for disease resistance have been unsuccessful. For example, major gene resistance has effectively controlled certain monocyclic soil borne diseases, e.g. Panama disease of banana (*Fusarium oxysporum* f. sp. *cubense*) (Knott 1972), wart disease (*Synchytrium endobioticum*) of potato (Lupton 1974), whilst loose smut (*Ustilago nuda*) of barley and North American maize rust (*Puccinia sorghi*) have been effectively controlled by polygenic resistance (Simons 1972, Lupton 1974). The success in controlling the latter disease strongly suggests much greater effort should be given to exploiting polygenic resistance in breeding programmes for disease resistance. Indeed rapid recovery of polygenic resistance can occur in the face of disease epidemics following introduction of pathogens previously separated geographically for several hundreds of years as shown in the field with tropical maize rust (*Puccinia polysora*) (van der Plank 1968, 1975) and experimentally with potatoes (Simmonds & Malcolmson 1967). These examples show how quickly cross pollination and selection pressure can accumulate resistance in a heterogeneous population provided that it has an adequately broad genetic base.

Another possibility of manipulation of the host to increase and broaden the genetic basis of resistance is the use of F_1-hybrid cultivars (Macer 1975).

Whilst many disease problems have originated from the misuse of monogenic resistance which has forced the evolution of parasites to greater virulence (Wolfe 1972b) the widespread use of highly specific systemic fungicides is leading to tolerant and resistant strains rising in the populations of pathogens and thereby increasing the problems of control (Wolfe 1971, Dekker 1973, Day 1974).

Conclusion

Much more intensive research into the population biology of plant pathogens is required. Macer (1975) has stated that 'plant pathology . . . as a subject has generated a voluminous literature much, however, concerned with observation, less with experimentation and relatively little with theory with the notable exceptions of van der Plank (1968) and Day (1974)'. Pathologists should be directed through teaching and stimulated through research to understand the factors that affect shifts in the size and structure of pathogen populations. The population dynamics of plant pathogens have been neglected by ecologists. The techniques of ecological genetics used principally in studies on populations of plants and animals should be employed to study the host and pathogen and the forces of selection acting on them at different stages of their life histories and in different environments.

References

ANON. (1975) Recommended varieties of cereals 1975. Farmers' Leaflet No. 8, 1975. *Nat. Inst. Agri. Bot., Camb.*

BAKER K.F. & COOK R.J. (1974) *Biological Control of Plant Pathogens.* W.H. Freeman & Co., San Francisco.

BEAUMONT A. (1947) The dependence on the weather of the dates of outbreak of potato blight epidemics. *Trans. Br. mycol. Soc.* 31, 45–53.

BLACK W., MASTENBROEK C., MILLS W.R. & PETERSON L.C. (1953) A proposal for an international nomenclature of races of *Phytophthora infestans* and of genes controlling immunity in *Solanum demissum* derivatives. *Euphytica* 2, 173–9.

BLUMER S. (1967) *Echte Mehltaupilze (Erysiphaceae).* Gustav Fischer Verlag, Jena.

BOURKE P.M.A. (1964) Emergence of potato blight 1843–46. *Nature, Lond.* 203, 805–8.

BRASIER C.M. & GIBBS J.N. (1973) Origin of the Dutch elm disease epidemic in Britain. *Nature, Lond.* 242, 607–9.

BRASIER C.M. & GIBBS J.N. (1975) Variation in *Ceratocystis ulmi*: significance of the aggressive and non-aggressive strains. *Proc. I.U.F.R.O. Conf. on Dutch Elm Disease* (1973), 53–66.

BRASIER C.M. & GIBBS J.N. (1976) Inheritance of pathogenicity and cultural characters in *Ceratocystis ulmi*: hybridization of aggressive and non-aggressive strains. *Ann. appl. Biol.* 83, 31–7.

BROWN J.K. (1975) Factors affecting the relative ability of strains of fungal pathogens to survive in populations. *J. Aust. Inst. agri. Sci.* **41**, 3–11.

BROWNING J.A. & FREY K.J. (1969) Multiline cultivars as a means of disease control. *A. Rev. Phytopathol.* **7**, 355–82.

BURDON J.J. & CHILVERS G.A. (1974) Fungal and insect parasites contributing to niche differentiation in mixed species stands of eucalypt saplings. *Aust. J. Bot.* **22**, 103–14.

BURDON J.J. & CHILVERS G.A. (1975a) A comparison between host density and inoculum density effects on the frequency of primary infection foci in *Pythium* induced damping-off disease. *Aust. J. Bot.* **23**, 899–904.

BURDON J.J. & CHILVERS G.A. (1975b) Epidemiology of damping-off disease (*Pythium irregulare*) in relation to density of *Lepidium sativum* seedlings. *Ann. appl. Biol.* **81**, 135–43.

BURDON J.J. & CHILVERS G.A. (1976) Epidemiology of *Pythium*-induced damping-off in mixed species seedling stands. *Ann. appl. Biol.* **82**, 233–40.

BUTT D.J. & ROYLE D.J. (1974) Multiple regression analysis in the epidemiology of plant diseases. In *Epidemics of Plant Diseases (Mathematical Analysis and Modeling)* (Ed. by J. Kranz), pp. 78–114. Chapman & Hall, London.

CATEN C.E. (1974) Intra-racial variation in *Phytophthora infestans* and adaptation to field resistance for potato blight. *Ann. appl. Biol.* **77**, 259–70.

CHILVERS G.A. & BRITTAIN E.G. (1972) Plant competition mediated by host specific parasites—a simple model. *Aust. J. biol. Sci.* **25**, 749–56.

COOPER A.J. (1975) Nutrient-film techniques. *Rep. Glasshouse Crops Res. Inst.* 1974, 78–9.

CRUTE I.R. & DAVIS A.A. (1976) New virulence gene combinations in British isolates of *Bremia lactucae*. *Ann. appl. Biol.* **83**, 173–5.

CRUTE I.R. & JOHNSON A.G. (1976) The genetic relationship between races of *Bremia lactucae* and cultivars of *Lactuca sativa*. *Ann. appl. Biol.* **83**, 125–37.

DAY P.R. (1960) Variation in phytopathogenic fungi. *A. Rev. Microbiol.* **14**, 1–16.

DAY P.R. (1973) Genetic variability of crops. *A. Rev. Phytopathol.* **11**, 293–312.

DAY P.R. (1974) *Genetics of Host–Parasite Interaction.* W.H. Freeman & Co., San Francisco.

DEKKER J. (1973) Selectivity of and resistance against systemic fungicides. *Bull. Org. Europ. Medit. Prot. Plant.* **10**, 47–57.

EMARA Y.A. & SIDHU G. (1974) Polygenic inheritance of aggressiveness in *Ustilago hordei*. *Heredity* **32**, 219–24.

ESHED N. & WAHL I. (1970) Host ranges and interrelations of *Erysiphe graminis hordei*; *E. graminis tritici* and *E. graminis avenae*. *Phytopathology* **60**, 628–34.

FLOR H.H. (1955) Host–parasite interaction in flax rust: its genetics and other implications. *Phytopathology* **45**, 680–5.

FLOR H.H. (1956) The complementary genic systems in flax and flax rust. *Adv. Genet.* **8**, 29–54.

FLOR H.H. (1971) Current status of the gene-for-gene concept. *A. Rev. Phytopathol.* **9**, 275–96.

GIBBS J.N. & BRASIER C.M. (1973) Correlation between cultural characters and pathogenicity in *Ceratocystis ulmi* from Britain, Europe and America. *Nature, Lond.* **241**, 381–3.

GIBBS J.N., HEYBROEK H.M. & HOLMES F.W. (1972) Aggressive strain of *Ceratocystis ulmi* in Britain. *Nature, Lond.* **236**, 121–2.

GOLDSCHMIDT V. (1928) Verebungsuersuche mit den biologischen Arten des Autheren brandes (*Ustilago violacea*). Ein Beitrag zur Frage per parasitaren spezialisierung. *Z. Bot.* **21**, 1–90.

HABGOOD R.M. (1976) Differential aggressiveness of *Rhyncosporium secalis* isolates towards specified barley genotypes. *Trans. Br. mycol. Soc.* **66**, 201–4.

HARLAN J.R. (1972) Genetics of disaster. *J. environ. Qual.* **1**, 212–15.

HASAN S. (1972) Specificity and host specialisation of *Puccinia chondrillina*. *Ann. appl. Biol.* **72**, 257–63.

HASAN S. (1974) Recent advances in the use of plant pathogens as biocontrol agents of weeds. *PANS* **20**, 437–43.

HORSFALL J.G. *et al.* (1972) *Genetic Vulnerability of Major Crops*. National Academy of Sciences, Washington, D.C.

HOWES N.K., SAMBORSKI D.J. & RÖHRINGER R. (1974) Production and bioassay of gene specific RNA determining resistance of wheat to stem rust. *Can. J. Bot.* **52**, 2489–97.

JAMES W.C., SHIH C.S., HODGSON W.A. & CALLBECK L.C. (1972) The quantitative relationship between late blight of potato and loss in tuber yield. *Phytopathology* **62**, 92–6.

JANSSEN B.D. (1973) *Studies of host-specific variation in the late blight fungus of potato*. Ph.D. thesis, University of Wales.

JEFFREY S.I.B., JINKS J.L. & GRINDLE M. (1962) Intraracial variation in *Phytophthora infestans* and field resistance to potato blight. *Genetica* **32**, 323–38.

JOHNSON R. & TAYLOR A.J. (1972) Isolates of *Puccinia striiformis* collected in England from the wheat varieties Maris Beacon and Joss Cambier. *Nature, Lond.* **238**, 105–6.

JOHNSON T. (1961) Man guided evolution in plant rusts. *Science* **133**, 357–62.

KATSUYA K. & GREEN G.J. (1967) Reproductive potentials of races 15B and 56 of wheat stem rust. *Can. J. Bot.* **45**, 1077–91.

KERR E.A. & BAILY D.L. (1966) Breeding for resistance to *Cladosporium fulvum* Cke. in tomato. *Acta Hort.* **4**, 145–7.

KNOTT D.R. (1968) The inheritance of resistance to stem rust races 56 and 15B-11 (Can.) in the wheat varieties Hope and H-44. *Can. J. Genet. Cytol.* **10**, 311–20.

KNOTT D.R. (1972) Using race specific resistance to manage the evolution of plant pathogens. *J. environ. Qual.* **1**, 227–31.

KRANZ J. (1974a) *Epidemics of Plant Diseases (Mathematical Analysis and Modeling)* (Ed. by J. Kranz) Chapman & Hall, London.

KRANZ J. (1974b) The role and scope of mathematical analysis and modeling in epidemiology. In *Epidemics of Plant Diseases (Mathematical Analysis and Modeling)* (Ed. by J. Kranz) pp. 7–54. Chapman & Hall, London.

LARGE E. C. (1940) *The Advance of the Fungi*. Jonathan Cape, London.

LEONARD K.J. (1969) Factors affecting rates of stem rust increase in mixed plantings of susceptible and resistance oat cultivars. *Phytopathology* **59**, 1851–7.

LEPPIK E.E. (1970) Gene centres of plants as sources of disease resistance. *A. Rev. Phytopathol.* **8**, 323–44.

LEVIN D.A. (1975) Pest pressure and recombination systems in plants. *Amer. Nat.* **109**, 437–51.

LOEGERING W.Q. & POWERS H.R. (1962) Inheritance of pathogenicity in a cross of physiological races 111 and 36 of *Puccinia graminis* f. sp. *tritici*. *Phytopathology* **52**, 547–54.

LUIG N.H. & WATSON I.A. (1970) The effect of complex genetic resistance in wheat on the variability of *Puccinia graminis* f. sp. *tritici*. *Proc. Linn. Soc. N.S.W.* **95**, 22–45.

LUPTON F.G.H. (1974) Plant breeding for disease resistance. In *Biology in Pest and*

Disease Control (Ed. by D. Price Jones and M.E. Solomon), pp. 87–96. Blackwell Scientific Publications, Oxford.

McIntosh R.A., Luig N.H. & Baker F.F. (1967) Disease resistance in Hope wheat. *Aust. J. biol. Sci.* **20**, 1181–92.

Macer R.C.F. (1975) Plant pathology in a changing world. *Trans. Br. mycol. Soc.* **65**, 351–67.

Malcolmson J.F. (1969) Races of *Phytophthora infestans* occurring in Great Britain. *Trans. Br. mycol. Soc.* **53**, 417–23.

Malcolmson J.F. & Black W. (1966) New R genes in *Solanum demissum* Lindl. and their complementary races of *Phytophthora infestans* (Mont) de Bary. *Euphytica* **15**, 199–203.

Martens J.W. (1975) Virulence dynamics in the wheat stem rust population of Kenya. *Pl. Dis. Reptr.* **59**, 763–7.

Martin W.J. (1949) Strains of *Phytophthora infestans* capable of surviving high temperatures. *Phytopathology* **39**, 14 (Abstract).

Mills W.D. & La Plante A.A. (1954) Diseases and insects in the orchards. *Ext. Bull. Cornell agric. Exp. Stn.*, No. 711, 20–2.

Moseman J.G. (1966) Genetics of powdery mildews. *A. Rev. Phytopathol.* **4**, 269–90.

Nelson R.R. (1972) Stabilizing racial populations of plant pathogens by use of resistance genes. *J. environ. Qual.* **1**, 220–7.

Niederhauser J.S., Cervantes J. & Servin L. (1954) Late blight in Mexico and its implications. *Phytopathology* **44**, 406–8.

Niederhauser J.S. & Cobb W.C. (1959) The late blight of potatoes. *Sci. Amer.* **5**, 100–12.

Ogle H.J., Taylor N.W. & Brown J.F. (1973) A mathematical approach to the prediction of differences in the relative ability of races of *Puccinia graminis tritici* to survive when mixed. *Aust. J. biol. Sci.* **26**, 1137–43.

Paxman G.J. (1963) Variation in *Phytophthora infestans. Eur. Potato J.* **6**, 14–23.

Person C. (1959) Gene-for-gene relationships in host:parasite systems. *Can. J. Bot.* **37**, 1101–30.

Person C. (1966) Genetic polymorphism in parasitic systems. *Nature, Lond.* **212**, 266–7.

Person C., Samborski D.J. & Röhringer R. (1962) The gene-for-gene concept. *Nature, Lond.* **194**, 561–2.

Richards A.J. (1973) The origin of *Taraxacum* agamospecies. *Bot. J. Linn. Soc.* **66**, 189–211.

Roelfs A.P. & Rothman P.G. (1974) Races of *Puccinia graminis* f. sp. *avenae* in the U.S.A. during 1973. *Pl. Dis. Reptr.* **58**, 605–7.

Röhringer R., Howes N.K., Kim W.K. & Samborski D.L. (1974) Evidence for a gene-specific RNA determining resistance in wheat to stem rust. *Nature, Lond.* **249**, 585–8.

Rotem J. & Palti J. (1969) Irrigation and plant diseases. *A. Rev. Phytopathol.* **7**, 267–288.

Sansome E. & Brasier C.M. (1973) Diploidy and chromosomal structural hybridity in *Phytophthora infestans. Nature, Lond.* **241**, 344–5.

Schrodter H. & Ullrich J. (1965) Untersuchungen zur Biometerologie und Epidemiologie von *Phytophthora infestans* (Mont.) de By. auf mathematisch-statistischer Grundlage. *Phytopathol. Z.* **54**, 87–103.

Shattock R.C. (1974) *Variation and its origins in* Phythophthora infestans (*Mont.*) *de Bary*. Ph.D. thesis, University of Wales.

Shattock R.C. (1976a) Variation in *Phytophthora infestans* on potatoes grown in walk-in polyethylene tunnels. *Ann. appl. Biol.* **82**, 227–32.

SHATTOCK R.C. (1976b) Winter survival of field isolates of *Phytophthora infestans* in seed tubers and development of primarily infected plants. *Ann. appl. Biol.* 84, 273–4.

SHATTOCK R.C. & SHAW D.S. (1975) Mutants of *Phytophthora infestans* resistant to, and dependent upon, antibiotics. *Trans. Br. mycol. Soc.* 64, 29–41.

SHATTOCK R.C. & SHAW D.S. (1976) Novel phenotypes of *Phytophthora infestans* from mixed cultures of antibiotic resistant mutants. *Trans. Br. mycol. Soc.* (in press).

SIDHU G. & PERSON C. (1971) Genetic control of virulence in *Ustilago hordei*. II. Segregations for higher levels of virulence. *Can. J. Genet. Cytol.* 13, 173–8.

SIDHU G. & PERSON C. (1972) Genetic control of virulence in *Ustilago hordei*. III Identification of genes for host resistance and demonstration of gene-for-gene relations. *Can. J. Genet. Cytol.* 14, 209–13.

SIMMONDS N.W. & MALCOLMSON J.F. (1967) Resistance to late blight in Andigena potatoes. *Eur. Potato J.* 10, 161–6.

SIMONS M.D. (1972) Polygenic resistance to plant disease and its use in breeding resistant cultivars. *J. environ. Qual.* 1, 232–40.

SMITH L.P. (1956) Potato blight forecasting by 90 per cent humidity criteria. *Pl. Path.* 5, 83–7.

SUMNER D.R. & LITTRELL R.H. (1974) Influence of tillage, planting date, inoculum survival and mixed plantings on epidemiology of southern corn leaf blight. *Phytopathology* 64, 168–73.

THRESH J.M. (1974) Vector relationships and the development of epidemics: the epidemiology of plant viruses. *Phytopathology* 64, 1050–6.

TOXOPEUS H.J. (1956) Reflections on the origin of new physiologic races of *Phytophthora infestans* and the breeding of resistance in potatoes. *Euphytica* 5, 221–37.

UPSHALL A. (1969) Intraracial variation in *Phytophthora infestans* and its relationship to different host varieties. *Can. J. Bot.* 47, 863–7.

VAN DER PLANK J.E. (1960) Analysis of epidemics. In *Plant Pathology* (Ed. by J.G. Horsfall and A.E. Dimond), Vol. 3, pp. 229–89. Academic Press, New York.

VAN DER PLANK J.E. (1963) *Plant Diseases: Epidemics and Control.* Academic Press, New York.

VAN DER PLANK J.E. (1965) Dynamics of epidemics of plant disease. *Science* 147, 120–4.

VAN DER PLANK J.E. (1968) *Disease Resistance in Plants.* Academic Press, New York.

VAN DER PLANK J.E. (1971) Stability of resistance to *Phytophthora infestans* in cultivars without R genes. *Potato Res.* 14, 263–70.

VAN DER PLANK J.E. (1975) *Principles of Plant Infection.* Academic Press, New York.

WAGGONER P.E. (1974). Simulation of epidemics. In *Epidemics of Plant Diseases (Mathematical Analysis and Modeling)* (Ed. by J. Kranz), pp. 137–60. Chapman & Hall, London.

WAGGONER P.E. & HORSFALL J.G. (1969) EPIDEM: a simulator of plant disease written for a computer. *Conn. Agri. Exp. Stn., Bull.* No. 698.

WAGGONER P.E., HORSFALL J.G. & LUKENS R.J. (1972) EPIMAY: a simulator of southern corn leaf blight. *Conn. Agri. Exp. Stn. Bull.* No. 729, 84 pp.

WATSON I.A. (1970) Changes in virulence and population shifts in plant pathogens. *A. Rev. Phytopathol.* 8, 209–30.

WHEELER H. (1975) *Plant Pathogenesis.* Springer-Verlag, New York.

WILSON M. & HENDERSON D.M. (1966) *British Rust Fungi.* Cambridge University Press.

WOLFE M.S. (1971) Fungicides and the fungus population problem. *Proc. 6th Br. Insect and Fungicide Conf.*, 724–34.

WOLFE M.S. (1972a) Genetics of barley powdery mildew. *Rev. Pl. Path.* 51, 507–22.

WOLFE M.S. (1972b) The forced evolution of cereal disease. *Outlook on Agriculture* 7, 27–31.

WOLFE M.S. & MINCHIN P.N. (1976). Quantitative assessment of variation in field populations of *Erysiphe graminis* f. sp. *hordei* using mobile nurseries. *Trans. Br. mycol. Soc.* 66, 332–4.

WOLFE M.S. & SCHWARZBACH E. (1975) The use of virulence analysis in cereal mildews. *Phytopathol. Z.* 82, 297–307.

WOLFE M.S., WRIGHT S.E., BARRETT J.A. & O'DONALD P. (1973) Population genetics of cereal mildews. *2nd Int. Cong. Pl. Pathol., Minneapolis, U.S.A.* (Abstract 0882).

WOLFE M.S., BARRETT J.A., SHATTOCK R.C., SHAW D.S. & WHITBREAD R. (1976). Phenotype–phenotype analysis: field application of the gene-for-gene hypothesis in host–pathogen relations. *Ann. appl. Biol.* 82, 369–74.

WOODHAM-SMITH C. (1963) *The Great Hunger, Ireland 1845–9.* Hamish Hamilton, London.

ZADOKS J.C. (1971) Systems analysis and the dynamics of epidemics. *Phytopathology* 61, 600–10.

ZADOKS J.C. (1972) Methodology in epidemiological research. *A. Rev. Phytopathol.* 10, 253–76.

Introductions and pest and weed problems

F. J. SIMMONDS and D. J. GREATHEAD
*Commonwealth Institute of Biological Control, Curepe,
Trinidad, W.I.*

The term pest is to an extent vague in that it can include any organism
detrimental to man or his activities, including weeds, and will be used in
this sense where the term is not qualified, as for example 'insect pest'.
Except for parasites of man himself all pests have become such as a result
of human manipulation of the environment. In the early days of the evolution
of mankind, pests would have been of entirly local origin but once trade and
exploration began, an artificial means of dispersal was provided. As human
mobility has increased the risks of inadvertent introductions has increased.
On the other hand many species are highly mobile even unaided by man;
many birds and insects migrate long distances and organisms are dispersed
across the sea and by air currents and on flotsam. Given these natural means
of dispersal and with the antiquity of human travel and trade, it is now often
impossible to determine the origin and means of spread of a particular pest.
Thus human ectoparasites, many insect pests (particularly of stored products), and weeds of cultivation had become cosmopolitan before geographical
studies began and the significance of distribution was appreciated.

Bearing in mind these limitations, in this paper we seek to examine the
importance of introduced species as pests, the means and mechanisms of
introductions and to touch on remedial measures. Biological control, in its
classical form, is essentially a means of combating the ravages of introduced
pests and will be treated in some detail. Biological control experience in the
deliberate introduction of species can also contribute to an understanding of
what is required for the establishment of a species in a new area.

Aspects of the problem have been treated at length by Elton (1958) in
his study of the ecology of invasions, and genetic aspects are examined by
the contributors to Baker & Stebbins (1965). Numerical data have proved
hard to come by so that much of the treatment will perforce be of an anecdotal
nature. It is hoped that it will stimulate study of some of the problems raised.

The importance of introduced pests

It is striking how few of the many species of plants and animals deliberately or accidentally introduced through human agency have become pests or even become established permanently in the wild. One needs to think only of the number of species of plants carried around the world to adorn gardens; few are able to survive without being tended and few of those that have escaped cultivation have become important weeds. Mayr (1965) has assembled data showing that among birds relatively few introduced species have become successful colonists (i.e. abundant and widespread) in the wild in continental areas (only the pheasant in Europe and four species in North America) in contrast to islands where in many instances introduced species now outnumber endemic species and are also more abundant. Equally, many species of invertebrates must have been transported unknowingly by man and only those few which have become pests are normally even noticed. Some examples of insects established in England include *Epiphyas* (*Tortrix*) *postvittana*, an Australian species known elsewhere as a pest of fruit trees, which has been established a long while in Cornwall but has not spread or become of great importance, and several species of stick insects which have escaped from captivity and are now established in the southwest. It is therefore impossible to estimate what proportion of introduced species either become established or having done so attract attention as pests. Elton's (1958) opinion that through human agency eventually all species would become distributed throughout the regions suited to them seems somewhat pessimistic.

Table 1.   Pest complexes of crops to show the part played by introduced species.

| Crop | No. of pest spp. recorded | No. of pest spp. introduced in some part of their present range | | | Source |
		Total	Sterno-rhyncha	Coccoidea	
Sugar cane	1,300	77 (5.9%)	16	12	Box (1953)
Cotton	1,360	32 (2.4%)	10	8	Hargreaves (1948)
Coffee	838	30 (3.5%)	17	15	Le Pelley (1968)
Cocoa	1,400	36 (2.6%)	17	16	Entwistle (1972)

On the other hand it is possible to give some indication of what proportion of insects associated with particular crops are introductions. Table 1 shows the numbers of species associated with certain widespread crops. In preparing this table only species quite clearly introduced—in that they occur in disjunct geographical regions—have been classed as introduced. It has not been possible to determine in each case which pests have spread by

Table 2. Analysis of the distributions of the pests mapped by the Commonwealth Institute of Entomology (Anon. 1951–1975)

Taxon	Total no. pest spp.	No. pest spp. discontinuously distributed	No. of biogeographical regions in which present					
			6	5	4	3	2	1
Acari	16	5	7	5	0	2	1	1
Collembola	1	1	0	0	1	0	0	0
Orthoptera	6	3	0	0	1	1	2	2
Dermaptera	1	1	0	0	0	1	0	0
Isoptera	2	2	0	0	0	1	1	0
Heteroptera	17	4	2	0	0	1	4	10
Homoptera								
Auchenorhyncha	18	10	0	0	2	5	7	4
Sternorhyncha								
Psyllidae/Aleyrodidae	10	6	1	2	2	2	1	2
Aphididae	27	27	15	6	2	3	1	0
Coccoidea	48	47	25	11	4	5	2	1
All Homoptera	103	90	41	19	10	15	11	7
Thysanoptera	9	6	3	2	0	0	2	2
Lepidoptera								
Tortricidae	13	9	1	1	0	2	5	4
Pyralidae	21	11	1	1	1	6	10	2
Other Microlepidoptera	16	9	3	0	1	4	4	4
Noctuidae	28	12	1	2	5	8	7	5
Other Macrolepidoptera	8	4	1	0	1	0	5	1
All Lepidoptera	86	45	7	4	8	20	31	16
Diptera								
Cecidomyidae	6	5	1	0	0	3	1	1
Tephritidae	20	6	0	1	0	1	7	11
Others	11	9	1	1	1	3	4	1
All Diptera	37	20	2	2	1	7	12	13
Coleoptera								
Chrysomelidae	7	3	0	0	0	0	4	3
Scolytidae	10	7	1	1	4	0	1	3
Curculionidae*	24	16	1	3	4	3	9	4
Scarabaeidae	11	7	0	0	1	1	5	4
Others	6	2	0	1	1	0	0	4
All Coleoptera	58	35	2	5	10	4	19	18
Hymenoptera								
Tenthredinoidea	11	6	1	0	0	0	5	5
Formicidae	3	3	1	1	0	0	1	0
All Hymenoptera	13	9	2	1	0	0	6	6
Total	350	221	66	36	31	52	89	74
%		63.3	18.9	10.3	8.9	14.9	25.4	21.1
% less Sternorhyncha		53.2	9.4	6.4	8.7	15.8	32.0	26.8

* Field pests only.

dispersal into contiguous areas. Thus the estimates in the table are minimal and suggest an overall percentage of less than 10%. However, many of the species listed from a crop are of little consequence and one suspects that the proportion of serious pests introduced in a particular area will be much higher. Taking the 350 agricultural pests of sufficient importance to have been mapped by the Commonwealth Institute of Entomology (Table 2) and using the same method as before, the percentage of introduced species is 63·3%. Although this estimate is likely to be biased in the opposite direction it does indicate that the proportion of introduced species is higher amongst 'serious' pests. A measure of frequency of introduction is provided by the number of zoogeographical regions in which the pest occurs, thus the proportion of cosmopolitan Acari and Homoptera Sternorhyncha is striking while in other orders the tendency is for fewer regions to be invaded; fully a third have entered only one (or straddle two) biogeographical regions and another third have spread little if at all.

This crude analysis suggests that the mode of dispersal is important in the origin of introduced pests, thus the mites and bugs are all small and can readily be transported on plant material or passively by air currents. Other very widespread pests are notably those that can easily be transported in produce such as fruit (fruitflies, codling moth [*Cydia pomonella*]), tubers (weevils, potato tuber moth [*Phthorimaea operculella*]), produce (pink bollworm [*Pectinophora gossypiella*]), and timber (termites, timber beetles). Others included in the analysis such as Acrididae and Noctuidae are migrants and were undoubtedly already widespread before trade began.

Species which are not polyphagous could only be introduced after their host plants had already been grown as crops in new areas. Notable examples are the potato tuber moth, a pest of potatoes and tobacco, the coffee berry borer (*Hypothenemus hampei*), and the eucalyptus weevils (*Gonipterus* spp.).

Another important source of introduced pests is oligophagous species, e.g. graminaceous stemborers, which have become pests of introduced crops, e.g. maize and sugarcane, and then have been introduced to other areas. A particularly interesting example is the European corn borer (*Ostrinia nubilalis*) which originally fed on *Artemesia* spp. (Thompson & Parker 1928), became a pest of maize in Europe and was introduced into North America! A similar example is provided by the Colorado beetle (*Leptinotarsa decemlineata*) which became a pest of potatoes when they were grown in North America and was later introduced into Europe (Trouvelot 1936). A further analysis of the smaps comparing oligophagous pests (including pests of Graminae) with polyphagous pests (no taxonomic pattern to host list) suggests that polyphagous species are more likely to become widespread introduced pests than those with a restricted host range (Table 3). To an extent oligophagous pests will be limited by the presence of suitable crops on which to feed but most groups of crop plants are now so widespread that this limitation is unlikely to account for the whole of the discrepancy.

Table 3. Further analysis of data in Table 2 in relation to host plant specificity

	Oligophagous pest	Polyphagous pest
Distribution not extended	40	20
Spread to contiguous areas	45	31
Introduced to disjunct areas	93	121

The proportion of introduced species which have become pests also shows differences from region to region. Thus Clausen (1956) comments that more than 60% of the important insect pests in North America are introduced species mostly from Europe, whereas in Europe the majority of pests are either indigenous or derived from adjacent regions of Asia. Outside the Mediterranean area there are so few important pests of North American origin on outdoor crops that they can be readily enumerated—phylloxera (*Viteus vitifoliae*), apple woolly aphid (*Eriosoma lanigerum*), Colorado beetle (*Leptinotarsa decemlineata*). This has been discussed by Pschorn-Walcher (Greathead 1976), who attributes it and the relative abundance of pests of European origin in certain other areas especially Australia, New Zealand, and South Africa, to the effects of colonization by Europeans who took with them their animals, crops, ornamentals, and baggage. Thus the pattern of human settlement has had an effect on the numbers of introduced pests in a particular area.

Islands, as opposed to continental areas, seem to have a higher proportion of introduced pests. Pemberton (1964) estimated that of the 2,000 introduced species of insects in the Hawaiian Islands about 296 are pests; Pemberton & Williams (1969) note that in Mauritius thirty-six of the forty-three cane insects are introduced including all the important ones (for comparison, in East Africa Le Pelley (1959) records thirty-two cane insects of which six (all Aphididae and Coccoidea) are introduced).

Differences are also apparent between crops. For example oil palm pests in Malaysia (Wood 1968) are all indigenous except for some cosmopolitan Coccoidea, although the palm is of West African origin, but of the citrus pests in the Mediterranean area listed by Bodenheimer (1951) fifty (forty of which are Coccoidea) out of 107 are introduced.

A similar situation is found with weeds. Salisbury (1961) notes that weeds of European origin predominate in North America (Canada over 60%), in Australia 60% of introduced plants are weeds, in New Zealand 80%, whereas in Europe there are few important alien species from other continents. Only the composites, *Solidago* spp. and *Conyza canadensis* from North America and *Ambrosia* spp. from the USSR, have been sufficiently invasive of uncultivated land to warrant the consideration of biological control.

In the tropics there are a number of well-known weeds now almost pan-tropical, including the shrub *Lantana camara* and prickly pears, *Opuntia* spp., spread as ornamentals; *Tridax procumbens*, *Tagytes minuta* and many other annuals in fields and waste places; and several floating waterweeds notably the water hyacinth (*Eichhornia crassipes*), the water fern (*Salvinia molesta*) and the Nile cabbage (*Pistia stratiotes*). Many others of more limited distribution are serious invasive pests.

Thus introduced species are often among the more important pests but their proportion varies with, among other factors, the taxonomic group, geographical area, crop, and means of dispersal. In discussing the mechanisms of introduction examples will be given which illustrate these points and reasons for the varying proportions will be suggested.

Mechanisms of introduction

Man, through travel and trade, has greatly increased the movement of species into new areas but chance and its own powers of dispersal have in the past, and probably still continue to be the major means of colonization. Thus were oceanic islands colonized before the advent of man. MacArthur & Wilson (1967) and MacArthur (1972) have used a mathematical approach to show that the number of species present depends chiefly on the area of land, distance from source and the balance between immigration and extinction. Their examples and analysis explain to a great extent the unbalanced nature of island biotas and that chance accounts for the majority of successful colonizations. Natural colonization may result from transport on flotsam, from storms, or be the result of good dispersal powers. Thus in many instances a widespread pest cannot with certainty be classed as introduced by man unless its introduction has occurred recently and its arrival observed. How, for example, did the red locust (*Nomadacris septemfasciata*) reach Mauritius? There is no record of its arrival and it is one of the first pests recorded on the island. Did it reach Mauritius aided by a cyclone or was it inadvertently introduced with plant material by man? Many pests widespread in the Mediterranean and western Asia; in eastern Africa and India; and in the western Pacific, can thus not with certainty be classified as indigenous or immigrant as man has been journeying back and forth within these three regions since the dawn of history. Consequently examples will be drawn mainly from recent introductions where man is known or suspected to have been responsible.

Means of dispersal provided by man are listed below.

Transport on the host

By this means fleas, lice and ticks have been carried around the world on the bodies of man and domestic animals. Likewise live plants have provided a mechanism for the dispersal of a number of plant pests. Aphids, mealybugs and scale insects are readily dispersed in this way, the notorious *Icerya purchasi* spreading from one country to another around the Mediterranean on nursery stock (Greathead 1976). In the days of lengthy sea voyages, shipping of growing plants in Wardian cases doubtless provided easy transport for many widespread pests. There are no documented examples, but such insects as the chafer (*Clemora smithii*), only known from Barbados and recorded from Mauritius for the first time in 1911, almost certainly travelled as larvae among the roots of sugarcane transported in Wardian cases.

Dispersal through trade

This has undoubtedly been the major source of immigrant pests. The pests that travel most readily by this means are those feeding on the produce. The many cosmopolitan pests of stored products are continually transported as contaminants. Other pests may be transported incidentally in packing materials, timber or in furniture. A notorious example of the last are the dry wood termites (*Cryptotermes* spp.) which first appeared in many tropical countries at seaports, where many pests have appeared in a country, for the first time. Even in recent times, since the enforcement of quarantine and inspection of cargoes, pests still enter by this route, and the establishment of such a large insect as the rhinoceros beetle (*Oryctes rhinoceros*) in Mauritius in 1962 is a striking example. Once established, it did severe damage to palms until brought under control by an imported virus. Likewise weed seeds are well known contaminants of grain cargoes and bales of fibres such as wool as well as of seed imported for planting (see Salisbury (1961) and Baker (1965) for reviews). Ballast has provided transport for many insects and plants (Lindroth 1957).

Chance transport as stowaways

Although insects travelling by sea and taken on board in packaging and ballast can be classed as stowaways, others travel accidentally on ships or in aircraft. As frequent air transport developed, a number of studies of the problem were made. These were reviewed by David (1949) who stated that 'so far as is known there is no record of a major pest of any kind being introduced to any country by an aircraft and becoming established'. Rainwater (1963) analysed the statistics for insects intercepted on aircraft landing in Hawaii and concluded that the chances of establishment of pests as a result of transport in aircraft were slight as the numbers were small in relation to the total transported in cargoes (between 1952 and 1961 only 269 out of a total of 74,134 individuals intercepted), most were dead, very few were pests, very few were intercepted repeatedly, almost two-thirds were males and less than one-third of the females were gravid. However, there is strong circumstantial evidence that pests have become established by this means. Hall (1958) reporting the establishment of *Aphis craccivora* in Fiji noted that large numbers were found around the New Zealand air force base. It is believed that the eucalyptus weevil (*Gonipterus scutellatus*) must have entered Kenya on flying boats landing at Kisumu as it first became established in the district near the town (see Greathead 1971). J. Monty (personal communication) suggests that the banana skipper (*Erionota thrax*) may have entered Mauritius on RAF transport aircraft from Malaya during 1968 when troops were flown in to assist in checking riots.

Weeds may also be dispersed accidentally as seeds adhering to animals or clothing. This is presumably how the Ngoora burr (*Xanthium pungens*) reached Australia.

Deliberate introduction

Important weeds have been introduced as ornamentals and escaped from gardens; prickly pear cactus and *Lantana camara*, notorious examples on land and the water hyacinth and the water fern in fresh water have already been noted. The pigs and cats which have destroyed the fauna and flora of many islands, grey squirrels in Britain, rabbits in Australia and deer in New Zealand are examples of deliberately introduced mammals which have become pests. Mongooses introduced into many islands, particularly in the West Indies for control of rats, are now pests. Introduced birds too have become pests, e.g. the starling in the United States of America. Even insects, particularly butterflies and moths, have been deliberately introduced. Naturalists have been responsible for the establishment of a number of species in new areas. The most notorious example is the gypsy moth (*Lymantria dispar*) imported into the United States in about 1870 by an amateur lepidopterist to test its value for silk production. Some moths escaped and it became established as a pest in the north eastern United States of America.

Likelihood of establishment

Given that physical conditions are suitable and food is present, the likelihood of establishment will depend on many factors. Except where the potential pest has escaped from a garden or farm, been introduced in numbers in produce or deliberately colonized, the initial numbers must be very small, often a single gravid female or seed. In these circumstances parthenogenetic reproduction among animals and selfing or vegetative reproduction among plants will be an advantage. Many important weeds have these properties (Baker 1965) but Harper (1965) lays emphasis on the importance of germination and seeding behaviour in the establishment of plant aliens. Among animals aphids are frequently parthenogenetic and this, as well as their ready transport on plants, may account for the large number of widely distributed pest species.

Lack of competition will favour establishment. This may account for the higher proportion of introduced pests on islands. Greathead (1971) seeking an explanation for the apparent higher proportion of biological control successes on the islands of the West Indian Ocean as compared with continental Africa suggested that it might be easier for species to become established owing to their lower density on islands compared with equivalent continental areas or even so-called ecological islands within continents. Pimentel (1966) has drawn attention to the relative lack of species diversity in coastal areas which may provide a means for invading species to gain a foothold on continents. Reduced competition can also occur inland on continents where large areas are planted with exotic species. Thus the eucalyptus weevil has become established in a number of African countries and on adjacent islands because eucalypt plantations are unattractive to most indigenous insects leaving a vacant niche. Equally its egg parasite (*Patasson nitens*) was readily established and has provided a high degree of control (Greathead 1971).

Preadaptation is also a characteristic of invasive pests; most important is the ability to feed on species not previously encountered. Thus, generalists rather than specialists stand a better chance, which explains why so many widespread pests are polyphagous. For example, most widely dispersed scale insects and mealybugs are pests of a number of diverse plants and those notorious widespread migratory pests, locusts and armyworms, have a wide range of food plants.

Among obligate sexually reproducing species a high fecundity and effective long range attraction of the sexes will be an advantage.

In spite of advantages such as those discussed, establishment may fail through accident, a sudden storm or flood, a drought, fire or clearance of vegetation by man destroying an incipient colony.

Successful colonizers are generally classed as aggressive species. Baker (1965) has set out characteristics of the ideal weed which sum up what is meant by this term. His ideal weed lacks specialized physical requirements and is thus able to survive under a wide range of conditions, it is capable of a high seed production, is not specialized in its pollination mechanism, has vigorous growth especially in the seedling stage and is otherwise capable of competitive growth. If perennial, it has vigorous vegetative reproduction and readily regenerates from severed parts. A similar list could be drawn up for the successful animal colonist, viz. tolerance of a wide range of physical conditions, high reproductive rate, efficient long-range attraction between the sexes, or the ability to produce parthenogenetically, non-specialized feeding or polyphagy, aggressiveness to competing species and so on.

Processes of colonization

Following the successful establishment of a species in a new area, there often appears to be an explosive phase of colonization. Unfortunately it is usually only at this point that the invader is noticed and seldom are any quantitative data available to substantiate these claims.

Elton (1958) cites data on the spread of the Japanese beetle (*Popillia japonica*) in the USA. When plotted this shows, after an initial acceleration, a steady progress with a linear relationship between the square root of the area covered and time. A similar relationship is shown for the spread of the white grub (*Clemora smithii*) through Mauritius (Moutia & Mamet 1946). With the invasion of a sugar estate in Tanzania by the scale insect *Aulacaspis tegalensis*, the relationship was between the square root of the area and the logarithm of time, so that in this instance the rate of spread declined with time. The population density was also measured and a density was achieved more than twice that attained subsequently during the colonization phase (Greathead 1975). In this instance the acquisition of a complex of natural enemies, including one species deliberately introduced, may have been responsible. The importance of *C. smithii* also declined after it completed colonization and here a number of factors including establishment of introduced parasites, introduction of improved and possibly resistant cane varieties and the recognition of other causes of crop loss which had been attributed to white grubs are held to be responsible (see Greathead 1971).

A further recent example is provided by Lewis *et al.* (1976) who show that the crazy ant (*Anoplolepis longipes*) is showing signs of decline in its original outbreak area on the Island of Mahé in the Seychelles, as has happened on Rodriguez and in the Agalega Islands, and suggest depletion of food resources as a cause.

Explosive growth of water weeds on first colonization of a new habitat has also been noted, e.g. Salisbury (1961) discusses the rise and decline of the Canadian pondweed (*Elodea canadensis*) and other water weeds in Britain. A similar eruption and decline of the water fern on Lake Kariba has been attributed to the exhaustion of nutrients following the decay of vegetation flooded by the filling of the dam.

Not all colonizations proceed at a steady rate following establishment. Lewis *et al.* (1976) note that there was a period of some ten years before the crazy ant began to spread rapidly through Mahé. This has often been noticed by biological control workers after the deliberate introduction of a control agent. An instructive example is that of the chrysomelid beetles (*Chrysolina* spp.) introduced in various areas for the control of St. John's wort (*Hypericum perforatum*). While progress was rapid following colonization in California, in Australia and in Canada, progress was slow and it is postulated that a period of genetic adaptation took place before the population density could increase rapidly and control the weed in areas where the insects were at the limit of their climatic tolerance (Harris *et al.* 1969).

A similar period of adaptation seems to have been required by the gypsy moth which was confined to a limited area until recently when a rapid extension began into Florida to the south, Canada to the north and more slowly westwards. Many examples of the delayed spread and population increase of introduced weeds in Britain are cited by Salisbury (1961) who suggests that ecological changes, sometimes after several hundreds of years, were necessary before increase was possible. Biological control experience suggests that genetic adaptation proceeds slowly and examples of changed status due to adaptations are rare within the time scale covered (Wilson 1965). However, it is still less than one hundred years since the first successful introduction, which is a short time even in the history of human impact on the environment. During this period there have undoubtedly been many genetic changes allowing successful colonization by introduced species not preadapted to the new environment. The examples of the European corn borer and Colorado beetle have already been noted. The sporadic appearance of the stem borer, *Eldana saccharina* as a pest of sugarcane and cereals, in different parts of Africa (Girling 1972) is an example of a species extending its range of host plants to include crops, although it has not yet been introduced beyond its natural distribution area.

Another consequence of colonization may be competitive displacement by a more aggressive immigrant species. A familiar British example is the displacement over the greater part of its range, of the native red squirrel

E

by the North American grey squirrel. This has frequently happened among introduced insect pests. Thus the citrus scale insect *Aonidiella citrina* has been replaced in parts of California by *A. aurantii* (DeBach & Sundby 1963) and in Hawaii the Mediterranean fruitfly (*Ceratitis capitata*) by the later invading Oriental fruitfly (*Dacus dorsalis*) (Christenson & Foote 1960).

Remedial measures

The obvious action to be taken against future introductions of potential pests is the institution of quarantine measures. These are now almost universal although not always rigorously enforced. They take the form of regulation of imports by prohibition of potentially dangerous material and authorization of imports only of material free of pests or from areas known to be free, fumigation of produce cargoes and vehicles, importation of live material for propagation only through quarantine stations where it may be observed before release and so on.

Even in areas where quarantine is rigorously applied new introductions of important pests periodically occur. The appearance of the pine woolly aphid *Pineus pini* in Kenya during the 1960's is an important example as are the steady stream of immigrant pests reported from the Hawaiian Islands which, since 1968, have included such diverse important insect pests as the grasshopper, *Schistocerca vaga*, the bean fly *Ophiomyia phaseoli*, citrus whitefly (*Aleurocanthus spiniferus*) and the banana skipper. Once detected, eradication campaigns are frequently mounted but are seldom successful as in the instance of *Pineus pini* in Kenya.

When established, the pest is often more important than it is in its country of origin, and this can frequently be attributed to escape from its natural control factors. Where these are natural enemies, biological control provides an attractive means of establishing a more favourable balance, by introducing some of the more important missing natural enemies without their own natural enemies. Thus freed of their own burden of checks, it is hoped that they will be even more effective than in their original home.

However, in practice it is seldom so simple. The ecology of both pest and natural enemy may be somewhat different so that the latter may be unpredictably less or more effective. The tolerance of the two species to other factors may not coincide, so that control is only effective in part of the affected area. This has happened with control of the cottony cushion scale, *Icerya purchasi*, in the Mediterranean where the scale is more tolerant of cold than its natural enemy, *Rodolia cardinalis*, and remains a menace in the foothills of the Alps (see Greathead 1976). The natural enemy may find conditions highly favourable and a population explosion may result, as discussed in relation to pests, which leads to a degree of control.

While research may indicate which natural enemies should in theory be the most successful, the actual introduction is the final crucial experiment. The new environment may, in some subtle way, be different so that it is extremely difficult, if not impossible, to forecast exactly what will happen just as it is virtually impossible to predict accurately the degree to which a species may become a pest on introduction to a new environment (Simmonds 1972). On the other hand progress has been made both in the field of biological control, and in forecasting the danger from pests; in defining the physical requirements of species from laboratory studies (Baker 1972); and in experimental releases such as the attempt to colonize the thistle feeding beetle, *Altica carduorum*, in Britain (Baker *et al.* 1972). More such studies are required to reduce the area of uncertainty surrounding introductions both accidental and deliberate.

In spite of the uncertainties biological control by introduction of natural enemies has been successfully practised since the well known success when the ladybird, *Rodolia cardinalis*, and the dipterous parasite, *Cryptochetum iceryae*, were introduced into California in 1888 for the control of the cottony cushion scale. This success has been repeated round the world usually with the same success—some thirty-seven introductions of *R. cardinalis* have been traced. Other biological controls which have been repeated many times are the control of the apple woolly aphid by the parasite *Aphelinus mali* first carried out in France in 1919 it has been repeated at least thirty times although not with uniform success. Control of the citrus blackfly, *Aleurocanthus woglumi*, by parasites first in Cuba in 1930 has been attempted, usually very successfully, in fourteen countries.

The control of prickly pears in Australia is the first and most famous success in the biological control of weeds. Similar success was achieved in South Africa, India and on many islands. Another repeated success in weed control has been that of St. John's wort by *Chrysolina* spp. first in Australia where a long period of acclimatization ensued before control was achieved, later with rapid success in California and other Pacific seaboard states of the USA, and since then in other areas including Canada, New Zealand, Chile and South Africa.

There have now been many successes in biological control. DeBach (1964) lists 113 more or less successful original attempts at classical biological control of insect pests and his list is not exhaustive. Since then there have been other successes. Not all attempts have provided a degree of control and this is a disadvantage of the method.

It has at times been claimed that biological control in the classical sense has already achieved most of the possible successes but with the accelerating mobility of man and even with increasingly strict quarantine measures introductions continue to be made, there is no shortage of opportunities to attempt biological control of introduced species.

An area in which little has yet been achieved but in which much research has been carried out is that of control of tropical waterweeds. The most ubiquitous are the water hyacinth and the waterfern. Control agents have been selected and screened against both of these (e.g. Bennett 1972, 1975) and preliminary releases are promising.

There are opportunities in using introduced natural enemies against native pests, a recent dramatic example is the control of the sugarcane moth borer, *Diatraea saccharalis* in Barbados by a parasite, *Apanteles flavipes* originally a natural enemy of pyralid stemborers in Asia (Alam *et al.* 1971). The changing status of pests also opens further opportunity for biological control. The recent spread of the gypsy moth in North America has stimulated a new attempt. There is therefore much scope for future application of the classic approach to biological control but there is an even greater potential in the use of its methods in developing integrated control or pest management programmes of which there are as yet few examples.

Conclusions

In spite of the large number of species moved around the world by human agency, few have become pests. Although perhaps less than 10% of the insect pest fauna are introduced species the percentage of important pests which are introduced is much higher and in extreme circumstances might include all important pests of a crop. Even though the dangers have been recognized and increasingly widespread and strict quarantine measures are applied, important pests are still establishing themselves in new areas. The method of classical biological control provides an appropriate means of combating such pests and has achieved much although it is still and perhaps always will be impossible to predict the results.

References

ALAM M.M., BENNETT F.D. & CARL K.P. (1971) Biological control of *Diatraea saccharalis* (F.) in Barbados by *Apanteles flavipes* Cam. and *Lixophaga diatraeae* T.T. *Entomophaga* 16, 151–8.
ANON. (1951–1975) Distribution maps of pests. *Commonw. Inst. Ent., London*, Nos. 1–351.
BAKER C.R.B. (1972) An approach to determining potential pest distribution. *EPPO Bull.* no. 3, 5–22.

BAKER C.R.B., BLACKMAN R.L. & CLARIDGE M.F. (1972) Studies on *Haltica carduorum* Guerin (Coleoptera: Chrysomelidae), an alien beetle released in Britain as a contribution to the biological control of creeping thistle, *Cirsium arevense* (L.) Scop. *J. appl. Ecol.* 9, 819–30.

BAKER H.G. (1965) Characteristics and modes of origin of weeds. In *The Genetics of Colonising Species* (Ed. by H.G. Baker & G.L. Stebbins), pp. 147–68. Academic Press, New York and London.

BAKER H.G. & STEBBINS G.L. (Eds.) (1965) *The Genetics of Colonising Species.* Academic Press, New York and London.

BENNETT F.D. (1972) Survey and assessment of the natural enemies of the water hyacinth, *Eichhornia crassipes. PANS* 18, 310–11.

BENNETT F.D. (1975) Insects and plant pathogens for the control of *Salvinia* and *Pistia. Proc. Symp. Water Quality Management biol. Contr., Florida,* pp. 28–35.

BODENHEIMER F.S. (1951) *Citrus Entomology in the Middle East.* Junk, The Hague.

BOX H.E. (1953) *List of sugar-cane insects.* Commonw. Inst. Ent., London.

CHRISTENSON L.D. & FOOTE R.H. (1960) Biology of fruitflies. *A. Rev. Ent.* 5, 171–92.

CLAUSEN C.P. (1956) Biological control of insect pests in the continental United States. *Tech. Bull U.S. Dep. Agric.* No. 1139.

DAVID W.A.L. (1949) Air transport and insects of agricultural importance. Commonw. Inst. Ent., London.

DEBACH P. (1964) *Biological Control of Insect Pests and Weeds.* Reinhold, New York.

DEBACH P & SUNDBY R.A. (1963) Competitive displacement between ecological homologues. *Hilgardia* 34, 105–66.

ELTON C.S. (1958) *The Ecology of Invasions by Animals and Plants.* Methuen, London.

ENTWISTLE P.F. (1972) *Pests of Cocoa.* Longmans, London.

GIRLING D.J. (1972) *Eldana saccharina* Wlk. (Lepidoptera: Pyralidae) a pest of sugar-cane in East Africa. *Proc. 14th Congr. Int. Soc. Sug. Cane Technol., Baton Rouge,* pp. 429–34.

GREATHEAD D.J. (1971) A review of biological control in the Ethiopian Region. *Tech. Commun. Commonw. Inst. biol. Contr.,* no. 5.

GREATHEAD D.J. (1975) The ecology of a scale insect, *Aulacaspis tegalensis,* on sugar cane in East Africa. *Trans. R. ent. Soc. Lond.,* 127, 101–14.

GREATHEAD D.J. (Ed.) (1976) A review of biological control in western and southern Europe. *Tech. Commun. Commonw. Inst. biol. Contr.,* no. 7.

HALL J.W. (1958) Outbreaks and new records. *Pl. Prot. Bull. FAO* 6, 159.

HARGREAVES H. (1948) List of recorded cotton insects of the world. Commonw. Inst. Ent. London.

HARPER J.L. (1965) Establishment, aggression, and cohabitation in weedy species. In *The Genetics of Colonising Species* (Ed. by H.G. Baker & G.L. Stebbins), pp. 243–65. Academic Press, New York and London.

HARRIS P., PESCHKEN D. & MILROY J. (1969) The status of biological control of the weed *Hypericum perforatum* in British Columbia. *Canad. Ent.* 101, 1–15.

LE PELLEY R.H. (1959) Agricultural insects of East Africa. E. Afr. High Comm., Nairobi.

LE PELLEY R.H. (1968) *Pests of Coffee.* Longmans, London

LEWIS T., CHERRETT J.M., HAINES I., HAINES J.B. & MATHIAS P.L. (1976) The crazy ant (*Anoplolepis longipes* (Jerd.) (Hymenoptera, Formicidae)) in Seychelles, and its chemical control. *Bull. ent. Res.* 66, 97–111.

LINDROTH C.H. (1957) *The Faunal Connections between Europe and North America.* John Wiley & Sons, New York.

MAC ARTHUR R.H. (1972) *Geographical Ecology. Patterns in the distribution of species.* Harper & Row, New York.

MAC ARTHUR R.H. & WILSON E.O. (1967) *The Theory of Island Biogeography.* Princeton University Press, Princeton.

MAYR E. (1965) The nature of colonisation in birds. In *The Genetics of Colonising Species* (Ed. by H.G. Baker & G.L. Stebbins), pp. 29–43. Academic Press, New York & London.

MOUTIA L.A. & MAMET R. (1946). A review of twenty-five years of economic entomology in the island of Mauritius. *Bull. ent. Res.* 36, 439–72.

PEMBERTON C.E. (1964) Highlights in the history of entomology in Hawaii 1778–1963. *Pacif. Insects* 6, 689–729.

PEMBERTON C.E. & WILLIAMS J.R. (1969) Distribution origin and spread of sugar cane pests. In *Pests of Sugar Cane* (Ed. by J.R. Williams, J.R. Metcalfe, R.W. Mungomery & R. Mathes), pp. 1–9. Elsevier, Amsterdam.

PIMENTEL D. (1966) Population ecology of insect invaders of the Maritime Provinces. *Canad. Ent.* 98, 887–94.

RAINWATER H.I. (1963) Agricultural insect pest hitchhikers on aircraft. *Proc. Hawaii ent. Soc.* 18, 303–9.

SALISBURY E.J. (1961) *Weeds and Aliens.* Collins, London.

SIMMONDS F.J. (1972). Approaches to biological control problems. *Entomophaga* 17, 251–64.

THOMPSON W.R. & PARKER H.L. (1928) Host selection in *Pyrausta nubilalis* Hübner. *Bull. ent. Res.* 18, 359–64.

TROUVELOT B. (1936) Le doryphore de la pomme de terre (*Leptinotarsa decemlineata* Say) en Amerique du Nord. *Annls. Epiphyt.* (*NS*) 1, 277–336.

WILSON F. (1960) A review of the biological control of insects and weeds in Australia and Australian New Guinea. *Tech. Commun. Commonw. Inst. biol. Contr.* no. 1.

WILSON F. (1965) Biological control and the genetics of colonising species. In *The Genetics of Colonising Species* (Ed. by H.G. Baker & G.L. Stebbins), pp. 307–25. Academic Press, New York and London.

WOOD B.J. (1968) *Pests of Oil Palms in Malaysia and their Control.* The Incorporated Society of Planters, Kuala Lumpur.

Monoculture

Pest and disease status in mixed stands vs. monocultures; the relevance of ecosystem stability

M. J. WAY *Department of Zoology and Applied Entomology, Imperial College, London*

Introduction

It is widely accepted that the stability of a community and of its constituent species is positively related to its diversity. Few people would dispute this as a broad generalization yet, everywhere, successful agriculture with its increasing emphasis on decreased diversity seems to present a direct challenge to this principle. Questions have therefore been posed, such as—is the diversity/stability relationship a functional one, and can it apply meaningfully to arable crops equivalent to early stages in ecological succession that are naturally very unstable (i.e. impermanent) irrespective of their diversity? In this respect, man-manipulated annual crops are undoubtedly more stable than their equivalent in nature except in some arid regions where natural succession is seasonally halted by drought at the annual plant stage. The question must nevertheless be asked whether the more unnatural agro-ecosystems obey different laws from those which apply to natural ecosystems. Some implications for pest incidence and control have been recently discussed by van Emden & Williams (1974) and Murdoch (1975) whose views in general are supported here.

In this paper, diversity is taken to cover the richness and variety of plant and animal species and occasionally of abiotic habitat components. The definition of stability is more elusive (van Emden & Williams 1974, Orians 1975, Southwood 1975). In the present discussion, it is used to cover the maintenance of a fairly constant mean state, either of a species mix in a community (community stability) or of a particular species population (species stability) irrespective of periodic displacements from the mean. A community or population is, however, more or less stable depending on the amplitude and frequency of the fluctuations. Pest damage will tend to be less in more stable populations because the peaks in fluctuations and probably the overall means are lower than in less stable populations. However, very stable conditions, whilst conferring freedom from periodic pest outbreaks, do not ensure freedom from pest damage. There are, for example, many low-density pests that maintain relatively very stable populations yet are severely damaging, e.g. tse-tse fly (*Glossina* spp.), rhinoceros beetle (*Orcytes* spp.) and various insect pests of tropical trees.

In this paper, the time scale for stability is taken as variable, perhaps only a few months for an annual plant community or centuries for a forest community. Man is usually envisaged as seeking to maintain or impose artificial forms of community and species stability. In agriculture, for example, he usually seeks the highest levels of productivity compatible with his ability to afford time and expense on artificial stabilizing mechanisms such as cultural practices and pesticides.

Diversity in relation to the incidence of pests and diseases—the regional situation

It is well known but not always remembered that, fundamentally, most of our pests occur because there is too much diversity in the form of alternate food and refuges that are essential at some stage in the life cycle of the pest. Some of the evidence has been outlined by van Emden (1965); it includes the classic example of the Rocky Mountain locust, *Melanoplus spretus*, which in the 19th century devastated crops in parts of the United States of America. This species is now extinct, or eradicated as the damaging migratory phase, because advancing agriculture destroyed the diverse natural grasslands where the swarming populations built up (Uvarov 1928). Pest problems, in general, seem to be most acute at the interface between two kinds of habitat where diversity is greatest and where the diverse elements of both habitats are exploited. Thus it is well recognized that many arthropod transmitted diseases such as yellow fever, leishmaniasis and tick typhus, are most serious at the boundary between different ecosystems, e.g. forest and savannah or forest and agro-ecosystems (Garnham 1971). Similarly, there are serious crop pest problems in the Sudan Savannah, for example, where the main pests of sorghum—birds, grasshoppers, sorghum bug, shoot fly and midge—all originate from wild tree and herbaceous vegetation of the natural savannah. The characteristic patchwork of 'bush' and cropped land throughout much of the tropics also creates great diversity that exacerbates pest problems; thus, many pests depend upon and originate from the bush, such as the cotton stainers, *Dysdercus* spp., whose pest status in a particular region is entirely related to presence and abundance of alternate hosts in the bush (Pearson 1958).

The above evidence indicates that severe pest problems are exacerbated at intermediate stages in the process of simplification of the ecosystem but can diminish in the ultimately simple system. In other words, the first stages in breakdown of the delicate natural 'climax' are particularly deleterious. Further simplification can lead to another form of stability when pest problems recede. This last stage is well exemplified by the vast areas of large wheat monocultures in the Canadian prairies which Turnbull & Chant (1961) recognize as being highly stable compared with many natural ecosystems.

The above evidence implies that within a region or group of ecosystems, diversity does not prevent pest problems; on the contrary, it more often seems to create them. However, only within smaller ecological units such as a forest or farm ecosystem can intricate relationships be expected to evolve, and it is at this level that diversity/stability relationships and their implications for pest and disease problems need also to be examined.

The reality of the diversity/stability concept in an ecosystem context

It is significant that field observations supporting the diversity/stability concept come overwhelmingly from forest ecosystems. This is no doubt partly because they have provided obvious long-term comparisons between the more diverse natural systems and the less diverse plantation monocultures. Moreover, forests are also climax-type ecosystems and hence by definition are fundamentally stable compared with earlier stages in succession.

The many examples of pest action in temperate forests support Harper's (1969) emphasis on the importance of herbivores in maintaining vegetation diversity. Outbreaks are the symptom of species' instability yet they are said to maintain community diversity and stability through periodic devastation of any tree species which tends to become dominant (Turnbull 1969). There are numerous examples of such community stabilizing action, especially in North America, ranging from spruce budworm in parts of Canada and the United States of America, to situations as in California where, in places, a five or six species mix of major conifers is said to be maintained by specific insects, together with diseases, killing back populations of the species which begins to predominate (Graham 1956, Dahlsten, personal communication). Such action explains the greater frequency of pest outbreaks in simplified forests and some plantations (Pimentel 1961, Gibson & Jones 1977). Catastrophic species-destabilizing but community-stabilizing outbreaks are therefore associated with the tendency of a particular tree species to predominate.

The question now arises as to the role of species diversity *per se* in maintaining species stability (i.e. minimizing outbreaks) as distinct from the importance of mere sparsity of the tree species at risk. In forests, there is little evidence of the reality of complex balances and checks acting through the trophic web structure of a diverse, as distinct from a simple, system. Most often, there seems to be a basically simple interaction between the tree and a specific herbivore—usually an insect or an insect associated with a disease. When the plant population tends to rise above the equilibrium level, there is feedback of more conspicuous herbivore action and vice-versa (Harper 1969). This maintains an equilibrium density of each tree species which is determined primarily by the dispersive qualities and the searching ability of its specific herbivore and by the nature of the herbivore's impact on the tree. In temperate conditions, the herbivore action is frequently delayed until the particular plant population has risen well beyond the equilibrium density, so creating delayed density-dependent over-compensating outbreaks. Man's aim since the 18th century has been to log the crop before it reaches the over-mature outbreak-inducing stage and to use cultural control practices which help extend the time-lag (Vité 1971). This has proved surprisingly successful for timber forest monocultures in Europe and in parts of the United States of America (Gibson & Jones 1977). It is remarkable that, given good husbandry, pest outbreaks in many artificially established temperate forest monocultures are as scarce as they are. The most intractable conditions seem to be where the natural forest has been altered, perhaps relatively little, leaving a dislocated climax system which has lost natural stability but is poorly sited and designed for the cultural practices that become feasible in a planned forest.

The tropical rain forest is always epitomized as the quintessence of the diversity-stability relationship; but as pointed out by Gray (1972), it is not so free of pest outbreaks as is popularly imagined nor is there significant evidence of stability created by the complex trophic web that is usually assumed to exist in these circumstances. On the contrary, the work of Janzen (1971), for example, indicates that diversity and stability are maintained by straightforward herbivore-host relationships comparable to, but operating more elegantly and directly than in temperate environments. Janzen (1971) points to intense destruction by specific fruit and seed eating beetles as the cause of the widely spaced dispersion of certain trees in Central American forests. A comparable situation exists in the equatorial forests in Africa where valuable timber trees such as the mahoganies and *Chlorophora* spp. are seemingly each kept widely dispersed in the forest by a single species of highly specific insect—a lepidopteran, *Hypsipila* spp., on mahoganies, and a psyllid, *Phytolyma* spp., attacking *Chlorophora*. These severely stunt growth and mostly kill young trees indirectly by making them non-competitive. They begin to fail to do this only where the host becomes sufficiently widely dispersed in the forest. Control can be envisaged as by mutual density dependent action of the herbivore and its tree host on each other. The crippling action of the pest has created a seemingly intractable problem for monoculture production of these species in plantations (White 1966, Gibson & Jones 1977). The answer no doubt lies in mimicking hazards for the pests that operate in the natural forest. Proponents of diversity might advocate the need to maintain or create a complex species mix, but this must be questioned on the basis that all that may be required is a matrix of other trees that need not be species diverse. Foresters and forest entomologists (Entwistle 1967) have, for example, recommended the culture of mahogany suitably spaced out in a monoculture of a useful exotic species (which would be free of major pests in an exotic environment) or in a monoculture of an indigenous successional tree species like *Triplochiton scleroxylon* in West Africa which, unlike the climax species, is characteristically not debilitated in monoculture-like conditions. It should be emphasized, however, that it is not yet known whether a single non-host species provides a better or worse hazard than a complex mix. This is a sign of lack of understanding of searching behaviour of major insect pests. Thus Kennedy's (1966, 1974) ideas on 'rebound' from non-hosts need to be examined to test whether a species is more, or less, likely to get lost when faced with a diverse mix of non-host species than with a non-host monoculture.

This conclusion suggests that critical forms of species diversity creating hazards or camouflage in a diverse system may be more important than is realized for stabilizing populations and minimizing pest problems, not only in perennial systems for also for arable crops. Such effects may have been largely responsible for the differences in pest infestation recorded by Pimentel (1961) between a mixed and a pure stand of a brassica crop. The camouflaging effect of weeds was the prime factor responsible for much smaller infestations of weedy than weed-free brassica crops by the aphid, *Brevicoryne brassicae* (Smith 1969, 1976), and undercover crops can have similar effects on several different pests (O'Donnell & Coaker 1975). It has long been recognized that field beans are less infested by *Aphis fabae* when mixed with oats than in a monoculture, but a similar result can be obtained in a field bean monoculture sown sufficiently densely for the crop to meet across the rows before the aphid colonists arrive (Way 1971). So, the diversity may act merely as a camouflage, filling spaces between the plants at risk and preventing them from being highlighted against bare soil. Nevertheless, self-camouflage is probably of limited significance, as is the crude quantity of diversity. Most often, it is the right quality of diversity appropriately dispersed that is likely to be critically important. Unfortunately, as pointed out by Orians (1975), ecological theory still has relatively little to offer on the role of pattern and diversity in space. Yet everywhere very important practical questions are being posed. For example, there seems little doubt that the empirically developed mixed cropping systems which provide the livelihood of about 1,000 million people in peasant families in the tropics contain important elements of plant protection from pests and diseases. The reasons for decreased damage are not understood and, until they are, it will be impossible to manipulate diversity to improve the system. As already indicated, other crops can act as a barrier or camouflage; but it is also known that, if they are alternative hosts, they can divert pests from the crop at-risk. Paradoxically, however, as already indicated, other hosts can be a source of pests. It is well known, for example, that the same plant can sometimes be a damaging source of a cotton pest such as the bollworm, *Heliothis armigera*, and sometimes a valuable diversionary host for it, depending on the nature of the cropping sequence (Pearson 1958). Whether to simplify in order to dislocate the pests' host plant sequence or, conversely, to diversify to provide hazards or diversionary hosts, is the kind of quandary to which at present we can rarely offer a useful answer.

Diversity in relation to the stabilizing action of natural enemies

Many published data relate to this problem. They can be summarized as follows:

1. There is no evidence that there is any relationship between the diversity of different natural enemy species in a system and the success of the complex in keeping pests in control. Thus most of the approximately forty-five species of arthropods predacious on red spider mites in an English apple orchard are seemingly irrelevant (Collyer 1953). That a very few predator species in a complex can be of key importance is apparent from the success of a single pesticide-resistant species in controlling spider mites in some orchards in the United States of America where the other natural enemies have been virtually obliterated by a pesticide (Crofts 1975a, 1975b).

2. There is evidence that other species in a natural enemy complex can upset control by one or a few key species. In English apple orchards, it seems probable that one of the few important natural enemies of the fruit tree red spider, a capsid, may sometimes be detrimental because, when spider mites are scarce, it can prey on the crucially important predacious mites (M. E. Solomon, personal communication). This suggestion is borne out by the thirty-eight species of arthropods recorded in Germany preying on predacious mites as well as on the pest mites (Kramer 1961). The best documented examples are perhaps Taylor's (1937) work on biological control of the coconut leaf miner in Fiji, and Varley, Gradwell & Hassell's

(1973) studies on dynamics of the winter moth in England together with Embree's (1966) work on its biological control in Canada. They and others provide striking support for Varley's (1959) conclusion that some insects are pests because of, rather than in spite of, a large diversity of natural enemies.

3. In all the successful examples of classical biological control, a single natural enemy species is responsible in most instances and never more than two or perhaps three. In the latter cases, there is complementary action, e.g. one parasite takes over in winter, the other in summer (Huffaker & Kennett 1966) or one or two parasites act at low density and a pathogen at higher density (Turnbull & Chant 1961). Furthermore, virtually all the successful examples of classical biological control are in very simple ecosystems such as monoculture crops of coconuts, coffee, tea, citrus, olives, walnuts, grapes and glasshouse crops.

In summary, it seems likely that in many circumstances, a few species of natural enemy are important, the rest being irrelevant or detrimental. Therefore diversity of a natural enemy complex is not necessarily desirable in pest management. On the contrary, the available evidence suggests that we could appropriately discourage it at least in some perennial habitats. Thus Newsom (1974) recommends augmenting good predators and parasites by using selective pesticides to release them from pressures by their natural enemies and competitors. However, the situation in ephemeral agro-ecosystems needs to be much more carefully examined. Here, as in cereal ecosystems (Potts 1977), a diverse group of generalized natural enemies can kill many pest individuals in circumstances where elegant density-dependent control is impossible. Such crude mortality may often be very important (Hagen, van den Bosch & Dahlsten 1971, Rabb 1971) and is an essential component of some integrated control programmes. Maintenance or enhancement of diversity may sometimes be desirable in these circumstances but there is, no doubt, still a maximum beyond which mutual interference or predator/predator predation becomes detrimental.

Summary and conclusions

Southwood & Way (1970) used diagrammatic models to indicate biotic links important in relation to pest problems. Levins' (1975) loop analysis is a notable step forward in defining biotic links both diagrammatically and mathematically but neither method adequately illustrates the important time and space features that have been emphasized in this paper. In an elaboration of Southwood & Way's (1970) approach, Fig. 1a indicates the situation where a key natural enemy (III) can be distracted from the pest

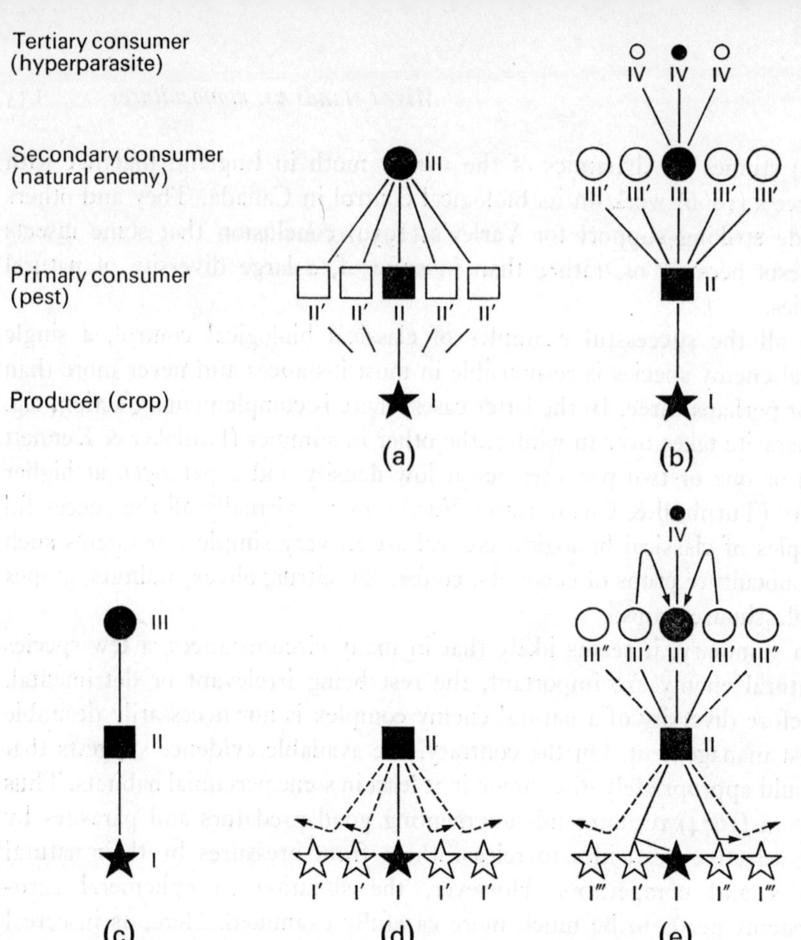

Figure 1. Diagrams indicating some effects of diversity at different trophic levels.
(a) A potentially important natural enemy (III) may be diverted to alternate prey or hosts (II′) and so fail periodically or consistently to control the pest (II) at below damaging levels. Alternatively, one or more of the alternate prey/hosts (II′) may beneficially bridge a gap in prey/host (II) availability.
(b) Competing natural enemies (III′) may, with hyperparasites (IV), upset the delicate control that the key natural enemy (III) could otherwise exert on the pest (II).
(c) A system which describes most examples of successful classical biological control. An introduced specific natural enemy (III) free of hyperparasites exerts elegant density dependent control on the key pest (II) without being diverted or interfered with by competitors.
(d) The system that may regulate distribution and abundance of some tropical forest tree species. A severely damaging pest (II) is specific to a particular tree (I), the two species mutually regulating each other in a system where a preponderance of non-host trees (I′) provides hazards for the pest, the hazard increasing with increase in abundance of other trees relative to the tree at-risk.
(e) A composite of important elements of many situations. A key natural enemy (III) is killed or interfered with by other natural enemies (III′) either incidentally or when normal prey is scarce. Crop I′ acts as an alternate host on which the pest can survive or build up before dispersing in damaging numbers to the crop at-risk (I). Crop I″ is an alternative diversionary host which is attractive to pest II at the same time as the crop at-risk (I). Crops I‴ are not hosts of the key pest but in a suitably diverse system may provide a barrier or camouflage decreasing numbers of the pest that reach I.

(II) by other prey/hosts (II'). Fig. 1b indicates that competing natural enemies (III') like hyperparasites (IV) can interfere with control by the key species (III). In both these situations, diversity may be said to be destabilizing. In contrast to undesirable effects of such diversity at the primary and secondary consumer levels, a very simple system, as in Fig. 1c, gives stability as represented by many successful examples of classical biological control. Figure 1d also illustrates a basically even less diverse relationship which is highly stable. Mutual direct density dependent action between a herbivore (II) and its plant host (I) is buffered by hazards (I'). Finally, Fig. 1e is a composite of several real-life situations; III is a key natural enemy of pest II. III' are other natural enemies which are destabilizing because they can prey on III. III'' are natural enemies of II but are insignificant in regulating it. Further along the trophic chain, the pest (II) may depend on host I' in order to move damagingly to host I. In contrast, I'' acts as a beneficial diversionary host and plants I''' as hazards.

In terms of pest incidence and control, possible components of monoculture strength may be summarized as follows:

1. Decreased diversity can dislocate the pests' life cycle:
   (a) the pest may be deprived of alternate plant food and refuges.
   (b) pest attack is diluted by the sheer abundance of a plant host.
   (c) with large uniform units of monoculture, pests that concentrate on field edges cause less overall damage.
   (d) with annual crops, standardizing the sowing date can put crops out of phase with a pest.

2. Decreased diversity can benefit natural enemies:
   (a) appropriate decrease in diversity of natural enemy species can decrease harm from interfering natural enemies and hyperparasites.
   (b) decreased plant diversity decreases the hazards to natural enemies searching for their prey/hosts on the crop at risk.
   (c) decreased prey/host diversity can decrease likelihood of key natural enemies being diverted to other hosts.
   (d) decreased diversity as indicated by a continuous sequence of annual-crop monoculture creates opportunities for perennial crop-type equilibria between natural enemies and their host/prey.

In contrast appropriate plant diversity might be expected to decrease pest attack as follows:

1. by providing camouflage making the at-risk crop less visible to the pest.
2. by acting as a barrier or hazard to the pest.
3. by providing alternative hosts which divert the pest from the at-risk crop.

4. by benefiting natural enemy action:
(a) providing food for non-predatory/parasitic stages of natural enemies.
(b) providing food for alternate prey/hosts needed to perpetuate predatory or parasitic stages of natural enemies.
(c) providing refuges for natural enemies.

In conclusion:

1. The general statement that diversity and stability are related must be examined very critically in relation to any particular situation; the key functional links conferring stability are likely to be few.

2. That usually only a few factors seem to be crucially important in creating or ameliorating pest problems makes population ecology as a basis for pest control both meaningful and practical (Way 1973).

3. More attention needs to be paid to the ways in which spatial diversity works and to the nature of spatial diversity that best hampers pests.

References

COLLYER E. (1953) Biology of some predatory insects and mites associated with the fruit tree red spider mite *Metatetranychus ulmi* Koch. in south-eastern England. *J. hort. Sci.* 28, 85–97.

CROFTS B.A. (1975a) Integrated control of orchard pests in the USA. *C.r. 5th Sym. Lutte intégrée en vergers.* OILB/SROP, 109–24.

CROFTS B.A. (1975b) Tree fruit pest management. In *Introduction to Insect Pest Management* (Ed. by R.L. Metcalf & W. Luckmann), pp. 471–507. Wiley Interscience, New York.

EMBREE D.G. (1966) The role of introduced parasites in the control of winter moth in Nova Scotia. *Canad. Ent.* 98, 1159–68.

EMDEN H.F. van (1965) The role of uncultivated land in the biology of crop pests and beneficial insects. *Sci. Hort.* 17, 121–36.

EMDEN H.F. van & WILLIAMS G.F. (1974) Insect stability and diversity in agro-ecosystems. *A. Rev. Ent.* 19, 455–75.

ENTWISTLE P.F. (1967) The current situation on shoot, fruit and collar borers of the Meliaceae. *Proc. 9th Br. Commonw. For. Conf.* New Delhi, *1968.* Comm. For. Instit. Oxford.

GARNHAM P.C.C. (1971) *Progress in Parasitology.* Athlone Press, London.

GIBSON I.A.S. & JONES T. (1977) Monoculture as the origin of major forest pests and diseases, especially in the tropics and southern hemisphere. In *Origins of Pest, Parasite, Disease and Weed Problems* (Ed. by J.M. Cherrett & G.R. Sagar), pp. 139–61. Blackwell Scientific Publications, Oxford.

GRAHAM S.A. (1956) Forest insects and the law of natural compensations. *Canad. Ent.* 88, 45–55.

GRAY B. (1972) Economic tropical forest entomology. *A. Rev. Ent.* 17, 313–53.

HAGEN K.S., VAN DEN BOSCH R. & DAHLSTEN D.L. (1971) The importance of naturally-occurring biological control in the western United States. In *Biological Control* (Ed. by C.B. Huffaker), pp. 253–93. Plenum, New York.

HARPER J.L. (1969) The role of predation in vegetational diversity. In *Diversity and Stability in Ecological Systems*. Brookhaven Symp. Biol. **22**, 48–61.

HUFFAKER C.B. & KENNETT C.E. (1966) Biological control of *Parlatoria oleae* (Colvée) by the compensatory action of two introduced parasites. *Hilgardia* **37**, 283–335.

JANZEN D.H. (1971) Seed predation by animals. *A. Rev. Ecol. and Systematics* **2**, 465–92.

KENNEDY J.S. (1966) The balance between antagonistic induction and depression of flight activity in *Aphis fabae* Scopoli. *J. exp. Biol.* **45**, 215–28.

KENNEDY J.S. (1974) Changes of responsiveness in the patterning of behavioural sequences. In *Experimental Analysis of Insect Behaviour* (Ed. by L. Barton Browne), pp. 1–6. Springer-Verlag, Berlin.

KRAMER P. (1961) Untersuchungen über den Einfluss einiger Arthropoden auf Raubmilben. *Z. angew. Ent.* **48**, 257–311.

LEVINS R. (1975) Evolution in communities near equilibrium. In *Ecology and Evolution of Communities* (Ed. by M.J. Cody & J.M. Diamond), pp. 16–50. Belknap, Harvard.

MURDOCH W.W. (1975) Diversity, complexity, stability and pest control. *J. appl. Ecol.* **12**, 795–807.

NEWSOM L.D. (1974) Predator–insecticide relationships. *Entomophaga* **7**, 13–23.

O'DONNELL M.S. & COAKER T.H. (1975) Potential of intra-crop diversity for the control of brassica pests. *Proc. 8th Brit. Insecticide & Fungicide Conf. 1975* **1**, 101–7.

ORIANS G.H. (1975) Diversity, stability and maturity in natural ecosystems. *Proc. 1st Int. Cong. Ecol.*, The Hague, *1974*, 139–50.

PEARSON E.O. (1958) *The Insect Pests of Cotton in Tropical Africa*. Commonw. Inst. Ent. London.

PIMENTEL D. (1961) Species diversity and insect population outbreaks. *Ann. ent. Soc. Amer.* **54**, 76–86.

POTTS G.R. (1977) Some effects of increasing the monoculture of cereals. In *Origins of Pest, Parasite, Disease and Weed Problems* (Ed. by J.M. Cherrett & G.R. Sagar), pp. 183–202. Blackwell Scientific Publications, Oxford.

RABB R.L. (1971) Naturally occurring biological control in the eastern United States, with particular reference to tobacco insects. In *Biological Control* (Ed. by C.B. Huffaker), pp. 294–311. Plenum, New York.

SMITH J.G. (1969) Some effects of crop background on populations of aphids and their natural enemies on brussels sprouts. *Ann. appl. Biol.* **63**, 326–30.

SMITH J.G. (1976) Influence of crop background on aphids and other phytophagous insects on brussels sprouts. *Ann. appl. Biol.* **83**, 1–13.

SOUTHWOOD T.R.E. (1975) The dynamics of insect populations. In *Insects, Science and Society* (Ed. by D. Pimentel), pp. 151–99. Academic Press, New York and London.

SOUTHWOOD T.R.E. & WAY M.J. (1970) Ecological background to pest management. In *Concepts of Pest Management* (Ed. by R.L. Rabb & F.E. Guthrie), pp. 6–28. North Carolina State Univ., Raleigh.

TAYLOR T.H.C. (1937) *The Biological Control of an Insect in Fiji: an Account of the Coconut Leaf-mining Beetle and its Parasite Complex*. Imperial Institute of Entomology, London.

TURNBULL A.L. (1969) The ecological role of pest populations. *Proc. Tall Timb. Conf. Ecol. Anim. Contr. Habit. Manage. No. 1, Tallahassee, Fla. 1969*, 219–32.

TURNBULL A.L. & CHANT D.A. (1961) The practice and theory of biological control of insects in Canada. *Canad. J. Zool.* **39**, 698–753.

UVAROV B.P. (1928) *Locusts and Grasshoppers*. Imperial Bureau of Entomology, London.

VARLEY G. (1959) The biological control of agricultural pests. *J. R. Soc. Arts* **107**, 475–90.

VARLEY G., GRADWELL G.R. & HASSELL M.P. (1973) *Insect Population Ecology, an Analytical Approach.* Blackwell Scientific Publications, Oxford.

VITÉ J.P. (1971) Silviculture and the management of bark beetle pests. *Proc. Tall Timb. Conf. Ecol. Anim. Contr. Habit. Manage. No. 3, Tallahassee, Fla. 1971,* 155–68.

WAY M.J. (1971) *A prospect of pest control.* Inaugural lectures, Imperial College, University of London. 127–62.

WAY M.J. (1973) Applied ecological research needs in relation to population dynamics and control of insect pests. *Proc. FAO Conference ecol. in relation to plant pest control,* 283–97.

WHITE M.G. (1966) The problem of the *Phytolyma* Gall Bug in the establishment of *Chlorophora. Comm. For. Instit. Oxford.* Paper No. 37.

Monoculture as the origin of major forest pests and diseases

I. A. S. GIBSON *Commonwealth Mycological Institute, Kew*
T. JONES *Centre for Overseas Pest Research, London*

A discussion of the relation between the adoption of monoculture* in forestry and the emergence of serious forest pests and diseases has a dual attraction. It is a subject that is likely to produce interesting contrasts with other papers in this symposium and it is also one on which foresters hold strong views which need evaluation.

There are two ways in which the adoption of monocultures may modify the pest and disease hazards of a crop, which may be considered separately although they are related to some extent. There are the immediate effects of the environmental change which may alter the rate of spread and severity of attack by insects, fungi and other pathogens, and the longer-term selection pressures of the cultural systems associated with monocultures, which modify the genetic basis of the crop and alter its resistance to these agencies. The results of the latter course of events are most likely to be found in agricultural field crops, where monocultures have been practised for a very long time. In contrast to this, artificial monocrop systems in forestry have much more recent origin and their effects on the emergence of new pests and diseases are most likely to be the direct result of environmental change. It is this relationship that we shall mainly discuss in this paper.

Before the middle of the last century, foresters were concerned with the management of natural forests so that a balance between exploitation and regeneration was maintained. While this may not always have been successful, prior consideration was given to the role of the forest in the conservation of land and water resources, and to its amenity value. This has meant that the training of foresters has always been oriented towards the preservation

* We have used the terms monocrop and monoculture throughout the text, usually as synonyms, but where the need for a distinction arises, monocrop is used to refer to stands of trees consisting entirely or very largely of a single species which may have natural or artificial origins while monoculture refers to artificial even-aged stands of a single tree species.

of natural communities to a much greater extent than the training of farmers. Indeed, even today, when the objectives of forestry are changing, the first remit of most foresters is to conserve natural forest resources and the forester's outlook still has much in common with that of the ecologist. However, during the last hundred years a rising world population with increasingly sophisticated tastes, combined with the demands of two world wars, have created a drain on our forest resources which can only be met by the adoption of intensive systems of forest management. As the accessible natural forests of the world have been depleted, so they have been largely replaced by even-aged plantations of select fast-growing species. In Europe and America the first major impetus to this was provided by World War I, leading to a steady increase in the area of forest plantation established on natural forest sites or on sites of low agricultural worth. Introduced and local conifers were largely used in this development, with an important element of fast-growing hardwoods, such as poplars. At about the same time, certain territories in the southern hemisphere sought to remedy local shortages of natural timber by planting exotic tree species. Australia, South Africa, New Zealand and Chile were forerunners in this respect, with extensive plantations of exotic softwoods and hardwoods in which pines and eucalypts figured prominently. These ventures were remarkably successful, so that many of the new and developing nations that emerged after World War II, particularly in Africa and parts of Latin America, set up similar industrial forestry schemes, also based on pines and eucalypts but with a significant proportion of other conifers and hardwoods including teak, mahogany, *Acacia*, *Cupressus* and *Araucaria*. This trend is not so apparent in many parts of Asia where there are still appreciable reserves of natural forest in many regions. However, in the Indian sub-continent and (to a lesser extent) Malaysia, there is an increasing interest in plantations of fast-growing exotics, and there are important areas of plantations of native tree species. The area commanded by this type of forestry continues to expand, by 1968 it amounted to a global total of 80 million ha, it increased to an estimated 100 million ha by 1972 and is expected to exceed 160 million ha by 1985. Although much of this represents investment by the developed world, it is in the less developed nations that the rate of expansion is now at its highest (Fraser 1976).

The risks of serious loss from pests and diseases that may accompany the adoption of forest monocrops of this kind have not gone unmarked. Foresters will argue, rightly, that the substitution of the balanced plant communities of the natural forest by even-aged plantations of single species may remove many of the established constraints on local tree pests and pathogens, leading to damaging attacks from these agencies, if not immediately, at any rate in the course of time. While this is essentially true, such arguments have frequently been overstated to the point of asserting that all attempts to develop forest crops using intensive management systems on a large scale are doomed to failure. Indeed, it is surprising how these views have persisted despite the evident success of 'tree farming' in most places where it has been tried. It will be important, therefore, for this reason, to examine how far the establishment of monocultures have led directly to the emergence of serious pests and diseases in these crops and the circumstances under which this may be expected to occur. For such studies we shall require ideally a number of situations in which we can compare the pest and disease regimes of a tree species when it is grown in a heterogeneous environment, with that occurring when it is grown uniformly as an even-aged crop. Such circumstances are hard to find and comparable disease and pest assessments have rarely been made using them.

Most work on forest pathology and entomology is related to intensively managed stands and we know little of the part played by pests and pathogens in the ecology of most natural forest communities. We know that these agencies may be present in relative abundance at times and there is little doubt that the analytical principles, such as those developed by van der Plank (1960, 1963) and others for the interpretation of epidemics, could be applied here. However, this will only be possible when the present lack of field study is made good. While we are better informed on the pests and pathogens of forest plantations, we still lag seriously behind our counterparts in agriculture in this respect. This is partly because of the difficulties posed by the size and longevity of trees in making a study of most kinds of pests and pathogens, and also because the economy of most forest crops still leaves little margin for special protection studies or control methods, despite the improved returns that have accrued to forestry since the end of World War II. It is almost as if to compensate for the forester's limited prospects for effective pest and disease control, that he often has the advantage of avoiding these problems by a choice from a relatively wide range of different tree species for a given quality and quantity of product. Thus, serious risks of loss from insects and fungi that have become evident at an early stage of forest plantation development have often been averted by a timely switch to a more resistant alternative species. However, this has also had the drawback of being too easy an expedient to use. Until recently, diagnoses have often not been properly made under these circumstances and we may be left with too little information on the basis for rejection of one species, or for the final choice of another, for an afforestation scheme. In this way, in particular, we have probably lost information on those diseases especially favoured by nursery monocultures. Despite these limitations we still have sufficient information to determine the origins of a number of our most important forest pests and diseases.

The origins of some pest and disease problems

White pine blister rust, chestnut blight and Dutch elm disease are probably the three most widely quoted examples of important forest diseases and all of these achieved their status by the recent rapid increase in the speed and efficiency of international transport that permitted the introduction of exotic pathogens to regions containing appreciable populations of highly susceptible host trees. The existence of monocultures had little to do with the subsequent damage that was caused.

Similar circumstances, but on a different scale, led to the spread of Jarrah blight in Western Australia, where the extensive losses caused by attack of native *Eucalyptus marginata* by *Phytophthora cinnamomi* was probably directly due to the parochial movement of the pathogen into a natural monocrop by the intensive mechanization of logging in these forests. Here, again, monoculture (as opposed to monocrop) has not been involved (Newhook & Podger 1972). The more recent spread of *Dothistroma* blight into exotic pine plantations of the tropics and southern hemisphere and *Melampsora* rusts into exotic poplars in Australia and New Zealand are examples of two more serious forest diseases which cannot be said to be due to the practice of monoculture. Again, these have been due primarily to the introduction of pathogens from elsewhere and in both cases the damage caused does not appear to have been appreciably modified where the host has been growing in plantations. Certainly, it is our experience in East Africa that *Dothistroma* blight may often be equally severe in young plantations of *Pinus radiata* as it is in isolated trees of the same age growing in the same region. The accounts recently published of the spread of poplar rust in Australia and New Zealand imply the same observations can be made there with regard to poplars (Walker *et al.* 1972, Kraayenoord *et al.* 1974).

Soil-borne diseases caused by wood-destroying basidiomycetes such as *Armillariella mellea, Heterobasidion annosum* (*Fomes annosus*), *Rigidoporus lignosus* and others are also sometimes quoted as examples of diseases that have their origins in forest monoculture systems. However, as these pathogens rely for their activity on the presence of stumps and root residues in the soil, the true origin of these diseases lies more properly in the site history previous to planting rather than its subsequent use. Nevertheless, the fact that monocultures may well have profound effects on the subsequent development of these diseases will be referred to later.

As with diseases, there are many insects whose origins as forest pests have been wrongly attributed to monoculture. The Nun moth, *Lymantria monacha*, is perhaps surprisingly a classic example. The devastation caused by this pest to central European conifer plantations in the 1920's and earlier has been widely cited since then as an example of the kind of disaster likely to follow the widespread adoption of monoculture in forestry. However, this view ignores previous records of *L. monacha* as a major pest of natural conifer stands (as well as a minority of plantations) in central Sweden and southern Germany, long before monocultures were widely established (Hopf, personal communication). Many other notable forest pests can also be shown to have origins unconnected with monoculture. *Adelges abietis* and other species, *Diprion pini*, *Neodiprion* spp. and *Lymantria dispar* in temperate North America (Anon. 1969), *Dendroctonus frontalis* in Honduras and other Central American countries (Beal 1965, Becker 1952). *Ips* spp. and *Paropsis obsoleta* in Australasia (Jones & Gibson 1966) and *Pineus pini* in Africa (Brown 1969, Odera 1974) are all examples of insects which have found natural and artificial forest stands equally suitable for their population build-up and assumption of pest status. Several of these insects are known to have emerged as exotic pests where monocultures of their favoured hosts are virtually non-existent and others (e.g. *Pineus pini*) were major pests in their native habitats before their dispersal elsewhere. Indeed, with *P. pini* in East Africa, we have observed that infestations were often as severe on isolated *Pinus patula* trees as they were in plantations of the same species. In none of the cases noted briefly above can serious infestations be attributed to monoculture practices. Instead, like many of the more notorious forest pathogens, their origins as forest pests can be often ascribed to the efficiency of modern transport systems or conditions preceding the establishment of the host crop. In the latter context, there are species such as the cerambycid borer, *Oemida gahani*, which, like the root diseases caused by cellulose-destroying basidiomycetes, owe as much to previous site history as to practices associated with monoculture for their pest status. We shall discuss *O. gahani* later on, but it is appropriate to note at this point that this pest achieved notoriety when it caused serious losses to recently established *Cupressus* plantations in East Africa (Jones & Curry 1964). The sequence of these events strongly suggested that the insect had been favoured by the forest monoculture that had been adopted. However, it transpired later that the first cause of this pest lay in poor site-clearing practices, which left large quantities of woody residues open to colonization by the pest. Once large populations of *O. gahani* had been built up in these sites, the depletion of available nutrient led to invasion of the standing crop through wounds and pruning scars. As the latter are a general result of plantation practice, there is no doubt that the severity of the outbreaks owed much to the establishment of monocultures. However, this was essentially secondary to the origin of the problem, which lay in poor site preparation. Improvements in site sanitation and control of pruning operations have since gone far to control this pest which is now greatly reduced in significance.

These examples should be sufficient to show that recent outbreaks of serious forest pests and diseases have been by no means always due to the adoption of even-aged plantations of single species and the practices associated with this, and frequently these have had nothing at all to do with these or similar systems. Nevertheless, there is no doubt that monocultures can and do contribute to the variety and severity of forest pests and diseases.

A forest monoculture consists of a crop of a single species, of the same age, planted at uniform spacing with stock derived from a nursery where similar but much more crowded conditions prevail—even making allowance for the smaller size of the individual plants. In the course of its growth the plantation may be subject to thinning and pruning operations which modify its density and expose internal tissues to infection. The most important change that this involves is a large reduction in the average distance between individuals of the same species and age, compared with that in the native state. This will favour parasites with limited radius of effective spread and increase risks from those with a wider dispersal range. Furthermore, in the case of most diseases this will be accompanied by an increase in the birth-rate of the pathogen, which will be dependent on the rate of lesion formation, while its death-rate will remain unchanged (unless special circumstances intervene). When the birth-rate of the pathogen exceeds the death-rate there will be a discontinuous change from sporadic disease to an epidemic state and disease explosion conditions (van der Plank 1960). The latter rule will not apply to most insect infestations, however, as their rate of reproduction depends on other factors and is not usually so closely linked to the number of attack sites or the damage that is caused. Nevertheless, a close relationship between the readily available food source provided by plantations and the rapid build-up of pest populations can often be found. On this basis we may expect the adoption of forest monocultures to enhance risks from diseases to a greater extent than those from pests, but that both of these will be at their highest in the nursery and young plantation stages. However, as the nature of the crop changes with maturity and with the intervention of thinning and pruning operations later on, fresh conditions may be imposed to alter susceptibility of the crop to attack. In addition to this, we may also expect the degree of risk to differ between monocultures of native and exotic species.

Where native species are planted, risks will accrue from host-specialized agencies as well as those of wide host range and facultative habit. By contrast, the planting of exotic species in regions far removed from their native habitats will expose them only to unspecialized pests and pathogens. This distinction is only valid for exotic plantations which have been established for a limited period, but it is sound for most forest plantations at the present day, most of which have been in existence for no more than a single rotation and which consist of species from families and genera limited to certain parts of the world (i.e. tropical pines and *Eucalyptus* spp.). It is, however, a temporary state which will disappear as specialized pests and pathogens catch up with their hosts in exotic locations.

In the following sections we shall examine our information on important forest pests and diseases in the light of these rules to evaluate, as far as possible, the role of monoculture in their origin. In this we shall confine ourselves to pests and diseases which have caused direct economic loss. Problems that have not been assessed objectively or in which increased loss is due to enhanced value of the crop rather than increased activity of pest or disease, will be avoided.

Nursery diseases and pests

DISEASES

There seems to be relatively little reference in the literature to specific examples of forest diseases favoured by nursery systems. Yarwood & Gardner (1972) in discussing *Erysiphe cichoracearum* on *Eucalyptus* spp. in the USA note that powdery mildew infections, while very rare in natural communities, are frequent in cultivated crops, and Peterson & Jewell (1968) quote other authors supporting the view that monoculture of pines in the USA has been associated with increases of attack by stem rusts (*Cronartium* spp.). Neither of these authorities suggest mechanisms by which this relationship occurs but it seems likely that in the case of powdery mildews it is through a direct increase of the infection rate of the pathogen from crowding in the host. The *Cronartium* spp. present a more complex situation, as the cycle for pines calls for infection of an obligate alternate host. Thus, while the effective radius of infection of sporal inoculum from *Cronartium* spp. on pines is often very limited, increasing the population density of pines will only be likely to affect the rate of reinfection by rust fungi if a minimal population of the alternate host is included. Failing this, the density and uniformity of the pine crop will have little effect on its liability to stem rust. de Kam (1973) has recently described *Septotinia podophyllina* as an exotic pathogen of poplar nurseries; this fungus is not important in its native regions where nurseries are infrequent.

If monocultures provide conditions leading to increased pest and disease losses through an increased proximity between host units, then we may expect the greatest incidence and diversity of these problems, particularly diseases, to occur in nurseries where plant crowding is greatest. This appears to be the case, as the lists of pests and diseases of most forest species show a disproportionately large number of them for the short nursery period, compared with those for the subsequent years of growth in the field. Further evidence that these nursery problems are largely favoured by crowded conditions is provided by the fact that many of them are much reduced, or disappear altogether, after the stock is transplanted to the field. This appears as true of diseases of leaves and stems as it is for less specialized root pathogens. In both types it would appear that increasing spacing between nursery and plantation conditions often reduces the rate of reproduction of the pathogen to a point where sporadic rather than epidemic spread occurs.

Three shoot and foliage diseases of pines, terminal crook, *Cercospora* blight and *Lophodermium* needle cast, have been selected to illustrate varying degrees of this decline in disease incidence coincident with out-planting of forest nursery stock.

Terminal crook is caused by *Colletotrichum acutatum* f.sp. *pinea*, a pathogenic form of a widely distributed fungus which can attack pines in the nursery causing a severe stunting and die-back of transplanted seedlings (Dingley & Gilmour 1971). *Pinus radiata* is particularly susceptible and the disease is known in New Zealand, Queensland and Kenya. If severely infected nursery stock is transplanted to the field at a spacing of 1·8 m or more, the disease ceases to spread and the affected plants will recover completely in the course of a few months. The failure to spread is certainly due to decrease in proximity between host plants and we suggest that the recovery occurs because the amount of tissue susceptible to infection (the growing point) is small and requires regular reinfection to create a state of chronic disease. This is possible under crowded nursery conditions but ceases to occur in the plantation.

Cercospora pini-densiflorae causes *Cercospora* blight which was first described from pines native to Japan and achieved importance later when these and other pine species were propagated in nurseries on a large scale in that country. The fungus causes severe defoliation of nursery stock but ceases to spread when infected plants are transferred to the field (Ito 1972). Recovery of individual plants may take place after this, but it is a lengthy process as spread of the pathogen within the infected plant will continue. During the last twenty years *C. pini-densiflorae* has spread to other parts of Asia and to Africa where its epidemiology is rather different. In East Africa, in particular, it is not only a serious nursery pathogen of pines such as *P. radiata* but may also cause severe defoliation in young plantations. There appears to be little to explain this change in the epidemiology under exotic conditions, but it is more likely to be due to the different environment rather than any variation in virulence of the fungus.

A third nursery pathogen, *Lophodermium pinastri*, has become increasingly important in recent years to nurseries in North America and Europe where local pine species are raised and the fungus occurs naturally (Skilling

& Nicholls 1975). The disease often poses sufficient threat to stock to justify regular fungicide spray routines and research continues to try to find more effective compounds for this. While attack by *L. pinastri* is always most severe in nursery populations it may also cause serious defoliation in young plantations. It is a widespread fungus and has been recorded in most places where pines have been planted as an exotic, but under these conditions it is rarely important as a pathogen. This curious feature has yet to be explained and could be due equally to environmental factors or because the more virulent pathotypes of this fungus are still only found where pines are native. There are many other examples of leaf and shoot pathogens achieving prominence when their hosts have been subject to the intensive production methods of the nursery, but which have little importance after the host is planted in the field. *Pleiochaeta albiziae* on *Albizia falcata* in Sri Lanka and Indonesia (Webster 1952), *Phyllachora pterocarpi* (Boaler 1966) in nurseries of *Pterocarpus angolensis* in Africa, *Phyllosticta swieteniae* on *Swietenia mahogani* in the West Indies (Garcia 1939) and *Puccinia psidii* on *Eucalyptus* spp. in Brazil (Joffily 1944, Reis & Hodges 1975) are four random examples illustrating this trend from three continents.

It is tempting to suggest that the majority of root diseases in forest nurseries may also find their origins in the practice of monoculture and relate to crowded host conditions in the same simple way as stem and leaf diseases. However, soil-borne pathogens are generally facultative organisms which spread by vegetative growth through the extremely complex soil environment, often presenting circumstances which are extremely difficult to interpret. It is likely that under these circumstances increases in nutrient that accompany the adoption of uniform crops may often contribute as much to the activity of pathogens as increased proximity between host units. There is evidence that this may occur in damping-off in pine seed beds where the activity of the pathogens has been found to relate to the mass of seeds rather than the numbers sown per unit area (Gibson 1956). In general, therefore, while more root diseases can be found in the nursery than in the plantation, it is generally difficult to establish the direct role of a uniform host population in bringing this about. It is true also, that few of these diseases seem to survive transplantation of their host to the field and, where survival occurs, as in crown gall (*Agrobacterium tumefaciens*) charcoal root disease (*Macrophomina phaseolina*) or *Fusarium* root disease of pines (Smith 1975) there is a marked decline in their importance. However, it is seldom clear whether this is due to a direct effect on the ability of the pathogen to spread or indirectly caused by changes in environmental and host factors.

PESTS

Within the Insecta there is a wide range of phytophagous pests including cutworms, chafers, grasshoppers, ants, aphids, coccids, weevils and termites which have been particularly troublesome in forest nurseries. However, most of these, like a few of the nursery pathogens, are long-established agricultural pests that have taken advantage of the additional source of food that the nursery provides. Apart from this, the monoculture of the forest nursery has made little difference to the intensity of their attack. There are, however, other nursery pests that have clearly found their origin through the special conditions present in nurseries. Infestations of the leaf-gall insect (*Pauropsylla tuberculata*) and leaf-roller (*Parotis marginata*) of *Alstonia scholaris* in Bangladesh (Chowdhury, personal communication), the scale insect *Eriococcus araucariae* on *Araucaria cunninghamii* in Kenya (Cole, personal communication) and *Phytolyma lata* (in the broad sense), the cause of leaf-gall of *Chlorophora excelsa* in East and West Africa (White 1966) all provide examples where nursery conditions have favoured the emergence of insects from relative obscurity to take their place as forest pests of major importance. The emergence of certain scolytid and cerambycid pests may be due, in part at least, to nursery conditions leading to host-stress from overcrowding, drought, insolation and other factors. These species would otherwise normally infect dead and dying trees or logs and are insignificant as pests of healthy trees of their favoured hosts. However, periodically some of these insects emerge as serious nursery pests. Examples include *Xyleborus mascarensis*, *X. sharpae* and *X. semiopacus* on *Khaya ivorensis* seedlings in Ghana (Jones & Roberts 1959, Browne 1962), the cerambycid *Phryneta leprosa* on young *Chlorophora excelsa* in Nigerian nurseries (Roberts 1969) and *Phoracantha semipunctata* on eucalypts in Israel (Bytinski-Salz & Neumark 1953). It appears, therefore, that nursery conditions have given rise to a significant range of insect pest problems, including all combinations of exotic and native hosts and pests at one location or another, many of which will be considered further in the context of the plantation.

GENERAL

This review has been necessarily brief, but it is perhaps sufficient to show the conditions provided by monoculture which have played their most important part in contributing to the origins of a number of forest nursery pests and pathogens. It is curious, therefore, that while many have caused serious loss, none has achieved the classic status of the species that were mentioned in the earlier part of this paper. There seems to be no simple explanation for this. However, it is very likely that any forest species showing high susceptibility to nursery pests or diseases would have been discarded at an early stage of its trials for plantation use, so that the maximum expression of the problem was avoided. As we have indicated already, under these circumstances we often have little detailed information on the basis for the rejection of one tree species or the retention of another. In addition, control of pests and diseases can be achieved with much more ease, speed and economy in nurseries than in the forest plantation, so that the impact of many problems may have been reduced at an early stage, if they were not avoided. This is particularly true for many pests which have been brought under such effective control by modern insecticides that the dangers they present have been virtually eliminated.

Plantation pests and diseases

While the transfer of nursery stock to the field for planting may be expected to be accompanied by a decline in the variety and impact of pests and diseases favoured by dense and uniform host crops, the converse may also occur as changes in host morphology and physiology with age make it susceptible to fresh parasites. In addition to this, further pests and diseases favoured by thinning and pruning practices may appear later on. It is our object in this section to determine how far these new problems may be attributed directly to the adoption of monoculture rather than to other factors.

PESTS

If we regard enrichment planting and trial plots as primitive precursors to more extensive monocultures we can find numerous examples where these systems can be correlated with, and have apparently favoured, outbreaks of serious forest pests, particularly in the tropics. Attacks by members of the Scolytidae, Cerambycidae, Pyralidae and Psyllidae are most frequently quoted in this context as causing severe losses and one example from each family will be considered here to determine how far the crop system was directly involved in the emergence of the pest.

The scolytid *Xyleborus mascarensis* and related species, already mentioned in the section on nursery pests, caused extensive damage and death to line-planted *Khaya ivorensis* and plantations of *Aucoumea klaineana* in Ghana in 1954 and later (Browne 1961). Damage was caused by girdling of the host and death either ensued directly from this or through secondary fungal infections. These attacks, which did not occur in the natural forest, were due to the accumulation of debris from the simultaneous elimination of weed species in the same forest region, rather than any move towards the intensive cultivation of the host. These woody substrates provided plentiful breeding and feeding sites for the insects, leading to a rapid population build-up. On the depletion of these resources the insects turned to the transplants as an alternative substrate.

Outbreaks of the cerambycid *Phryneta leprosa* were observed by Jones to cause severe losses to trial plots of *Chlorophora excelsa* in Tanzania between 1959 and 1965 and provide a contrasting example. This beetle lays its eggs under branches which are girdled by the larvae when they hatch. In this outbreak it was evident that exposure of the trees to insolation by establishing them in even-aged plots had caused a blistering of the bark and predisposition to attack by the insect. This pest had also been responsible for the destruction of thousands of hectares of *C. excelsa* plantations in the Belgian Congo (Zaire) between 1926 and 1936 (Mayné, personal communication), which was probably also connected with insolation damage to the host as a direct result of the adoption of monoculture.

The pyralid genus *Hypsipila*, which includes several species that are notorious for the difficulties that they have caused in the establishment of plantations of *Swietenia*, *Cedrela* and other species of the Meliaceae, present a more complex picture (Entwhistle 1968). Like *P. leprosa*, these pests are sensitive to shade and cause maximum damage when the host is exposed to direct sunlight. However there are also reports that close planting of the host will reduce *Hypsipila* attack, which to some extent runs counter to the previous observation on the effects of shade. Here it would seem, the effects of monoculture are likely to be complex, but in the early stages of plantation establishment of members of the Meliaceae, there seems to be little doubt that the conditions afforded by monoculture systems favour *Hypsipila* infestations in the majority of cases.

In Ghana, Nigeria and Zaire the psyllid, *Phytolyma lata*, as a pest of *Chlorophora excelsa* caused such damage in planted lines, trial plots and plantations that all attempts at monoculture of its host were completely abandoned (Roberts 1969, Mayné personal communication). *P. lata* has a very limited radius of dispersal and it is clear that these outbreaks were entirely due to the provision of an abundance of host plants in close proximity, facilitating spread of the pest.

These examples illustrate that even the relatively modest move towards artificial monoculture represented by enrichment practices and trial plots may be sufficient to favour the emergence of a new forest pest problem on some occasions, but that incorrect site preparation and management practice may be equally responsible for these outbreaks.

There are many other pest problems of plantation hardwoods in the tropics that have apparently been due to the practice of monoculture, but which prove, on closer scrutiny, to have originated from other causes. The termites that have caused extensive damage to eucalypts in East Africa have been found to be favoured by the weakening of the host by climatic and edaphic factors rather than the establishment of uniform host populations. Furthermore, the bostrychids, *Apate monachus* and *A. terebrans*, which devastated plantations of *Glyricidia* in Zaire (Mayné personal communication) and Ghana (Jones & Roberts 1959), also severely damaged isolated trees of the same species. In the latter example it seems that indiscriminate feeding on a number of different hosts, rather than the existence of monoculture, has led to the pest status of these insects. On the other hand, there are a number of good examples of pests emerging directly as the result of the establishment of plantations. *Hoplocerambyx spinicornis*, normally an inhabitant of logs and moribund trees of *Shorea robusta* in India and Pakistan, has now become a serious, albeit periodic, pest of healthy mature *S. robusta*, now that this species is frequently established in plantations (Beeson & Chatterjee 1925). This change is attributed to the improved opportunities for congregation and reproduction provided by plantations, leading to high populations and pressure to find new breeding and feeding sites. *Hecphora testator* has caused widespread damage to *Nauclea diderrichii* plantations in Nigeria while in the natural forest this insect is totally insignificant as a pest (Henry 1960). The borer, *Phoracantha semipunctata* and the snout beetle, *Gonipterus scutellatus*, both pests of *Eucalyptus* spp., appear to have had similar histories. Both these insects are natives of Australia, where they have only minor importance as pests. However, after introduction to regions in South and East Africa where extensive areas of *Eucalyptus* plantations had been established, they casued widespread damage in these crops (Kevan 1946, Tooke 1936, 1953). *P. semipunctata* also spread to similar plantation crops in the Mediterranean region where it caused serious damage in Israel, Egypt and Cyprus (Bytinski-Salz & Neumark 1953). While the origin of these outbreaks may be attributed to the efficient transport systems that enabled these pests to reach their host crops, their extensive and uniform nature certainly contributed significantly to their subsequent impact as exotic pests on exotic hosts. However, the appearance of *Analeptes trifasciata* on *Bombax rhodognaphalon* and *Tragocephala* spp. on *Cedrela odorata* observed by Jones in Tanzania are examples where monocultures of exotic species provided the basis for the assumption of pest status by two indigenous insects.

Softwood plantations have earned much the same reputation as hardwoods and many of the examples of pests emerging through the establishment of these species in monocultures are more justly attributable to other factors.

We have already noted that the importance of the borer of *Cupressus* spp., *Oemida gahani*, in East Africa owed more to the failure of foresters to observe simple hygiene in plantations than to the establishment of extensive monocrops of the host, and the same can be said of the infestation of young *Pinus patula* by *Hylastes angustatus* in South Africa (Bevan & Jones 1964). As with hardwoods, termite attack of softwood plantations is more dependent on a state of debility in the crop than on the existence of monoculture, *per se*. Site factors may also be responsible for outbreaks of insect pests in forest plantations.

However, there are still numerous examples of insect pests of softwoods which have derived significant advantage by monoculture systems, although in some cases it has apparently taken some thirty to fifty years for the transition to be completed to the new host and environment. The list of such insects is lengthy for Africa alone and consists largely of defoliating Lepidoptera including members of the Lasiocampidae, Lymantridae, Notodontidae, Geometridae and Saturniidae, all of which were established inhabitants of local forests which have migrated to pine plantations where they periodically cause serious but localized damage. An outstanding example of a non-lepidopterous insect pest of pines that falls into this class is *Plagiotriptus pinivora* which has caused sufficiently serious damage to plantations in parts of Malawi to justify chemical control by aerial spraying (Lee 1972). This is a clear example of a non-lepidopterous pest that has been favoured by the system of monoculture used for its new host. *Gonometa podocarpi* (Lepidoptera, Lasiocampidae) infects *Podocarpus gracilior* and other species of this genus native to southern and eastern Africa, while *Nudaurelia cytherea* (Lepidoptera, Saturniidae), *Orgya mixta* (Lepidoptera, Lymantriidae) and *Pachypasa capensis* (Lepidoptera, Lasiocampidae) are all long-established inhabitants of *Brachystegia* woodlands in the same region (Tooke 1936, Tooke & Hubbard 1941, Gardner 1957, Austara & Jones 1971). These have all adapted to exotic pine plantations and it is estimated that between them these four species cause a 50% loss in increment at each epidemic. Fortunately, such outbreaks rarely occur in successive years. However, there is evidence that the period of activity of one of these defoliators (*P. capensis*) on *Pinus patula* has recently extended from two months to six to eight months in Swaziland, so the impact of some at least, of these pests may not have reached their peak (Keat personal communication). *Buzura edwardsi* has emerged more recently as a very serious pest of *Pinus patula* in Uganda and Rhodesia)Brown 1962, Stubbings 1968) and in the latter country it has completely defoliated thousands of acres of the species. It seems reasonable to deduce from these examples that, in Africa at least, there is a host of indigenous lepidopterous pests capable of adapting to exotic conifers and hardwoods and of thriving on these new hosts. As we have noted already, these pests sometimes take a long time to emerge, but this makes little difference to their eventual impact on the crop.

While we have drawn the majority of our examples of forest pests in this section from African experience, similar examples can be found in abundance where forest plantations have been established in other continents. The reviews that are made of forest pest incidence at international conferences demonstrate how frequently populations of local pests will build up in association with plantations of their indigenous hosts and how local insects can adapt to new exotic hosts where these plantations have replaced their local source of food. There seems to be little doubt, therefore, that the most important and largest number of examples of insect pests finding their origins in the practice of forest monoculture are defoliators and, to a lesser extent, boring insects. This may be due, partially at least, to the delicate host–parasite balance that often exists in natural communities for pests depending exclusively on the living tissues of their hosts. In these circumstances host survival often depends on its low frequency in the community, making it relatively inaccessible to its more specialized pests, and it is these insects that are most likely to respond swiftly and positively to any increase in host availability. Even if these outbreaks are delayed by special circumstances, as may sometimes occur, it does not mean that the eventual impact on the crop is likely to be significantly reduced. In establishing this point we do not wish to imply that all serious loss through forest pests, particularly defoliators, is related simply to the population density of these insects. Other factors, including host morphology and the feeding habits of the insect, may significantly modify the net impact of the pests on the crop, as a comparison of infestations by *Buzura edwardsi* on *Eucalyptus* and pine crops shows. Thus, on the broad-leaved crop this defoliator will reduce the photosynthetic ability of the host in direct proportion to the amount of leaf tissue consumed. On pines, however, the insect travels mainly on the shoot, taking single bites from the base of each needle and leaving the rest to fall to the ground. The rain of needle tissue that results from this illustrates well the excessive amount of photosynthetic tissues destroyed by the feeding of these pests on conifers compared to the damage done by a similar infestation of hardwoods.

DISEASES

The reasons for the emergence of many new diseases of older forest crops are often less clear than they appear to be for insect pests. However, it is likely that the effects of monoculture on stem diseases at this stage will be more pronounced than on foliage pathogens. Canker diseases caused by fungi and bacteria are likely to be particularly favoured by a reduction in the distance between host units, as they are generally splash-dispersed and have a limited radius of effective spread, which can be further reduced if the centres of inoculum (the cankers) are sited close to the ground. The susceptibility of the host to these diseases often depends on the state of maturity of the bark, with the result that outbreaks may be confined to the older stages of the crop. This appears to have been the case when *Monochaetia* canker of cypress appeared a few years after the planting of *Cupressus* spp. on a large scale in the East African highlands. The pathogen (*Rhynchosphaeria cupressi*, imperfect state *Monochaetia unicornis*) has a limited radius of effective dispersal (Jones 1953) but, once established in plantations of a susceptible host species (*Cupressus macrocarpa*), it spread rapidly in the young crop with disastrous results until the introduction of *C. lusitanica* which was more resistant. The maximum susceptibility of the crop occurred between its second and sixth years. The origin of this pathogen is still in doubt, as it could have been derived from a mildly pathogenic form of the fungus found in association with local *Juniperus procera*, or it could equally well have been introduced (Jones 1953). Whatever its origins, conditions of monoculture favoured its spread. Two more recent and serious examples of stem diseases caused by local pathogens that have been apparently favoured by monoculture are those of *Corticium salmonicolor* on *Eucalyptus tereticornis* and other *Eucalyptus* spp. in the state of Kerala and elsewhere in India (Bakshi *et al.* 1970, 1972) and *Diaporthe cubensis* on *Eucalyptus* spp. in Central and South America (Reis & Hodges 1975). Both fungi have wide geographical distribution and host range, as would be expected with a local pathogen of a recently introduced crop. *C. salmonicolor* appears to have caused the more serious damage while, with the exception of Surinam, *D. cubensis* has had less severe effects overall (Hodges & Reis 1974). A further example of a stem pathogen in rather a different context, is given by the invasion of plantations of local species in the Chittagong Hill Tracts of Bangladesh by *Loranthus parasitica*. This mistletoe has spread slowly into plantations bordering the natural forest and is now causing extensive losses to stands of *Gmelina arborea* and *Tectona grandis* while threatening other species. The limited rate of spread characteristic of most angiospermic parasites has ensured that this problem has emerged many years after plantations of local species were started. Nevertheless, it now constitutes a serious source of loss to forest production in the region. A disease of unknown etiology, *Terminalia* dieback, also deserves mention here. This has caused extensive losses in young plantations of *Terminalia ivorensis* in Ghana and possibly elsewhere in West Africa (Ofosu-Asiedu & Cannon 1976). The disease is unknown in the natural forest, where the host is widely dispersed, but appears in young plantations a few years after establishment. It is possible that here too, the cause is a local pathogen with very limited range of dispersal to which the host becomes susceptible after a few years' growth. The emergence of serious leaf and stem diseases of poplars, such as those caused by *Marssonina brunnea* or *Aplanobacter populi*, has probably also been largely due to the adoption of monocultures, although in these cases enhanced susceptibility as a by-product of its intensive improvement through selection, hybridization and vegetative propagation may have contributed as much to the importance of these pathogens as modification of the environment.

It would be a mistake, of course, to assume that all stem diseases in plantations arise through factors associated with monoculture. *Diplodia* die-back (*Diplodia pinea*), although a widespread and well-known disease of exotic pine plantations, generally occurs in association with wounding or other circumstances reducing the resistance of the host. These range through hail damage in South Africa and Swaziland (Laughton 1937), poor pruning techniques in New Zealand (Gilmour 1964) to insect damage in North America (Haddow & Newman 1942). The host proximity factor provided by monoculture may contribute to the spread of the fungus from tree to tree, and to its build-up as a saprophyte in litter on the ground, but these are secondary in importance to the predisposition of the host to infection.

The relationship between the causes of many important shoot diseases of forest trees and environmental or host factors is still not fully understood. *Gremmeniella abietina* (*Scleroderris lagerbergii*) is an important example of these. This pathogen was first noted (but misidentified) by Rostrup in 1881 on Austrian pines in Denmark (Gremmen 1972) and later on it caused serious damage to *Pinus nigra* planted on the west coast of Norway (Brunchorst 1888, Roll-Hansen 1972). Since that time it has appeared as an important pathogen of pines and other conifers in many parts of Europe, in North America and, recently, in Japan (Kobayashi 1972, Saho & Takahashi 1972). The conditions favouring outbreaks of this disease are complex and are still not completely understood, although shading, frost and high humidity clearly play an important part (Donaubauer 1972). Roll-Hansen (1972) observed that introduced pines may be severely damaged by *G. abietina* if their new localities differ significantly from their native habitat, but as serious outbreaks of the disease have occurred in natural forests as well as in plantations (Gremmen 1972), it is extremely difficult to evaluate the role played by monoculture *per se* in its emergence. The silvicultural practices associated with nearly all systems of forest monoculture are likely to provide conditions favouring pests and diseases however carefully they are applied, by providing infection ports on host tissues and increasing available substrates for the saprophytic build-up of pests and pathogens. The outbreaks of *Xyleborus* spp. on *Terminalia ivorensis* in Ghana and

Oemida gahani in East Africa are examples already discussed. *Monochaetia* canker is another problem that can be much enhanced by pruning if this is done at times when inoculum is abundant, and there are many other similar canker diseases that can be favoured similarly. A recent outbreak of heart rot in young *Eucalyptus grandis* in Zambia provides a further interesting example of how early pruning and weeding operations can favour invasion of this crop by the rot fungi *Phellinus gilvus* and *Poria epimiltina* (Ivory 1975).

Soil-borne infestations of forest plantation trees by insects have seldom achieved significance but there can often be important relationships between root pathogens and forest monocultures especially where the wood-destroying basidiomycetes are involved. The pathogenicity of this group depends directly on their ability to build up inoculum potential by colonizing woody residues and in this they are favoured by clearing operations which leave plentiful stumps in the soil. It is because of this that we regard the origins of this class of disease to lie with the site history before plantation establishment, rather than in the nature of the plantations themselves. Nevertheless, once these root diseases are established, the uniform nature of the plantation can contribute very significantly to severity of attack and, in one case at least, thinning operations can lead to further infection centres in the crop. This occurs with *Heterobasidion annosum* (*Fomes annosus*) which sporulates profusely in the infected crop, providing airborne inoculum to colonize freshly cut stumps from which the fungus spreads by root contact. It has been suggested that other root pathogens may have similar cycles particularly *Armillariella mellea* and *Rigidoporus lignosus*, but no firm evidence for this has been found (Rishbeth 1972, Momoh 1976). A further example of a soil-borne disease apparently favoured by monoculture is a vascular wilt of *Pterocarpus angolensis*, associated with *Fusarium oxysporum*, that occurs in natural stands in Zambia. Attempts to establish *P. angolensis* in trial plots and plantations within its native areas have been followed by disastrous outbreaks of this disease (Piearce personal communication). A wilt disease of plantations of *Dalbergia latifolia* in Java, associated with *Fusarium solani* appears to have similar origins to the disease of *P. angolensis* (Suharti & Hadi 1974).

GENERAL

Overall, there appears to be a trend whereby the general impact of diseases due to monoculture declines in importance as the crop grows older. However, this does not apply to insect pests, particularly defoliators, many of which have their most serious impact when the crop is near maturity. Finally, there is some justice in the observation that the adoption of monocultures can lead to a significant reduction in loss from insect and fungal attack where this is associated with over-maturity. Invasion of the heartwood by borers, decay and stain fungi generally occurs through wounds and other damage when the host is at a relatively advanced age. This type of loss is brought to a minimum under plantation regimes, where the crop can be protected from wounding and is grown to a carefully controlled rotation period.

Conclusion

If we are now to answer the question 'Has the adoption of monoculture systems led directly to an increase in the number and severity of pests and diseases of forest crops?' the answer would have to be 'Yes'. We could add that there seems to be evidence that much of this is due to the uniform and crowded conditions that plantations provide and that the cultural operations associated with these crops have often accentuated problems that have had other origins. There appears to be some evidence also, to confirm our proposal that these effects are likely to be greatest for diseases at the nursery and early plantation stage, diminishing as the crop grows older. However this decline is not so evident for insect pests where the relationship between infestation and reproduction may be different from pathogens. Similarly, we can find some support for the proposal that diseases with their origins in monoculture are likely to be more numerous for species growing in their native regions than for exotics planted far from their original homes. This is likely to be a temporary situation and may not apply to the same extent to insect pests, which often have more flexible feeding habits than the specialized fungal, bacterial and viral pathogens. Overall, and despite the fact that certain pests and pathogens of over-maturity are avoided by plantation systems, it appears that the most pessimistic forecasts of traditional foresters on the dangers arising from forest monocultures have been fully vindicated.

However, this conclusion needs qualification. We have already observed that few if any, of the most disastrous outbreaks of forest pests and diseases, the classic cases, can be attributed to the introduction of monoculture systems. We think that this is partly due to the flexibility that the forester often enjoys in the choice of main species for his monoculture, which allows a good margin for avoidance of serious pest and disease problems when the crop is at an early stage of development. In addition, the economics of these intensive systems of forestry allow a much greater latitude for expenditure on protection measures than is usually available for other types of forestry. This, and the compact nature of plantation crops, provide clear advantages for the application of effective control measures against forest pests and diseases. Thus it is perhaps not surprising to find that the pest and disease problems engendered by forest monocultures have seldom reached the catastrophic levels of certain pests and diseases of natural forests or forest crops grown for non-commercial ends. It is with the latter type of tree crops that the prediction and early detection of new pests and diseases is most difficult, where there is also a general lack of funds for special protection measures and where physical circumstances make them difficult to apply. Briefly, while monoculture has provided conditions favouring the emergence of many new pest and disease problems, it has also gone far to facilitate their control.

These conclusions refer to the direct effect of environmental change due to monoculture on pests and pathogens, but do not take account of longer-term effects of selection on the genetic basis of the crop. The establishment of monoculture systems in agriculture has been followed by pressures, both planned and fortuitous, which have tended to alter the genetic base of the crop and at times has led to changes in its susceptibility to pests and diseases. This has been particularly evident in the present century, following intensive selection and breeding to improve quality and yield in staple crops. Similar methods have been applied more recently to improve those forest species selected for monocultures; so far with success. It is to be hoped that forest tree breeders and geneticists will learn from the experience of their counterparts in agriculture and ensure that the risk of new pest and disease problems is reduced to a minimum in their pursuit of improved quality and yield.

References

ANON. (1969) Biological control programme against insects and weeds in Canada, 1959–1968. *Tech. Comm. No. 4. Commonw. Inst. Biol. Control.*

AUSTARA O. & JONES T. (1971) Host list and distribution of lepidopterous defoliators of exotic softwoods in East Africa. *E. Afr. agric. For. J.* **36**, 401–13.

BAKSHI B.K., RAM REDDY M.A., SUJAN SINGH & PANDEY P.C. (1970) Disease situation in Indian forests. I. Stem diseases of exotics due to *Corticium salmonicolor* and *Monochaetia unicornis*. *Indian Forester* **96**, 828–9.

BAKSHI B.K., RAM REDDY M.A., PURI Y.N. & SUJAN SINGH (1972) *Forest Disease Survey (Final Technical Report) (1967–72).* Forest Research Institute and Colleges, Dehra Dun, India.

BEAL J.A. (1965) Bark beetles threaten destruction of Honduras pine forests. *Proc. FAO/IUFRO Symp. int. dang. Forest Dis. Insect*, Oxford, 1964.

BECKER G. (1952) Die Dendroctonus-Kalamital en Guatemala. *Proc. 9th int. Congr. Ent.*, Amsterdam, 1951, 1, 684–7.

BEESON C.F.C. & CHATTERJEE N.C. (1925) The economic importance and control of the Sal heartwood borer (*Hoplocerambyx spinicornis* Newm. fam. Cerambycidae). *Indian Forest Rec.* (ent.) **11**, 1047.

BEVAN D. & JONES T. (1964) *Hylastes angustatus (Herber) in pine plantations of Swaziland and adjacent areas of the Republic of South Africa.* Report of Ministry of Overseas Development to H.M. Commissioner, Swaziland.

BOALER S.B. (1966) The ecology of *Pterocarpus angolensis* DC in Tanzania. *Res. Publs Minist. Overseas Dev.* No. 12, HMSO, London.

BROWN K.W. (1962) Notes on the recent outbreaks of looper caterpillars in coniferous plantations in Uganda. *Tech. Notes Forest Dept. Uganda* No. 99.

BROWN K.W. (1969) The pine woolly aphid, *Pineus* sp. in East Africa. *Commonw. For. Rev.* 48, 194–5.

BROWNE F.G. (1961) Death of young transplants. *4th Rep. W. Afr. Timb. Borer. Res. Unit.*

BRUNCHORST J. (1888) Über einer neue verheerende Krankheit der Schwaarzföhre (*Pinus austriaca* Hoss.). *Bergens Mus. Aarsberetn. for. 1887* 6, 1–6.

BYTINSKI-SALZ H. & NEUMARK S. (1953) The Eucalyptus *borer* (*Phoracantha semipunctata* F.) in Israel. *Proc. 9th Int. Cong. Ent.* Amsterdam 1, 696–9.

DINGLEY J.M. & GILMOUR J.W. (1971) *Colletotrichum acutatum* Simmds. f sp. *pinea*, associated with Terminal Crook disease of *Pinus* spp. *N.Z. Jl Forest Sci.* 2, 192–201.

DONAUBAUER E. (1972) Environmental factors influencing outbreaks of *Scleroderris lagerbergii* Gremmen. *Eur. J. Forest Path.* 2, 21–5.

ENTWISTLE P.F. (1968) The current situation on shoot, fruit and collar borers of the Meliaceae. *9th Br. Commonw. For. Conf.* New Delhi 1968.

FRASER A.I. (1976) A manual on the planting of man-made forests. *FAO Working Paper* (unpubl.).

GARCIA L.A.A. (1939) A mahogany seedling blight in Puerto Rico. *Caribb. Forester* 1, 23–4.

GARDNER J.C.M. (1957) An annotated list of East African forest insects. *E. Afr. agric. For. Res. Org. Forest Tech. Notes* No. 7.

GIBSON I.A.S. (1956) Sowing density and damping-off in pine seedlings. *E. Afr. agric. J.* 21, 182–8.

GILMOUR J.W. (1964) Survey of *Diplodia* whorl canker in *Pinus radiata*. *Res. Leafl. N.Z. For. Serv.* No. 5.

GREMMEN J. (1972). *Scleroderris lagerbergii* Gr.—The pathogen and disease symptoms. *Eur. J. Forest Path.* 1, 1–5.

HADDOW W.R. & NEWMAN F.S. (1942) A disease of the Scots Pine caused by *Diplodia pinea* associated with pine spittle bug (*Aphrophora paralella* Say). *Trans. R. Can. Inst.* 24, 1–18.

HENRY P.W.T. (1960) *Hecphora testator* F., a pest attacking *Nauclea diderichii*. *Niger. For. Inf. Bull.* Nos. 4–7, pp. 18–20.

HODGES C.S. & REIS M.S. (1974) A influencia do cancro basal causado por *Diaporthe cubensis* Bruner na brotacao de *Eucalyptus saligna* Sm. *Bras. Florestal* 5, 25–8.

ITO K. (1972) *Cercospora* needle blight of pines in Japan. *Bull. Govt. Forest Exp. Stn. Meguro* No. 246, 21–33.

IVORY M.H. (1975) Heart rot of Eucalyptus in Zambia. *Proc. 2nd Wld tech. Consultation Forest Dis. Insect.* Delhi.

JOFFILY K. (1944) Ferrugem do Eucalipto. *Bragantia* 4, 475–87.

JONES D.R. (1953) Studies on a canker disease of *Cupressus* in East Africa caused by *Monochaetia unicornis* (Cooke & Ellis) Sacc. I. Observations on the pathology, spread and possible origins of the disease. *Ann. appl. Biol.* 40, 323–42.

JONES T. & CURRY S.J. (1964) *Oemida gahani* Desrair, its host plants, host range and distribution. *E. Afr. agric. For. J.* 30, 149–61.

JONES T. & GIBSON I.A.S. (1966) The present world situation in regard to the spread of internationally dangerous forest diseases and insects. *Proc. 6th Wld. For. Congr.* 2, 1897–908.

JONES T. & ROBERTS H. (1959) *2nd Rep. W. Afr. Timb. Borer Res. Unit.*

KAM M. de (1973) Life history, host range and distribution of *Septotinia podophyllina*. *Eur. J. Forest Path.* **3**, 1–6.

KEVAN D.K.Mc.E. (1946) The Eucalyptus weevil in East Africa. *E. Afr. agric. J.* **12**, 40–4.

KOBAYASHI T. (1972) Todomatsu edagarebyo—Micropera edagarebyo kara no byomei henko to sono gakumei ni tsuite. *Forest Prot. Tokyo* **21**, 208–9.

KRAAYENOORD C.W.S. van, LAUNDON G.F. & SPIERS A.G. (1974) Poplar rusts invade New Zealand. *Pl. Dis. Reptr* **58**, 423–7.

LAUGHTON E.M. (1937) The incidence of fungal disease on timber trees in South Africa. *S. Afr. J. Sci.* **33**, 377–82.

LEE R.F. (1972) A preliminary account of the biology and ecology of *Plagiotriptus* spp. (Orthoptera: Eumasthacidae). *Res. Rec. Malawi Forest Res. Inst.* No. 48.

MOMOH Z.O. (1976) Status of root rot disease of teak (*Tectona grandis* Linn. F.) in Nigeria. *PANS* **22**, 43–8.

NEWHOOK F.J. & PODGER F.D. (1972) The role of *Phytophthora cinnamomi* in Australian and New Zealand forests. *A. Rev. Phytopath.* **10**, 299–326.

ODERA J.A. (1974) The incidence and host trees of Pine woolly aphid, *Pineus pini* L. in East Africa. *Commonw. For. Rev.* **53**, 128–36.

OFOSU-ASIEDU A. & CANNON P. (1976) *Terminalia ivorensis* decline in Ghana. *PANS* **22**, (2) (in press).

PETERSON R.F. & JEWELL F.F. (1968) Status of American stem rusts of Pine. *A. Rev. Phytopath.* **6**, 23–40.

REIS M.S. & HODGES C.S. (1975) Status of forest diseases and insects in Latin America. *Proc. 2nd Wld Tech. Consultation Forest Dis. Insect*, Delhi.

RISHBETH J. (1972) The role of basidiospores in stump infection by *Armillaria mellea*. In *Root Diseases and Soil Borne Pathogens* (Ed. by T.A. Tousson, R.V. Bega & P.E. Nelson), pp. 141–6. University of California Press, Berkeley.

ROBERTS H. (1969) Forest insects of Nigeria with notes on their biology and distribution. *Commonw. For. Inst., Dept. For.*, Oxford *Pap.* No. 44, 1–206.

ROLL-HANSEN F. (1972) *Scleroderris lagerbergii:* Resistance and differences in attack between pine species and provenances. *Eur. J. Forest Path.* **2**, 26–39.

SAHO H. & TAKAHASHI I. (1972) Shinyoji ho byogenkin 3 shu *Scleroderris lagerbergii* Gremmen, *Lachnellula fuscosanguinea* (Rehm.) Dennis oyobe *Lachnellula suecica* (de By. ex Fr.) Nannf. *Forest Prot. Tokyo* **21**, 209–11.

SKILLING D.D. & NICHOLLS T.H. (1975) The development of *Lophodermium pinastri* in conifer nurseries and plantations in North America. *Eur. J. Forest Path.* **5**, 193–7.

SMITH R.S. (1975) In *Forest Nursery Diseases in the United States. U.S.D.A. Forest Serv. Agric. Handb.* No. 470, 9–13.

STUBBINGS J.A. (1968) Field observations of *Buzura edwardsi* Walker (Aeometridae): a defoliating looper on *Pinus patula* and *Eucalyptus* species in the Melsetter district of Rhodesia. *S. Afr. For. J.* **65**, 15–29.

SUHARTI M. & HADI S. (1974) Wilt disease of *Dalbergia latifolia* Roxb. in KPH Malang, East Java. *Laporan Departemen Pestanian, Lembago Penelitian Huten* No. 194.

TOOKE F.G.C. (1936) Insects injurious to forest and shade trees. *Bull. Dep. Agric. For. Un. S. Afr.* No. 142.

TOOKE F.G.C. (1953) The *Eucalyptus* snout beetle *Gonipterus scutellatus* Gyll., a study of its ecology and control by biological means. *Entomology Mem. Dep. Agric. Un. S. Afr.* No. 3.

Tooke F.G.C. & Hubbard C.S. (1941) The pine tree emperor moth, *Nudaurelia cytherea capensis* Stoll. *Sci. Bull. Dep. Agric. For. Un. S. Afr.* No. 210.

van der Plank J.E. (1960) Analysis of epidemics. In *Plant Pathology, An Advanced Treatise* 3, pp. 229–89. Academic Press, New York and London.

van der Plank J.E. (1963) *Plant Diseases: Epidemics and Control.* Academic Press, New York and London.

Walker J., Hartigan D. & Bertus A.L. (1972) Poplar rusts in Australia with comments on potential conifer rusts. *Eur. J. Forest Path.* 4, 110–18.

Webster B.N. (1952) A note on pathological matters. *Tea Q.* 23, 84–5.

White M.G. (1966) The problem of *Phytolyma* gall bug in the establishment of *Chlorophora. Commonw. For. Inst., Dept. For. Oxford, Inst. Pap.* No. 37, 1–52.

Yarwood C.E. & Gardner M.W. (1972) Powdery mildews favoured by agriculture. *Phytopathology* 62, 799.

Effects of different systems of monoculture on marine fish parasites

A. H. McVICAR and K. MacKENZIE *Department of Agriculture and Fisheries for Scotland, Marine Laboratory, P.O. Box 101, Aberdeen*

Introduction

For centuries man has regarded the seas as hunting grounds in which to capture wild prey, in much the same way as our early ancestors did on land. Only recently have we seriously entertained the idea that it may be to our advantage to tend marine fish with the same care we give terrestrial farm animals. The evolution of techniques for the cultivation and rearing of purely marine fish must rank as one of the most challenging and stimulating developments in animal husbandry in recent years. As a result of the short history of the subject, the associated disease and parasite problems are poorly documented even by comparison with those of cultivated fresh water fish. Indeed, marine parasitology in general is a sadly neglected subject, largely because in the past the formidable scale of the marine environment fostered the belief that the study of marine parasites was of little more than academic interest, since no control could be practised on such a scale. With the advent of marine fish cultivation this attitude is no longer valid, but present-day workers have inherited the legacy of an embarrassing lack of knowledge of the biology of even the most common parasite species. Consequently much of the current research into parasites encountered on marine cultivation units is necessarily of a very basic nature.

This paper deals with the effects on marine fish parasites of different monoculture systems. It is based largely on the results of our own investigations into parasites of flatfish, principally plaice *Pleuronectes platessa* and turbot *Scophthalmus maximus*, reared on the White Fish Authority's Scottish marine fish cultivation units; these have been supplemented with information gleaned from the relevant literature. From these data and from our knowledge of parasite biology, we have tried to draw practical conclusions, particularly in relation to preventing and controlling parasitic diseases. Up to now most marine fish cultivation has been based on monoculture and our experience is limited to this field. We have however, included a speculative discussion of the possible effects of polyculture on marine fish parasites. We have also tried to anticipate some problems which have not yet appeared in monoculture but which we consider likely to arise under certain circumstances. By doing so, we hope to alert fish farmers to potential dangers involving marine parasites.

Types of marine culture systems

Enclosures used for the rearing of marine fish are of six main types, as described and discussed by Milne (1972). We have carried out parasitological studies in four of these types of system.

The onshore system, represented by the Hunterston unit in south-west Scotland. The fish tanks, situated above sea level, are supplied with heated chlorinated sea-water, discharged from the adjacent nuclear power station.
The enclosed pond, with limited and controllable water exchange. This system has been used at Ardtoe, a fish cultivation unit situated in a remote area of the Scottish west coast. At this site a small inlet was walled off near its mouth so that the water level could be maintained as the tide ebbed.
Floating cages, moored in a sheltered situation for protection against adverse weather conditions. This is the system in use at Moidart, near the Ardtoe unit.
Sea-bed cages, which offer protection from bad weather and a near-natural environment for benthic species. This system was used in an experiment carried out in Loch Ewe in north-west Scotland and described by Steele (1966).

Of the two remaining systems, one using *mid-water cages* of adjustable buoyancy was reported by Hanson (1974) as being in the experimental stage in Japan. The other is the *tidal enclosure*, which is exposed to natural tidal rhythms and contains the stock within a permeable barrier, usually of netting. The latter is a long-established system in many parts of the world.

General comments on parasites and their control in the culture systems

A parasite may be defined as an organism which lives in or on its host, normally without conferring any advantage, but which may in some relationships be detrimental to the host. In general, if association between host and parasite is long established in an evolutionary sense, the relationship has usually evolved into a stable one. Such associations involve hosts which are incapable of avoiding or destroying the invading parasite, and parasites which do not usually cause serious damage to the host. The relationship is finely balanced and will remain so unless environmental conditions change abruptly and drastically. Such changes may be brought about when the host species is removed from its natural habitat and held in conditions of intensive culture. Where the new environment favours the host rather than the parasite, control may be achieved incidentally, but occasionally we find that parasites previously regarded as causing little harm can become pathogens in the new conditions. Increased success of reproduction or transmission of the parasite may account for this, but often the greater pathogenicity may be the result of a variable and sometimes elusive group of factors, which singly or in combination produce a 'stress' effect on the host. The source of these stress factors can often be traced to the quality of the husbandry. The relationship between stress and parasitism in general was critically examined by Esch *et al.* (1975), while the question of stress in relation to diseases of fish was discussed by Wedemeyer (1970) and Snieszko (1974).

Control of parasitic infections is best achieved by preventing the parasites from gaining access to cultivated stock in the first place. Measures aimed at prevention may be difficult or costly to carry out and may have limited success, particularly with parasites having direct single-host life-cycles. We have more opportunity to control parasites having multiple-host life-cycles, as these can be attacked at their most vulnerable stage, which must be determined for a given set of conditions and may not be a stage actually present in the cultivated stock. In ponds or tidal enclosures it may be practicable to remove the intermediate hosts of harmful parasites from the immediate area, but a few words of caution are in order here: one must beware of drastically altering the ecology of the enclosure and its environs by such measures. For example, elimination of herbivorous molluscs which are known to be hosts of harmful parasites may lead to uncontrollable algal blooms which can affect the water quality. It may be better to settle for a measure of control rather than to aim at complete eradication of the parasite. The next best form of control is through the natural defensive mechanisms of the fish. However, as Corbel (1975) noted, the immune response of only a few species of marine teleosts has been studied and only fragmentary information is available on species of economic importance. Even less is known about the reactions of marine fish to parasitic infections. In the face of this dearth of knowledge we frequently have to resort to tackling a parasitic infection after the disease symptoms have appeared and the parasite has become established in the culture system, by which time control may be extremely difficult. The behaviour in sea water of many of the chemical therapeutics used in fresh water parasite control is largely unknown and they should be used with caution in the marine medium. The use of reduced or increased salinity as a safer alternative control method for ectoparasites is discussed below under a separate heading. With tenacious infections it may be necessary to employ the most drastic method of control: the slaughter of all infected stock and the sterilization of the equipment.

The source of a parasitic infection is of prime concern to a fish culturist. The parasite may be locally endemic in wild populations of the same species as that being cultivated or, with parasites which show low host specificity, in other local wild hosts. The diet of the cultivated fish will have a major influence on their parasite fauna. When natural food is taken, the fish are exposed to infection by parasites with multiple-host life-cycles, whereas when only the food provided by the culturist is available the parasite range can be narrowed to a large extent to those with direct single-host life-cycles. It is the practice in marine fish cultivation to feed stock with either industrial fish or reclaimed tissue left after filleting. Although much of this material is frozen before feeding, fish culturists are understandably reluctant to treat it by freezing or heating, as these processes are expensive and apparently lower the nutritional value. This untreated material is undoubtedly a potential source of infection with some highly undesirable parasites.

At the Third International Congress of Parasitology in 1974 a resolution was passed to the effect that representations be made to governments urging them to introduce measures aimed at the control of biological contaminants (Baer 1974). The delegates were concerned about the transfer of dangerous parasites from one region to another in infected livestock generally. We are equally concerned that this point should be made in particular to those involved in marine fish cultivation. Not only can pathogens be introduced to a cultivation unit in this way, but they may have even more serious consequences if they leak into wild populations with little or no resistance to them. The dangers of this were discussed by Odum (1974) and emphasized, with cautionary examples, by Hoffman (1970) and Sindermann (1974). The last author set out a plan of action aimed at preventing the spread of pathogens, which should be followed before any proposed transfer of live animals is carried out.

By comparing the parasite faunas of wild fish populations with those of cultivated fish reared in several different systems, we hoped to be able to identify the principal features influencing the spread and pathogenicity of parasites within farm environments. It must be emphasized that a combination of factors usually operates to produce an observed disease condition, but the influence of particularly important single features of culture systems are discussed below with reference to certain parasites.

Crowding

Crowding is the most important single feature influencing parasite infections of cultivated fish and is one common to all culture systems. Its effect may be direct by facilitating the spread of the parasite or indirect through the lowering of host resistance due to a stress effect. The transmission and reproductive rates of a fish parasite in its natural environment are usually adapted to a relatively low host density. When the hosts are held in conditions of higher density, it may be anticipated that the parasite's transmission rate will increase, even if its reproductive rate remains the same. This point is of particular relevance to those parasites with direct single-host life-cycles, and especially to those which include a free-swimming stage, e.g. the ciliate protozoans.

Cryptocaryon irritans is the marine equivalent of *Ichthyophthirius*, a common and highly pathogenic parasite in fresh water fish farms. *Cryptocaryon* is not common in wild populations and some fish species appear to escape infection completely, whereas in marine aquaria most species become infected and many serious outbreaks have been reported (Lom 1970). Sindermann (1970) described 'white spot disease' of cultivated pompano due to *C. irritans*.

Species of *Trichodina* were the most common and widespread parasites of the cultivated flatfish we examined. In the context of fish cultivation, the particularly dangerous features of these ciliates are their apparently low host specificity and the wide geographical distribution of many species (Lom 1970). Gill infections of *Trichodina borealis* are common in both wild and cultivated plaice in Scotland (MacKenzie 1969, MacKenzie *et al.* 1976). Cultivated plaice may also have skin infections of a larger species, probably identical to that described but not named, by Pearse (1972) from hatchery plaice on the Isle of Man. At least two species of *Trichodina*, not yet identified or described, occur on the gills and body surfaces of turbot. The ease with which they spread has been demonstrated by their appearance in well-isolated hatchery turbot at a disturbingly early age. The host-specificities of these trichodinids from cultivated flatfish are unknown.

In the aquarium of the Marine Laboratory, Aberdeen, concurrent infections of *Trichodina* with monogeneans of the genus *Gyrodactylus* had more serious pathological effects on plaice than single infections with either parasite (unpublished). Typical symptoms of double infections were grey mucous-covered patches on the skin, which progressed to open lesions, with secondary infections of bacteria and free-living or saprophytic protozoans. Mortalities were attributed to loss of body fluids and osmotic imbalance resulting from parasitic infections. Lom (1973) pointed out that trichodinids are essentially ectocommensals which do not usually cause damage to the host, since they normally feed on water-dispersed particles. It is only under conditions adverse to the host, such as a concomitant or earlier invasion by other parasites, that the trichodinids themselves become real parasites, begin to feed on damaged host tissue and proliferate massively. In the situation described above the damage to plaice skin epithelium caused by the *Gyrodactylus* infection may have been the starting point of the pathological sequence.

Caligid copepods are another group of parasites of some importance to marine fish cultivation, which have a direct life-cycle with a free-swimming larval stage. Paperna & Lahav (1974) described an epizootic of *Pseudocaligus apodus* and *Caligus* sp. on grey mullet in Israel, which resulted in the deaths of all the fish in a pond. *Caligus spinosus* and *C. amplifurcus* were quoted by Ghittino (1973) as causing serious damage to cultivated yellowtail in Japan, and Bardach *et al.* (1972) reported *Pseudocaligus fugu* as a parasite of cultivated puffers, also in Japan. We have had no problems with parasitic copepods on marine flatfish, but Anderson & Conroy (1968) reported massive infections of *Lepeophtheirus nordmanni* on cultivated turbot. A related species, *Lepeophtheirus salmonis*, is well-known as a serious pathogen in the marine culture of salmon (Kabata 1970, Braaten 1975).

Gyrodactylus is a genus of viviparous monogeneans with a direct form of life-cycle, but no free-swimming stage. As the species depend largely on direct physical contact between hosts for transmission (Lester & Adams 1974), conditions of high host density would be expected to favour such parasites, and gyrodactylids are well-known pathogens of juvenile fish in fresh water farms. MacKenzie *et al.* (1976) described a heavy infection of plaice in floating cages with *Gyrodactylus unicopula*, a common parasite of wild juvenile plaice in Scottish waters. This particular infection built up slowly and then decreased rapidly without treatment of any kind. MacKenzie (1970) had previously suggested the existence of some mechanism controlling the population density of *G. unicopula* in wild juvenile plaice. The possibility of some form of immunity being involved in this host–parasite system in particular and in fish–monogenean relationships in general was discussed by MacKenzie *et al.* (1976).

Monogeneans in general are amongst the most serious pathogens of cultivated marine fish, mainly because of their direct form of life-cycle. In Japan, *Benedenia seriolae* and *Heteraxine heterocerca* can cause serious damage to yellowtail (Kubota & Takakuwa 1963, Hoshina 1968), *Microcotyle tai* can affect red sea bream (Ghittino 1973), and *Diclidophora tetrodontis* may cause serious problems on puffers (Bardach *et al.* 1972). In the United States, *Bicotylophora trachinoti* and *Benedenia* sp. may contribute to mortalities of pompano (Sindermann 1974), and in Israel *Benedenia* sp. is pathogenic to mullets and *Microcotyle* sp. to fingerling *Siganus* (Reiss & Paperna 1975).

We have found small numbers of *Entobdella soleae* on Dover sole under farm conditions. Although we have had no problems with this monogenean, Anderson & Conroy (1968) reported it in large numbers on breeding stock in an early experimental farm in Scotland. The virus lymphocystis occurred concurrently with *E. soleae* in this instance, and the authors speculated on the possible role of the monogenean in virus transmission.

Conditions of high population density also favour those parasites which spread by means of resistant infective spores, e.g. the microsporidian and myxosporidian protozoans. Lom (1970) discussed some highly pathogenic members of these groups which occur in wild populations of marine fish, but with few exceptions little is known of their behaviour in conditions of culture.

McVicar (1975) and MacKenzie *et al.* (1976) found that up to 80% of some groups of plaice in the Hunterston unit were infected with the microsporidian *Glugea stephani*. Only a small proportion of infected plaice had heavy infections and few deaths could be attributed directly to parasitism. Another microsporidian, tentatively identified from spore dimensions as *Nosema ovoideum*, was found in the livers of Hunterston turbot, but subsequent examinations revealed no further infections (unpublished). This parasite was treated as a potential hazard since Raabe (1936) attributed liver degeneration and deaths of aquarium mullet to *N. ovoideum*. Egusa (cited by Ghittino 1973) reported infections of the microsporidian *Plistophora* sp. and the myxosporidian *Kudoa* sp. as causing severe lesions in the body muscles and hearts of cultivated yellowtail in Japan. Sindermann (1974) reported on two myxosporidians of relevance to marine fish culture in the United States; *Henneguya* sp. in pompano and *Kudoa cerebralis* in striped bass. Paperna (1975) believed gill and tissue myxosporidians to be potential pathogens of mullet in marine and brackish water farms, and Raabe (1936) noted severe damage to the kidneys of aquarium mullet caused by a myxosporidian, probably a species of *Myxidium*, which resulted in emaciation and death of the host.

In Hunterston turbot, Anderson *et al.* (1976) found a correlation between the severity of liver and kidney lesions (hepato-renal syndrome), the intensity of liver infection with the myxosporidian *Myxidium incurvatum* and the frequency of occurrence of *Rhabdospora* cells in the kidney and liver. (Whether or not *Rhabdospora* is a parasite or a specialized host cell is still the subject of controversy.) Turbot with severe lesions did not grow well, were often dark in colour and had a high death rate. Although at first we suspected that the condition might have been parasite-induced, the eventual conclusion was that the parasites probably played a secondary opportunist role, but may still have contributed towards the death of the fish. It seems more likely that the original cause of the condition lay in the fundamentals of turbot husbandry, and Pearse (personal communication) has evidence that the primary factor may be present in food binders.

Maintaining a reasonable standard of sanitation is one of the major difficulties associated with high density fish rearing, and has stimulated considerable research into cage and tank design and feeding methods. In sea enclosures, water exchange is normally good provided weed growth is controlled, but onshore tanks are generally more vulnerable to pollution. Finucane (1970) referred to fungus and bacterial infections of cultivated pompano which were apparently the result of overcrowding and contamination with uneaten food. A serious outbreak of an unidentified coccidian protozoan which occurred in one tank of turbot at Hunterston (unpublished) was probably associated with the uneaten food and detritus which had accumulated in that particular tank. Clinical manifestations of the disease were the formation of cysts throughout the flesh and body cavity and, in severe cases, larger diffuse areas of infection. The lesions contained large numbers of leucocytes and there was a great increase in the numbers of circulating white blood cells, with up to four trophozoites infecting one cell. Ferguson & Roberts (1975) carried out a light and electron microscopy study of the disease, but the biology of the parasite remains largely unknown. Improvement in tank hygiene led to a recession of the condition, while turbot held in better conditions throughout never developed the disease. It is not known whether the spread of the parasite was a direct result of the improved transmission of infective stages through their accumulation amongst detritus on the tank bottom, or an indirect effect brought about by a decrease in fish resistance due to the stress of adverse living conditions. This experience demonstrates the value of maintaining high standards of husbandry.

Up to this point we have considered crowding with the individual fish as the basic unit; another aspect is the effect of crowding with the tank or bounded population as the unit. Even in an onshore tank system it is almost inevitable that some water or live organisms will be transferred from tank to tank. *Glugea stephani* was introduced with live plaice to the Hunterston unit (McVicar 1975). Although there was no apparent re-introduction during the following five years at least, and despite the absence of direct water exchange between tanks, the infection persisted and spread into several different years' hatchery stock. Boots, nets, measuring boards, handling tanks etc. were all considered as possible vectors of spores. Because of the high transfer risk in a farm situation, it is necessary to eradicate any highly pathogenic organism immediately on diagnosis. When the fungus *Ichthyophonus* was identified in a residual plaice stock at Hunterston, we advised the immediate slaughter of all suspect stocks, since we knew of no satisfactory method of controlling this parasite. The disease has not reappeared. With cages or enclosures in intertidal or sublittoral situations it is usually impracticable to control the interchange of water between adjacent cultivation units. Infective stages of parasites, together with carrier, intermediate and final hosts, are thus free to move to, or be carried from, one unit to another. Fortunately, we have encountered no problems of this nature, but Hempel (1970) gave an indication of the sort of situation which can arise under conditions of intensive culture. With reference to Japanese cultivation of large numbers of yellowtail in floating cages in sheltered bays, he stated that '. . . infestation by parasites and epidemic diseases are also increasing'.

Diet

Most forms of fish cultivation involve the provision of an enriched source of food aimed at increasing the growth rate. This may be supplemented with natural food to an extent varying with the type of culture system. Different parasites may be associated with the two food sources, and the parasite faunas of cultivated fish will reflect the amount and form of each taken in different culture systems.

In an experiment described by Steele (1966), tanks containing juvenile plaice were set on the sea-bed in two locations in Loch Ewe on the west coast of Scotland. As no additional food was provided, the fish were totally dependent on natural food. Almost the full range of parasite species found in local wild populations of the same age was recorded from the cage fish (MacKenzie 1968). At one location the plaice became exceptionally heavily infected with two helminth parasites, one of which, the larval nematode *Contracaecum aduncum*, is acquired through feeding on infected invertebrates. No plaice survived the eighteen weeks of the experiment in that location and MacKenzie (1968) considered that parasitism may have at least contributed to the mortality.

In contrast, most of the adult digeneans which occurred as gut parasites of wild plaice were absent from cultivated plaice held in onshore tanks and floating cages (MacKenzie *et al.* 1976). One exception was *Hemiurus communis*, which was found in Hunterston plaice as frequently as it occurred in a wild population. Calanoid copepods, the most common of the planktonic second intermediate hosts of *H. communis*, were frequently found in the stomach contents of plaice from Hunterston. Only one specimen of *H. communis* was found in cultivated plaice from the other sites, where there was no evidence that the fish had been feeding on copepods.

A fish cultivation unit will inevitably attract the attention of wild fish and can result in the unit becoming a focal point for locally occurring parasites. Shoals of small wild fish are frequently attracted to the vicinity of floating cages by spillage of food and by the weed growth around the cages.

The cultivated fish may become infected by eating either these wild fish or invertebrates which have been infected via eggs produced by mature helminths in the wild fish. Pipe fish, *Syngnathus* spp., appear to find the weed growth around the floating cages at Moidart a suitable habitat. Many of them carry an adult cestode of the genus *Bothriocephalus*, possibly the species *B. scorpii*, which we have found in turbot from these cages. We are concerned with the possibility that eggs produced by cestodes in the pipe-fish may infect copepods which may in turn be eaten by the cultivated turbot. Kändler (cited by Mann 1954) reported retardation of growth as a result of cestode infection of turbot, but did not name the species involved. Davey & Peachey (1968) reported that 29% of O-group turbot caught on Welsh beaches were infected with *B. scorpii*. Adult turbot are usually infected; we counted 148 *B. scorpii*, each approximately 9·5 cm long, from the gut of one 54 cm wild turbot. Williams (1972) recorded substantial increases in cestode infections of pompano, Atlantic croaker and sea catfish during a period of confinement in floating cages. No reference was made to feeding arrangements in this experiment, but it seems likely that the cestodes were ingested with natural food.

Food provided by the fish farmer for his stock may also be a source of parasitic infection. Reference was made earlier to the dangers of feeding untreated waste or industrial fish to cultivated stock. We suspect that this was the source of the outbreak of the fungus *Ichthyophonus* in plaice at Hunterston. The fungal infection is systemic and can infect a wide range of teleost species, having been recorded from haddock, plaice, herring, cottids and salmon in Scottish coastal waters (unpublished). On some fishing grounds haddock are commonly infected, and since the flesh is one of the main sites of infection, a large proportion of infected fish are rejected and can appear in the reclaimed material available to fish farmers. McVicar & MacKenzie (1972) showed that *Ichthyophonus* can be transmitted orally from one species of fish to another, and there is ample evidence of the high pathogenicity of the fungus in experimental aquarium infections (unpublished), in wild populations and on fish farms (Sindermann 1970). Most examples of the latter are from fresh water rainbow trout farms where infected marine or fresh water fish have been used as feed, but Egusa (cited by Ghittino 1973) found *Ichthyophonus* in cultivated yellowtail in Japan.

Larval anisakidid nematodes are among the most common helminth parasites of marine fish, sometimes occurring in large numbers in the viscera and musculature. The genus *Anisakis* is of particular importance to fish cultivation, since in some circumstances they can become a human health problem. They are also aesthetically unpleasant when visible in the musculature and can thereby affect the market value of the fish. Wootten & Smith (1975) showed experimentally that *Anisakis* can be transferred easily from one fish to another by feeding live larvae, and they concluded that infections of trout in fresh water fish farms may be acquired through feeding untreated marine fish offal. If similar material is fed to cultivated marine stock, cross-infection seems even more likely. Fortunately, this has not become regular practice in the United Kingdom and we have only recorded *Anisakis* on one occasion in cultivated plaice. *Anisakis* larvae are killed by freezing to −20°C for twenty-four hours (Gustafson 1953).

Nakajima & Syuzo (1969) described occasional infections of cultivated yellowtail in Japan with a larval cestode *Callotetrarhynchus* sp., which they showed was acquired through feeding on anchovy, one of the main supplied foodstuffs in yellowtail cultivation.

Particular care should be taken with the feeding of larval and juvenile cultivated fish, as they are usually more severely affected by parasitism than older fish. Rosenthal (1967) described about 10% mortality of herring larvae reared under aquarium conditions due to infection with larvae of the nematode *Contracaecum* sp. The infection was acquired through feeding the young herring with wild plankton. Some cestode larvae were picked up from the same source, but did not appear to have such harmful effects.

Siting of enclosures

When considering marine fish cultivation in a given area two questions should first be asked from the parasitological point of view: are there any local concentrations of parasites which are known to be infective and pathogenic to the proposed cultivated fish, and has one culture system any marked advantage over another in avoiding infection?

The importance of careful site selection for fish cultivation enclosures is well illustrated by the experiment described and discussed by Steele (1966) and MacKenzie (1968) and referred to in the previous section. Juvenile plaice were kept in sea-bed cages at a depth of about six feet sublittorally at two locations—an exposed beach with a sandy bottom and a sheltered bay with a muddy bottom. The cage floors were covered with the natural substrate. The parasite fauna of the plaice at the exposed sandy site was similar to that of a natural population from a nearby beach, but that of the plaice at the sheltered muddy site was markedly different. Eleven weeks after the start of the experiment, plaice at the muddy location had heavy infections of metacercariae of the digenean *Cryptocotyle lingua* and larvae of the nematode *Contracaecum aduncum*. No live plaice were recovered from this location after eighteen weeks and it was concluded that parasitism at least contributed to this mortality. Infections of both parasites were much lower and fish survival much better at the exposed sandy site, where juvenile plaice occur naturally. Rosenthal (1967) found that even single infections of larval *Contracaecum* could kill herring larvae and MacKenzie (1971) showed that heavy experimental infections of *Cryptocotyle lingua* could kill juvenile plaice.

Infections of *Cryptocotyle* metacercariae were invariably heavier in cultivated plaice and turbot than in wild populations (MacKenzie *et al.* 1976 and unpublished). This was probably due to the siting of the floating cages at Ardtoe and Moidart and of the water intake at Hunterston. Juvenile plaice and turbot occur naturally on sandy beaches, whereas the first intermediate host of *C. lingua*, the periwinkle *Littorina littorea*, is restricted to rocky areas. The shore-lines at Hunterston, Moidart and Ardtoe are all rocky and support periwinkle populations. Piscivorous birds, the final hosts of the digenean, are also common in these areas and are naturally attracted to the vicinity of fish enclosures. However, none of the infections of cultivated stock reached a serious level.

Variations in the occurrence of parasites in the wild and in different cultural conditions can yield valuable information for selecting the most suitable system for a particular area. In contrast to *C. lingua*, the metacercariae of two other digeneans, *Stephanostomum baccatum* and *Rhipidocotyle minimum*, were common in wild plaice but absent from cultivated stock. These variations can probably be explained in terms of the behaviour of the free-living cercariae. Those of *C. lingua* are active swimmers and positively phototactic (Rothschild 1939), which gives them a wide vertical range and brings them into contact with floating cages. The cercariae of *S. baccatum* on the other hand, is not an active swimmer but tends to lie or crawl on the substrate (Wolfgang 1955). The absence of *S. baccatum* and *R. minimum* from cultivated plaice was particularly rewarding, since we consider these species to be amongst the more undesirable of the indigenous parasites both from the point of view of fish health and the market value of the product. MacKenzie & Liversidge (1975) showed that juvenile plaice can be killed by exposure to large numbers of *S. baccatum* cercariae. Of the seven digenean species reported by MacKenzie (1971) as adults in the alimentary tract of wild plaice in the west of Scotland, only *Hemiurus communis* was found on one occasion in plaice held in floating cages. Two species, *H. communis* and *Lecithaster gibbosus*, were found at Hunterston. The significant point about these infections is that they are the only two species of this group to have planktonic second intermediate hosts; the others are found as metacercariae in benthic invertebrates. It is clear that holding flatfish in onshore tanks or floating cages has the advantage of raising the fish beyond the range of most of their common parasites, but this must be balanced against the possibility of increased exposure to other parasites such as *C. lingua*.

Temperature and chlorine

Most of the culture systems we studied have used natural unmodified sea water. The exception is the onshore system at Hunterston, where the water is both heated and chlorinated. A major reason for siting a cultivation unit on the coolant discharge route of a power station was to use the high temperature to enhance fish growth rates. Chlorine is added to combat mussel fouling in the pipes. Although neither modification was specifically designed to influence the parasite fauna of the cultivated fish, our observations showed that both are probably significant.

Bauer (1959) stated that 'For parasites of fish, temperature is the main factor, influencing them at all stages of their development'. The same author pointed out that fish parasites differ greatly from those of warm-blooded animals in that they are exposed to variations in temperature at all stages of their development. Each fish parasite has an optimum temperature at which the rates of growth, maturation and reproduction are at their maxima. When the water temperature in a fish cultivation unit tends towards the optimum for a pathogenic parasite already present in the system, in the absence of any barrier to the spread of infection, a disease problem is imminent.

The appearance of the microsporidian *Glugea stephani* in several age groups of plaice at Hunterston was at first surprising, since this parasite had not been recorded before in Scottish waters. McVicar (1975) demonstrated experimentally the total dependence of *G. stephani* on temperatures above those usually found in Scottish coastal waters for successful transmission. The parasite is apparently endemic in southern United Kingdom waters, where summer temperatures attain levels which favour its spread. We suspect that *G. stephani* was introduced to Hunterston with plaice from southern regions, and that the consistently high temperatures of the power station discharge enabled it to reproduce and spread to successive stocks hatched on the site. Although it was subsequently transferred with live stock to the floating cages at Moidart, there was no evidence of its further development or spread in the lower temperatures that prevail there (MacKenzie *et al.* 1976). None of the other differences between parasite infections of plaice at Hunterston and those of other cultivated and wild plaice could be attributed to the effects of temperature. Some effects of thermal effluents on fresh water and brackish water fish parasites were recorded by Eure & Esch (1974) and Markowski (1966).

The monogenean *Gyrodactylus unicupola* was absent from all Hunterston plaice stocks, but was a common parasite of both wild and cultivated plaice elsewhere (MacKenzie *et al.* 1976). There is no question of these Hunterston fish not being susceptible to infection, as *G. unicopula* appeared in plaice in the floating cages at Moidart about two months after transfer from Hunterston. According to Malmberg (1966), *Gyrodactylus* is sensitive to small amounts of chlorine, so it seems likely that the chlorination of the power station discharge may have provided an incidental means of control. In view of the severe pathological effects of *Gyrodactylus* infections already discussed, a detailed investigation of the effects of chlorine on these monogeneans might prove rewarding.

Salinity

Most techniques for the control of fish parasites relate to fresh water conditions (Hoffman & Meyer 1974). Oppenheimer (1962) pointed out that it may be difficult to extrapolate fresh water techniques to marine conditions and advised caution with the use of chemical therapeutics, as their activity in sea water may differ from that in fresh water. There is evidence now that the regular use of toxic chemicals such as formalin can have harmful effects on fish (Wedemeyer 1971, Smith & Piper 1972, Piper & Smith 1973, Wedemeyer & Yasutake 1974), and the efficacy of even high concentrations of formalin in controlling a parasitic copepod on a marine fish farm was questioned by Braaten (1975). Attention has therefore been focused on other methods of control, particularly the use of variations in salinity.

On the basis of their salinity tolerance, Bauer (1959) divided fish parasites into three groups: marine stenohaline, fresh water stenohaline and euryhaline. As far as the first two groups are concerned, control can be effected by siting fish enclosures in situations where the salinity is acceptable to the host but not to the parasite. Paperna & Lahav (1974) found that grey mullet held in sea water ponds in Israel developed heavy infections of the parasitic copepods *Pseudocaligus apodus* and *Caligus* sp., resulting in severe wounds from which the fish died. These parasites had not appeared in the fresh and brackish water mullet cultivation ponds used previously, but the difference in infection may not be directly related to salinity, since Reiss & Paperna (1975) reported that *P. apodus* can tolerate salinities ranging from 20‰ to 70‰ under laboratory conditions. Kabata (1970) described severe lesions on salmon caused by the parasitic copepod *Lepeophtheirus salmonis*, and Braaten (1975) attributed the loss of thousands of salmonids from Norwegian marine fish farms over a two-year period to the same parasite. Berger (1970) showed that *L. salmonis* does not survive in salinities below about 50% sea water, so that infection may be avoided by siting enclosures in low salinity locations.

The use of lowered salinity to control marine fish parasites may be extended from prophylaxis to the destruction of established infections. In most systems of marine fish cultivation the greatest problem with the latter approach lies in acquiring the necessary degree of control over the water within the enclosure. With floating cage systems it may be worth while periodically towing them to conditions of lower salinity, but in most situations treatment will probably involve dips and sprays of individual fish or groups of fish. These methods have so far been limited to small-scale operations. In the concurrent infection of aquarium plaice with *Trichodina* and *Gyrodactylus* described earlier, it was found that the tolerance limits of host and parasites to dilute solutions of formalin and malachite green did not differ greatly, but both parasites succumbed quickly to fresh water treatment which left the plaice apparently unharmed (unpublished). Kubota & Takakuwa (1963) and Hoshina (1968) reported the use of fresh water to control the monogenean *Benedenia seriolae*, and Kubota & Takakuwa (1963) reported the use of high salinity water to control another monogenean *Heteraxine heterocerca*, both on cultivated yellowtail in Japan.

Choice of stock species

It is often recommended that the fish selected for cultivation should be an indigenous species, since the ecology of the region must provide a suitable environment for growth. From the parasitological point of view this may have some disadvantages, as the cultivated stock would then be vulnerable to infection from local wild populations of the same species. Before setting up a cultivation unit an attempt should be made to determine if there are any local parasites which might represent hazards. A cultivated species which is not indigenous to the region might escape much parasitic infection because of the high degree of host specificity shown by many parasites. The other side of the coin is that natural specificity may break down in the artificial environment of a cultivation unit, resulting in infections which cross established specificity boundaries. Clearly a detailed knowledge of the biology and ecology of potential pathogens is necessary before the pros and cons of cultivating a particular species of fish in a given situation can be properly assessed.

In order to improve the economic viability of a cultivation unit it is logical to consider rearing more than one species in the same unit. We have not had the opportunity of making a parasitological study of a polyculture system so that we can only attempt to anticipate problems. We do not foresee many additional parasite problems arising from the cultivation of more than one species of teleost, provided the same precautions already discussed with reference to monoculture systems are observed. One action which must be particularly guarded against in polyculture is the introduction of a fish which may act as a carrier host for a parasite which has more serious pathological effects on another stock species.

The situation becomes more complicated when fish and invertebrates, particularly molluscs, are held in the same water column, even if they occupy separate tanks. Molluscs can act as intermediate hosts for a variety of fish parasites, but this type of system would particularly favour the gasterostome group of digeneans, in which the standard life-cycle involves a bivalve mollusc as first intermediate host, a teleost fish as second intermediate host and a piscivorous teleost as final host. There are a large number of gasterostome species and their bivalve hosts include many of commercial importance, such as oysters, cockles and mussels. Matthews (1973a) showed that turbot are suitable second intermediate hosts of *Prosorhynchus crucibulum*, with metacercariae encysting in the musculature and connective tissue. We found this species in mussels from the inlet pipes of the Hunterston unit (unpublished), but it did not appear in the cultivated turbot. The reason for the parasite's failure to continue its life-cycle in this instance remains unknown, but may be peculiar to the Hunterston system and possibly associated with chlorination. Matthews drew attention to the relevance of *P. crucibulum* to turbot cultivation and referred to economic loss in a commercial fish industry, described by Liston *et al.* (1960), due to the effects of *Prosorhynchus* sp. metacercariae. A particularly serious situation could arise if the life-cycle of a gasterostome could be completed within the confines of a cultivation unit. This may be possible with *P. crucibulum*, since immature forms of an unidentified species of *Prosorhynchus* have been found as gut parasites of cultivated plaice at Ardtoe and of wild turbot from a Hebridean beach (unpublished). Cannibalism occurs amongst cultivated turbot and could lead to completion of the life-cycle if turbot can act as final, as well as second intermediate hosts of *P. crucibulum*. (Digenea as a group, show low host specificity to their final hosts.) Introduction of another species of piscivorous teleost to the system would further increase the chances of a gasterostome completing its life-cycle. We are at present planning experiments to determine if *P. crucibulum* can develop to maturity in the gut of turbot.

Matthews (1973b) reported extensive damage to the livers of young plaice and gobies caused by migrating metacercariae of another gasterostome, *Bucephalus haimeanus*, the bivalve host of which is the cockle *Cardium edule*. Experimental exposure to fifty or more cercariae killed larval plaice within two days, despite the fact that plaice are not suitable hosts and only 2% of metacercariae survived more than two weeks. This result suggests that juveniles of fish species outside the normal range of intermediate hosts may be attacked and severely damaged by gasterostome cercariae. We strongly recommend that bivalves should be thoroughly checked for gasterostome infections before introduction to a mixed cultivation unit.

Concluding remarks

In this paper we have been concerned solely with the problems, actual and potential, resulting from parasitic infections in marine fish cultivation. If by this stage we may have left an impression of having struggled to overcome one disease problem after another in the cultivation units we have studied, we will now attempt to view these problems in their true perspective. The range of parasites in cultivated stocks is generally more restricted than that of wild populations, and the few serious pathogens which have appeared to date in marine fish cultivation have rarely presented problems which required extreme measures. An outbreak of a pathological condition may often be directly associated with inadequacies in husbandry techniques; the low prevalence of parasitic disease in White Fish Authority experimental fish cultivation units may be largely due to the good conditions in which the fish are held. Pressures on husbandry techniques will undoubtedly be increased in commercial units and it may be anticipated that parasite problems will become increasingly important. If a parasite were to restrict the profit margin in a fish cultivation project by, for example, forcing reduction of stocking density or by interfering with food conversion, growth rate or the condition of the fish, the infection could become a factor limiting the success of the project and a direct monetary value could be placed on its effect.

References

ANDERSON C.D., ROBERTS R.J., MACKENZIE K. & MCVICAR A.H. (1976) Hepato-renal syndrome in cultured turbot. *J. Fish. Biol.* 8, 331–41.

ANDERSON J.I.W. & CONROY D.A. (1968) The significance of disease in preliminary attempts to raise flatfish and salmonids in sea water. *Bull. Off. int. Epiz.* 69, 1129–37.

BAER J.G. (1974) Resolutions of the Third International Congress of Parasitology. *Environ. Conserv.* 1, 176.

BARDACH J.E., RYTHER J.H. & MCLARNEY W.O. (1972) *Aquaculture. The Farming and Husbandry of Freshwater and Marine Organisms.* Wiley-Interscience, New York, London, Sydney, Toronto.

BAUER O.N. (1959) The influence of environmental factors on reproduction of fish parasites. *Vop. Ekol. (Izdatelstvo kievskogo Univ.)* 3, 132–41. (English translation 1968 by Z. Kabata and L. Margolis, Fish. Res. Bd Can.)

BERGER V.YA. (1970) Effect of sea water of different salinity on *Lepeophtheirus salmonis* (Kroyer): an ectoparasite of salmon (in Russian). *Parazitologiya* 4, 136–8.

BRAATEN B.R. (1975) Recent Norwegian experience in fish farming. *Conf. 'Oceanology International 75',* Soc. Underw. Technol., Brighton, 168–72.

CORBEL M.J. (1975) The immune response in fish: a review. *J. Fish Biol.* 7, 539–63.

DAVEY J.T. & PEACHEY J.E. (1968) *Bothriocephalus scorpii* (Cestoda: Pseudophyllidea) in turbot and brill from British coastal waters. *J. mar. biol. Ass. U.K.* 48, 335–40.

Esch G.W., Gibbons J.W. & Bourque J.E. (1975) An analysis of the relationship between stress and parasitism. *Am. Midl. Nat.* 93, 339–53.

Eure H.E. & Esch G.W. (1974) The effects of thermal effluents on the population dynamics of helminth parasites in the largemouth bass, *Micropterus salmoides*. In *Thermal Ecology* (Ed. by J.W. Gibbons & R.R. Sharity), pp. 207–15. AEC. Symp. Ser. (CONF-730505).

Ferguson H.W. & Roberts R.J. (1975) Myeloid leucosis associated with sporozoan infection in cultured turbot (*Scopthalmus maximus* L.). *J. comp. Path.* 85, 317–26.

Finucane J.H. (1970) Pompano mariculture in Florida. *Proc. 2nd Conf. Food-drugs from the Sea* (Ed. by H.W. Youngken, Jnr.), pp. 135–43. Kingston, R.I.

Ghittino P. (1973) Present knowledge of the principal diseases of cultured marine fish. *Fourth Ann. Conf. Workshop Int. Ass. Aquatic Anim. Med.*, Victoria, B.C., Canada. 51–6.

Gustafson P.V. (1953) The effect of freezing on encysted *Anisakis* larvae. *J. Parasit.* 39, 585–8.

Hanson J.A. (1974) Concentrating and harvesting marine crops. In *Open Sea Mariculture* (Ed. by J.A. Hanson), pp. 237–60. Dowden, Hutchinson & Ross, Stroudsburg, Pa.

Hempel G. (1970) Fish-farming, including farming of other organisms of economic importance. *Helgoländer wiss. Meeresunters.* 21, 445–65.

Hoffman G.L. (1970) Intercontinental and transcontinental dissemination and transfaunation of fish parasites with emphasis on whirling disease (*Myxosoma cerebralis*). In *A Symposium on Diseases of Fishes and Shellfishes* (Ed. by S.F. Snieszko), pp. 69–81. *Spec. Publ. Am. Fish. Soc.* No. 5.

Hoffman G.L. & Meyer F.P. (1974) *Parasites of Freshwater Fishes: A Review of their Control and Treatment.* T.F.H. Publs. Inc., New Jersey.

Hoshina T. (1968) On the monogenetic trematode, *Benedenia seriolae*, parasitic on yellowtail. *Bull. Off. int. Epiz.* 69, 1179–91.

Kabata Z. (1970) *Diseases of Fishes, Book 1: Crustacea as Enemies of Fishes* (Ed. by S.F. Snieszko & H.R. Axelrod). T.F.H. Publs. Inc., New Jersey.

Kubota S.S. & Takakuwa M. (1963) Studies on the diseases of marine cultured fishes. I: General description and preliminary discussion on fish diseases in Mie Prefecture. *J. Fac. Fish. Prefect. Univ. Mie* 6, 107–24. (English Translation 1966 by Y.C. Pan, Fish. Res. Bd Can.)

Lester R.J.G. & Adams J.R. (1974) A simple model of a *Gyrodactylus* population. *Int. J. Parasit.* 4, 497–506.

Liston J., Peters J. & Stern J.A. (1960) Parasites in summer-caught rockfishes. *U.S. Fish Wildl. Serv. Spec. Sci. Rep., Fish.* No. 352.

Lom J. (1970) Protozoa causing diseases in marine fishes. In *A Symposium on Diseases of Fishes and Shellfishes* (Ed. by S.F. Snieszko), pp. 101–23. *Spec. Publ. Am. Fish. Soc.* No. 5.

Lom J. (1973) The adhesive disc of *Trichodinella epizootica*—ultrastructure and injury to the host tissue. *Folia parasitol.* 20, 193–202.

MacKenzie K. (1968) Some parasites of O-group plaice, *Pleuronectes platessa* L., under different environmental conditions. *Mar. Res.* No. 3.

MacKenzie K. (1969) *Scyphidia* (*Gerda*) *adunconucleata* n. sp. and *Trichodina borealis* (Dogiel, 1940) Shulman et Shulman-Albova, 1953 (Protozoa: Ciliata) from young plaice in Scottish waters. *J. Fish Biol.* 1, 239–47.

MacKenzie K. (1970) *Gyrodactylus unicopula* Glukhova, 1955, from young plaice *Pleuronectes platessa* L. with notes on the ecology of the parasite. *J. Fish Biol.* 2, 23–34.

MACKENZIE K. (1971) *Ecological studies of some parasites of juvenile plaice*, Pleuronectes platessa *L.* Unpublished PhD. thesis, University of Aberdeen.

MACKENZIE K. & LIVERSIDGE J.M. (1975) Some aspects of the biology of the cercaria and metacercaria of *Stephanostomum baccatum* (Nicoll, 1907) Manter, 1934 (Digenea: Acanthocolpidae). *J. Fish Biol.* 7, 247–56.

MACKENZIE K., MCVICAR A.H. & WADDELL I.F. (1976) Some parasites of plaice *Pleuronectes platessa* L. in three different farm environments. *Scott. Fish. Res. Rep.* No. 4.

MCVICAR A.H. (1975) Infection of plaice *Pleuronectes platessa* L. with *Glugea* (*Nosema*) *stephani* (Hagenmüller, 1899) (Protozoa: Microsporidia) in a fish farm and under experimental conditions. *J. Fish Biol.* 7, 611–20.

MCVICAR A.H. & MACKENZIE K. (1972) A fungus disease of fish. *Scott. Fish. Bull.* No. 37, 27–8.

MALMBERG G. (1966) Taxonomical and ecological problems in *Gyrodactylus* (Trematoda, Monogenea). *Proc. Symp. 'Parasitic Worms and Aquatic Conditions'*, Prague, pp. 203–30. Czechoslovak Academy of Sciences Publishing House, Prague.

MANN H. (1954) Die wirtschaftliche Bedeutung von Krankheiten bei Seefischen. *Fischwirtschaft, Bremerhaven* 6, 38–9.

MARKOWSKI S. (1966) The diet and infection of fishes in Cavendish Dock, Barrow-in-Furness. *J. Zool. Lond.* 150, 183–97.

MATTHEWS R.A. (1973a) The life-cycle of *Prosorhynchus crucibulum* (Rudolphi, 1819) Odhner, 1905, and a comparison of its cercaria with that of *Prosorhynchus squamatus* Odhner, 1905. *Parasitology* 66, 133–64.

MATTHEWS R.A. (1973b) The life-cycle of *Bucephalus haimeanus* Lacaze-Duthiers, 1854 from *Cardium edule* L. *Parasitology* 67, 341–50.

MILNE P.H. (1972) *Fish and Shellfish Farming in Coastal Waters*. Fishing News (Books) Ltd., London.

NAKAJIMA L. & SYUZO E. (1969) Studies on a new Trypanorhynchan larva, *Callotetrarhynchus* sp., parasitic on cultured yellowtail—II. On the source and route of infection. *Bull. Jap. Soc. scient. Fish.* 35, 351–7.

ODUM W.E. (1974) Potential effects of aquaculture on inshore coastal waters. *Environm. Conserv.* 1, 225–30.

OPPENHEIMER C.H. (1962) On marine fish diseases. In *Fish as Food. Volume 2. Nutrition, Sanitation and Utilization* (Ed. by G. Borgstrom), pp. 541–72. Academic Press, New York and London.

PAPERNA I. (1975) Parasites and diseases of the grey mullet (Mugilidae) with special reference to the seas of the Near East. *Aquaculture* 5, 65–80.

PAPERNA I. & LAHAV M. (1974) Mortality among grey mullets in a seawater pond due to caligid parasitic copepod epizootic. *Bamidgeh* 26, 12–15.

PEARSE L. (1972) A note on a marine trichodinid ciliate parasitic on the skin of captive flatfish. *Aquaculture* 1, 261–6.

PIPER R.G. & SMITH C.E. (1973) Factors influencing formalin toxicity in trout. *Progve Fish Cult.* 35, 78–81.

RAABE H. (1936) Etudes de microorganismes parasites des poissons de mer. 1. *Nosema ovoideum* Thel. dans le foie des rougets. *Bull. Inst. oceanogr. Monaco*, No. 696.

REISS Z. & PAPERNA I. (1975) Studies on diseases of marine fish. *Fourth Rep. H. Steinitz Mar. Biol. Lab. Elat, Israel*, 55 69.

ROSENTHAL H. (1967) Parasites in larvae of the herring (*Clupea harengus* L.) fed with wild plankton. *Mar. Biol.* 1, 10–15.

ROTHSCHILD, M. (1939) A note on the life cycle of *Cryptocotyle lingua* (Creplin, 1825). *Novit. zool.* 41, 178–80.

SINDERMANN C.J. (1970) *Principal Diseases of Marine Fish and Shellfish*. Academic Press, New York and London.

SINDERMANN C.J. (Ed.) (1974) Diagnosis and control of mariculture diseases in the United States. *Nat. Mar. Fish. Serv. Tech. Ser. Rep.* No. 2.

SMITH C.E. & PIPER R.G. (1972) Pathological effects in formalin-treated rainbow trout (*Salmo gairdneri*). *J. Fish. Res. Bd Can.* **29**, 328–9.

SNIESZKO S.F. (1974) The effects of environmental stress on outbreaks of infectious diseases of fishes. *J. Fish Biol.* **6**, 197–208.

STEELE J.H. (1966) Ecological problems of marine fish farming. *Proc. Nutr. Soc.* **25**, 126–9.

WEDEMEYER G. (1970) The role of stress in the disease resistance of fishes. In *A Symposium on Diseases of Fishes and Shellfishes* (Ed. by S.F. Snieszko), pp. 30–5. *Spec. Publ. Am. Fish. Soc.* No. 5.

WEDEMEYER G. (1971) The stress of formalin treatments in rainbow trout (*Salmo gairdneri*) and coho salmon (*Oncorhynchus kisutch*). *J. Fish. Res. Bd Can.* **28**, 1899–904.

WEDEMEYER G. & YASUTAKE W.T. (1974) Stress of formalin treatment in juvenile spring chinook salmon (*Oncorhynchus tshawytscha*) and steelhead trout (*Salmo gairdneri*). *J. Fish. Res. Bd Can.* **31**, 179–84.

WILLIAMS E.H. Jnr. (1972) Parasitic infestation of some marine fishes before and after confinement in feeding cages. *Alabama Marine Resources Bull.* No. 8, 25–31.

WOLFGANG R.W. (1955) Studies of the trematode *Stephanostomum baccatum* (Nicoll, 1907). III. Its life cycle. *Can. J. Zool.* **33**, 113–28.

WOOTTEN R. & SMITH J.W. (1975) Observational and experimental studies on the acquisition of *Anisakis* sp. larvae (Nematoda: Ascaridida) by trout in fresh water. *Int. J. Parasit.* **5**, 373–8.

Some effects of increasing the monoculture of cereals

G. R. POTTS *The Game Conservancy, North Farm,*
Washington, Pulborough, West Sussex; now at Burgate Manor,
Fordingbridge, Hampshire

Farmers normally aim to increase the quality, quantity and ease of harvesting of their produce and as they succeed, in part, by eliminating weeds, insect pests or diseases, their crops increasingly resemble true monocultures. Although pesticides were first extensively used in 1940, chemical control of many of the known pests, diseases and weeds was not possible until 1975.

The growing impact of new techniques on the agroecosystem has for some years caused concern not only to wildlife conservationists but also to farmers because increases in pesticide use are now justified by small and possibly short-term increases in yield. Indeed since studies began at North Farm on the Sussex Downs in 1968 it has become less clear that yields are increasing in the long term and more clear that modern techniques are having long-term repercussions on wildlife.

Since 1933, with a gap for World War II, the annual production of young partridges (*Perdix perdix* and *Alectoris rufa*) has been monitored over most of their range in the United Kingdom. There is clear evidence of a decrease in the chick survival rate of the grey partridge (*Perdix perdix*) beginning between 1952 and 1962. This decline appears to be due to a shortage of insect food in cereals (Potts 1970a, 1971). However, prior to this study, there was no comprehensive monitoring of arthropods in cereal fields, although there had been a considerable amount of research concerning some pest species (reviewed in Potts & Vickerman 1974) and some calculations of the overall effects of herbicides on the arthropods (Southwood & Cross 1969).

Hearsay evidence suggests that prior to 1950 cereal fields contained 'many grasshoppers' and far more 'butterflies' and 'ants' than is the case today and the basic aim of the arthropod monitoring programme at North Farm is to replace this type of evidence with an objective assessment and to establish the effects of modern agricultural techniques on the farmland fauna.

Cereals as monoculture

There are two different ways in which cereal growing approaches true monoculture.

Inter-field, inter-farm and inter-regional monoculture

In the United Kingdom mixed farming declined slowly from about 1870, when cheap grain began to be imported in quantity from North America, though there was a respite during both World Wars, and the decline only became marked in the early 1960's (Anon. 1969). Whole farms were given over to cereals (Britton 1969), farming systems became more specialized and regional cropping patterns became polarized (Southwood 1972); in particular the main cereal growing areas became relatively separated from the grassland areas (for England and Wales see Church *et al.* 1968).

Various economies of scale were made, especially towards the end of the 1960's, with the advent of bigger and better farm machinery. Hedges were removed; half the hedges in Norfolk were lost during the period 1946–70 (Baird & Tarrant 1973), and much the same happened elsewhere in the arable sectors of Europe (Terrason & Tendron 1975). These trends towards block farming were also carried through to species of cereals and to cultivars. The storage of several cultivars on the same farm, for example of malting barley, or cereals for seed, is difficult in modern grain silos and there is a tendency for farmers to sow much of the same variety. This is also true for the whole country and more than half the winter wheat sown in the United Kingdom in 1975 was *cv.* Maris Huntsman. It is well known that the genetic variability of cereals is often low, and even as early as 1939 85% of the commercial barley seed in the United Kingdom was derived from four individual plants (Beaven 1947).

Intra-field monoculture

In many regions such as East Anglia and the South Paris Basin cereals are no longer grown in rotation with grass and no longer used as a nurse crop to establish grasses and clovers (Potts 1970a) due to less need for grass on the farm and also to the fact that modern species are more likely to compete with their nurse. Fodder legumes, such as peas in dredge corn, are now rarely grown among cereals for similar reasons (Anon. 1969).

The abundance of dicotyledonous weeds has declined, due mainly but not entirely to the introduction of herbicides (Potts 1970a). The levels of abundance of these weeds are now stable and low (Table 1) reflecting the continued application of a range of herbicides (Table 2). By contrast with the dicotyledons, most of the monocotyledonous weeds could not be removed from cereals by chemical means in 1969. However, since then successive new herbicides have been developed to control grass weeds. The overall trend of these weeds in the study area would have been markedly upward, especially in autumn drilled barley, if it had not been for these new herbicides (Tables 1 and 2), though the incidence of wild oats (*Avena fatua*) and black grass (*Alopecurus myosuroides*) is far lower than in most of the United Kingdom.

Table 1.   Annual indices of weed abundance and fungal disease in cereal crops in late June 1970–1976, West Sussex study area.

| Year | No. of fields sampled | Mean indices for weeds and cereal diseases | | |
		Dicoty-ledons*	Monocoty-ledons*	Diseases†
1970	150	1.36	0.64	0.72
1971	115	1.16	0.80	0.91
1972	107	1.75	1.15	2.26
1973	143	1·22	1.54	3.32
1974	157	1.64	1.26	0.70
1975	176	1.17	1.31	1.29
1976	146	1.15	1.00	2.14

Sown monocotyledons and clovers in undersown fields are excluded.
* Weed indices are based on the ranking of each field into six categories of 0 (no weeds) to 5 (weeds smothering crop)
† Disease indices are based on the total for each of the rankings of the three disease groups mildew (*Erysiphe*), the rusts (*Puccinia*) and *Rhynchosporium* on a scale of 0 to 5, giving (theoretically!) a maximum score of 15 for each field.

Seed dressings have been used to control fungal diseases for about thirty-five years in our study area but in 1971 new systemic fungicides were introduced, the most important being ethirimol and tridemorph (Potts & Vickerman 1975). By 1976 most barleys were treated either with seed dressings or foliar sprays. The latter, not used at all until 1971, were used on more than half of all cereals in the study area in 1976 (Table 2) and nine different active ingredients were involved. By 1976 farmers in Sussex were barely considering genetic resistance to fungal attack when choosing new varieties, and there is no doubt that many cereals would have been drastically affected by mildew in the dry season of 1976 if it had not been for the new fungicides. However the overall annual incidence of cereal diseases has shown no significant trend since 1970 (Table 1).

Insecticides have been used for many years as seed dressings but only recently have they been applied to large acreages as foliar sprays, to protect crops from cereal aphids (Table 2). A national survey of aphicide use over the period 1966–75 has shown that it has increased from less than 1% of cereals sprayed between 1966 and 1972, to 3% in 1974 and 14% in 1975 (Vickerman, Potts & Sunderland unpublished). Aphicide use in 1976 was greater than in 1975. Part of the reason for this increase is that cereal aphids have increased (Baranyovits 1973, Potts & Vickerman 1974) and part that aerial application of pesticides is now much more feasible. The agricultural–aviation industry in the United Kingdom is not large with only seventy planes and helicopters though in 1975 over 400 000 ha were treated with pesticides or fertilizers from the air (Fuller 1976).

Table 2. Use of foliar applied pesticides on cereals in the West Sussex study area 1969–1976. Data exclude seed dressings (including ethirimol), stubble treatments and trials.

	Year							
	1969	1970	1971	1972	1973	1974	1975	1976
No. of fields in sample	110	145	115	105	148	149	141	133
Herbicides (% of fields treated)								
Narrow spectrum								
controlling dicotyledons	20	25	32	31	32	32	34	33
Broad spectrum								
controlling dicotyledons	70	57	58	59	57	55	46	45
Broad spectrum controlling								
dicotyledons and monocotyledons	0	3	3	4	7	10	16	17
Total	90	85	93	94	96	97	96	95
Fungicides (% of fields treated)	0	0	7	6	11	25	45	55
Aphicides (% of fields treated)	0	0	0	0	1	0	12	7

Between 1969 and 1973 stubble and straw burning was widely practised in the study area (Vickerman in preparation) and also in the country as a whole (Staniforth 1975). In part the intention was to reduce the possibility of carry-over of disease on crop residues and to aid stubble hygiene generally. More recent trends towards minimal cultivation, direct drilling, and reduction of straw burning, as opposed to stubble burning, have probably reduced the impact of post-harvest cultivation techniques on the soil fauna despite the use of desiccants and other pesticides.

Consequences of monoculture of cereals

METHODS

The main study area of 6,240 ha consists of 306 fields on and around North Farm, Washington, West Sussex. Most of the eighteen farms adopt crop rotations on a ley farming theme, but some did not have any stock from about 1968 to 1974 when grass was replaced with crops such as field-beans and oilseed rape.

Work started on the declining partridge population of this particular area in 1968—and the spatial and temporal availability of insect food to partridge chicks have been measured each year. Details of the area, including topography, background and methods for the work on arthropods and partridges have already been given (Potts & Vickerman 1974). Since 1969, the arthropod densities in cereal crops have been measured with the use of a 'backpack' petrol-driven Dietrick vacuum insect net. Each sample consists of five randomly placed sub samples of 0.09 m² and generally contains a mixture of arthropods, vegetation and soil. The samples are deep-frozen and then stored in alcohol for sorting under a microscope (for details see Potts & Vickerman 1974). About 900 species of insect inhabit cereal fields in the study area and eighty species of spiders have so far been identified, though the arthropod fauna is dominated by only about fifty common species. The routine work also involves the monitoring of the use of all agrochemicals and agricultural techniques on the farms in the study area.

CEREAL APHIDS

The role of predation in affecting the rate of change of cereal aphid numbers has long been given attention in the study area because it appeared to be important in determining their rates of increase (Potts 1970b). In Sussex there is a significant inverse relationship between the number of apterous cereal aphids present in samples taken from different fields and the proportion of total arthropods which are predatory (Potts & Vickerman 1975); many of these predators eat aphids in cereals (Table 3). The principal factor affecting peak numbers of aphids appears to operate during the early part of the logistic curve of aphid increase and therefore when density is low (Potts & Vickerman 1974). Many of the aphid predators are polyphagous (Sunderland 1975) and since they do not depend on aphids they can be present in cereal fields before the arrival of immigrants.

Aphid densities were measured and the proportion of predators containing aphid remains was determined by dissection in four fields in 1974 and

1975. When certain assumptions are made concerning rates of feeding it becomes virtually certain that predators controlled the cereal aphids in one field, and probable that they did so in another, but the evidence for effective predation in the other two fields is at best equivocal. Parasitism also appeared important at times, possibly in conjunction with predation. Work now in progress on our study area uses predator exclusion and predator boosting techniques in order to measure predation as a factor reducing the rate of increase of cereal aphids early in the season and in their decline later. Polyphagous species which appear in cereals early in the season and which are also aphid predators are given special attention in the Sussex work, *Agonum*

Table 3. Percentage of some predatory arthropods (collected from spring barley and winter wheat, May–July 1973–1975) which contained remains of aphids.

	Number dissected	%
Coccinellidae:		
Coccinellidae larvae	530	87·0
Coccinella 11-punctata	147	71·4
Coccinella 7-punctata	623	68·5
Propylea 14-punctata	137	46·7
Dermaptera:		
Forficula auricularia	794	41·0
Staphylinidae:		
Tachyporus chrysomelinus	497	7·2
T. obtusus	62	4·8
T. (callow)	118	3·4
T. hypnorum	759	1·7
Carabidae:		
Carabidae unid.	59	49·2
Calathus fuscipes	60	30.0
Feronia melanaria	418	29·9
Risophilus atricapillus	189	28·6
Notiophilus biguttatus	449	26·9
Agonum dorsale	723	25·0
Amara aenea	79	22·8
Amara plebeja	135	18·5
Amara similata	22	18·2
Nebria brevicollis	408	11·5
Asaphidion flavipes	104	10·6
Harpalus rufipes	166	10·3
Bembidion lampros	986	9·5
Loricera pilicornis	454	5·5
Amara familiaris	411	3·4

dorsale (Carabidae) is an example (see Table 3). Elsewhere it has also been noted that an early appearance of predators in May, is associated with lower rates of increase of cereal aphids in June and July (Jones 1972), whereas July populations are reduced by aphidophagous Syrphids (Dean 1974). Dean & Wilding (1971) also considered that fungal disease (*Entomophthora* spp.) may effectively lessen cereal aphid numbers throughout the summer in the United Kingdom when it appears early in the season.

In the United States of America Fenton & Fisher (1940) found that the Coccinellid *Hippodomia convergens* was of considerable value in checking infestations of the green bug *Toxoptera graminum* in some wheat fields, and current work in Sussex is leading to a similar conclusion for *Coccinella 7-punctata*. Walton (1921) considered that the green bug would be a pest every year if it was not kept in check by parasites, but little evidence was given. Cereal aphids cause damage directly, though the extent of this is uncertain (e.g. Vickerman *et al.* unpublished), and indirectly by transmitting viruses such as that causing barley yellow dwarf (BYDV). The percentage of fields with notable levels of BYDV has varied from 8% in 1974, the year of least incidence, to 40% in 1975, the year of highest incidence.

Much more work is required but the emphasis has now changed from establishing the fact of extensive predation, parasitism and disease to quantifying the significance of these factors to cereal growers.

CEREAL SAWFLIES

The mean peak density of sawfly larvae in cereals over the years 1970 to 1975 inclusive was 1·47 m^{-2}; 2·05 m^{-2} were found on traditional farms with undersowing and 1·39 m^{-2} on those farms with no undersowing. This is consistent with the view that pre-pupal and pupal mortality is increased by the cultivation of the soil in which they overwinter (Potts 1970a) although the difference is small (0·66 larvae m^{-2}) and not statistically significant in two of the six years. Moreover sawfly numbers have fluctuated much more violently than has the amount of undersowing, which has been fairly stable since 1968 in our study area.

Local outbreaks of sawflies appear to start with an emergence of adults from a field which has been undersown the previous year and which has a relatively high larval population, but such a set of conditions does not consistently give rise to an increase in sawfly numbers.

Dolerus haematodes, *D. puncticollis* and *D. gonager* have been successfully reared in the laboratory but of the three, *D. puncticollis* survived through the first instars only when grass as well as cereal was available. It is therefore of interest that the proportion of this species in the total of adult *Dolerus* species (Table 4) has increased consistently with the monocotyledon weed index for the previous year (Table 1), but this does not account for the overall fluctuations.

Table 4. Annual totals of sawfly adults trapped in cattle water troughs in the West Sussex study area; the genus *Dolerus* 1969–1976.

	Year							
	1969	1970	1971	1972	1973	1974	1975	1976
D. puncticollis	40*	66	52	18	119	188	42	50
D. gonager	90*	125	115	25	28	113	48	16
D. haematodes	75*	100	36	14	19	49	8	9
D. aeneus	—	9	2	1	1	9	3	2
D. sanguinicollis	—	12	2	0	0	4	3	2
D. nigratus	—	5	5	4	0	3	4	0
D. liogaster	—	1	2	6	4	3	0	7
D. niger	—	1	3	2	0	1	2	2
D. picipes	—	3	3	2	0	0	1	0
Total *Dolerus* spp.	205*	322	220	72	171	370	111	88

* Regular searches of the water troughs did not begin until 1970, the 1969 totals are therefore estimates based partly on the number of larvae; their density and species composition follow very closely that of adults in the other seven years.

However the number of adult *Dolerus* emerging in spring is related in a delayed density dependent manner to the number in the previous year (Fig. 1). This is the kind of relationship one would expect if pupal parasitism was involved and eggs of ichneumons have often been found attached to late instar larvae. Although research continues it does appear that parasitism may obscure the effect of undersowing or soil cultivation on sawflies.

OTHER ARTHROPODS

Interpretation of the arthropod data is hampered by a complete lack of base-line information about the 'normal' densities of the vast majority of species in cereals until 1970. For example, the density of *Coccinella 7-punctata* was very low in 1969, about $0.01 \ m^{-2}$ in barley, and work in Canada had shown that some of the herbicides used in cereals significantly reduced the survival rate of Coccinellidae larvae in the laboratory, even when used at normal field rates. Early in this study it was therefore thought that herbicides might be directly depressing Coccinellid densities in Sussex cereals (Potts 1970a) but subsequent monitoring revealed that this could be an erroneous conclusion; numbers were up three-hundred fold by 1974 (Table 5), and Coccinellids have remained common since then, with exceptionally high numbers (*ca.* 33m⁻²) in June 1976.

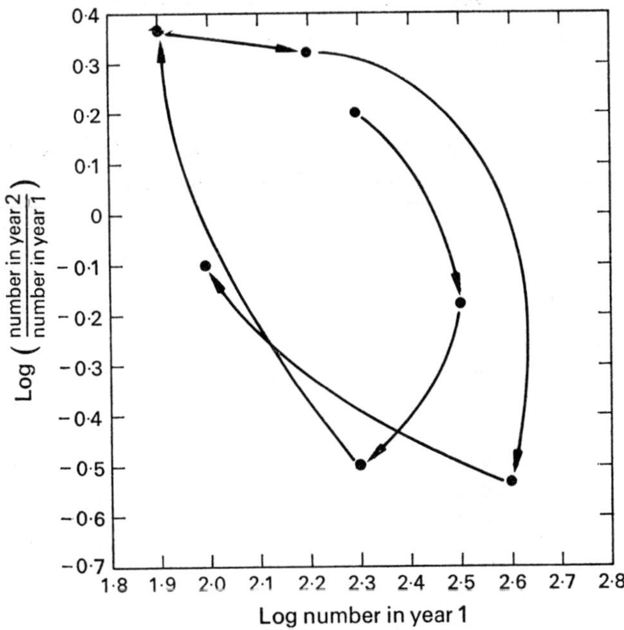

Figure 1. Annual mortality as log (density in year 2/density in year 1), plotted on density in year 1 for sawflies (Tenthredinidae) of the genus *Dolerus* showing delayed correlation with density. Points for successive years 1969–70 to 1975–76 are joined. West Sussex study area.

On the other hand, monitoring of the density of *Tachyporus* spp. (Staphylinidae) adults and larvae clearly indicates a decline since 1970, and probably since 1969 though the data for that year only concern barley (Fig. 2). *Tachyporus* adults and larvae are nocturnal predators of cereal aphids (Vickerman & Sunderland 1975) and also feed on fungi, including some pathogenic fungi of cereals. Field trials have shown that some of the recently introduced herbicides, for instance mixtures of metoxuron and simazine (used to control *Poa trivialis*), depress *Tachyporus* spp. larvae by about 91 % (Vickerman 1974). There is also evidence that the fungicide tridemorph reduces *Tachyporus* numbers, in this case by about 73% (Potts unpublished). The decline of *Tachyporus* and other mycetophagous species is marked when compared with the status of other sectors of the arthropod fauna (see Table 5), and coincides with the increased use of herbicides (against monocotyledons) and foliar applied fungicides (Table 2). This sort of situation demands further attention especially since it is the polyphagous predators of cereal aphids such as *Tachyporus* spp. which may be having a considerable impact on cereal aphid abundance.

Speight & Lawton (1976) have demonstrated that predatory beetles remove more artificial prey, *Drosophila* pupae, in wheat where it is infested with the weed *Poa annua*. Such predation could be one reason why fewer aphids are found in the weedy parts of cereal crops (Potts 1970b, Vickerman

Table 5. Changes in the late June density of the adults and larvae of some insect taxa in cereal crops over the period 1970–1975, West Sussex study area. Data are mean no. m^{-2}.

	Year					
	1970	1971	1972	1973	1974	1975
No. of fields sampled	135	110	98	148	149	141
Polyphagous predatory Coleoptera (adults)						
Risophilus atricapillus	0·56	0·98	0·80	0·63	0·55	0·18
Bembidion lampros	1·19	0·73	0·80	0·75	0·79	0·42
Agonum dorsale	0·26	0·11	0·03	0·10	0·06	0·05
Trechus quadristriatus	0·92	0·98	0·83	1·09	0·74	1·16
Notiophilus biguttatus	0·14	0·15	0·12	0·07	0·07	0·20
Paederus littoralis	0·06	0·06	0·07	0·10	0·04	0·03
Stenus spp.	0·39	0·32	0·35	0·41	0·07	0·23
Aphidophagous predators (adults & larvae)						
Neuroptera	1·00	1·06	0·06	1·44	0·97	0·37
Syrphidae	0·24	1·22	0·12	1·67	2·61	0·48
Coccinellidae	1·21	0·71	0·18	0·73	3·91	3·78
Phytophagous Coleoptera and Diptera (adults)						
Gastrophysa polygoni	1·26	0·38	0·00	0·28	0·64	0·94
Crepidodera transversa	0·43	0·58	0·06	0·37	0·16	0·06
Lema melanopa	0·47	0·07	0·09	0·09	0·38	0·57
Phyllotreta spp.	0·11	0·24	0·09	0·00	0·06	0·01
Chaetocnema spp.	0·27	1·29	0·12	0·04	0·04	0·06
Longitarsus spp.	0·05	0·15	0·12	0·04	0·04	0·03
Meligethes spp.	0·42	0·53	0·65	0·00	0·04	0·06
Nephrotoma spp.	1·60	1·10	0·51	0·54	0·66	0·62
Omnivorous Coleoptera and Dermaptera (adults)						
Tachinus rufipes	—	0·13	0·09	0·01	0·03	0·00
Forficula auricularia	0·39	0·78	0·33	0·40	1·25	0·17
Mycetophagous Coleoptera (adults)						
Atomaria spp.	43·60	43·58	34·32	25·50	14·46	10·17
Stilbus testaceous	4·35	3·82	2·42	0·94	1·55	0·35
Enicmus spp.	2·27	10·87	12·91	7·74	8·14	3·13
Lathridius spp.	20·33	7·67	8·76	6·23	24·45	2·23
Stephostethus lardarius	1·00	3·88	1·29	0·73	1·31	0·43
Tachyporus spp.	11·56	10·81	8·86	5·64	1·74	1·60

1974) and in fields with diverse faunas (Potts & Vickerman 1974), but the importance of polyphagous predators is not restricted to their role as aphid predators since they eat other pests (Sunderland 1975), and may be limiting to some extent the numbers of say, thrips or mites. As an example the larval

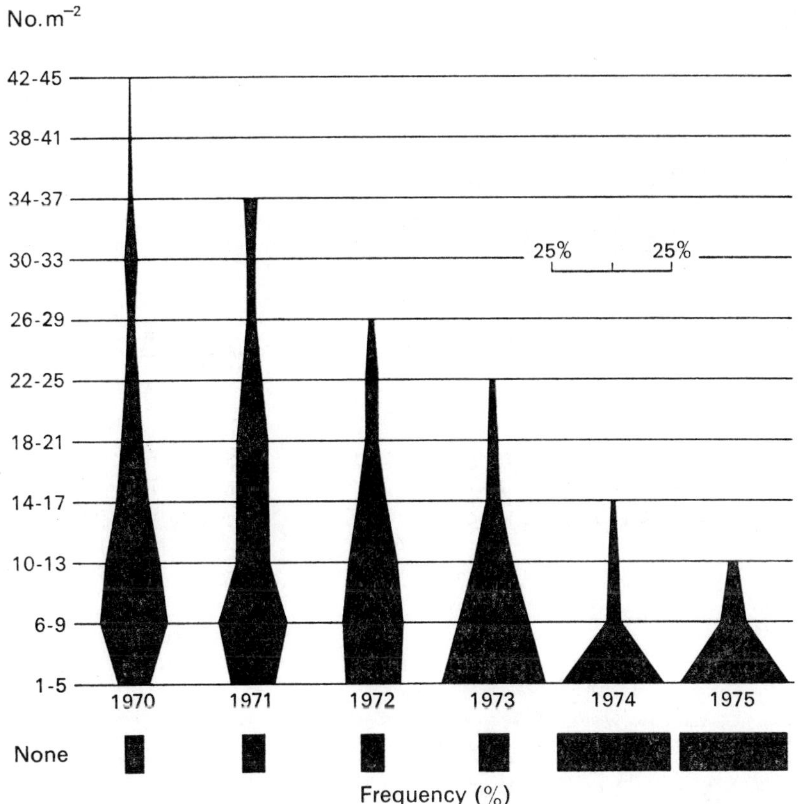

Figure 2. Six-year trend in the abundance of *Tachyporus* (Staphylinidae) adults in cereal fields in late June.

mortality of the wheat blossom midges (*Sitodiplosis mosellana* and *Contarinia tritici*) has been shown to be greatly increased by polyphagous predators in the Carabidae (Basedow 1975). Table 6 indicates the occurrence of some predatory species in a variety of crops and shows how often the predatory fauna of cereals can occur in other crops. There is clearly considerable movement of these species from crop to crop and between crops and other habitats (see for example Freeman (1945) and Lipkow (1966)). It is therefore possible that practices reducing the numbers of these predators in any one crop (such as the reduction in the predatory fauna of cereals after spraying

Table 6.   Some predatory insects recorded from cereals and a variety of other crops in the United Kingdom.

	Cereals	Permanent pasture	Root crops	Legumes	Brassicas	Lettuce and strawberries	Hops	Apples
Carabidae:								
Loricera pilicornis	2, 7, 10, 6, 33	1						7
Nebria brevicollis	33, 2, 3, 4, 7, 8, 6	10, 1	7, 22, 23	7	12, 13, 30	19	31	7
Harpalus rufipes	2, 7, 8, 6, 33	1	7, 22	7	12, 13, 30	19, *20		7
Harpalus aeneus	1, 7, 9, 6, 33	1	7, 22	7	12, 13, 30	19	31	7
Feronia melanaria	33, 2, 3, 4, 7, 8, 10, 6	28, 10, 1	7, 21, 22	7	12, 13, 30	19, *20	31	7
Calathus fuscipes	2, 7		7, 22		12, 13	19		7
Calathus melanocephalus	1, 7	28	7, 22					
Agonum dorsale	33, 2, 3, 4, 7, 8, 10, 6	1	7, 22	7	13, 30		31	7
Notiophilus biguttatus	33, 2, 3, 7, *8, 10, 6	10, 1	7, 21, 22	7	12		31	7
Bembidion lampros	33, 2, 3, 4, 7, *8, 6	1	7, 21, 22	7	12, 13, 30			7
Asaphidion flavipes	3, 33							7
Amara familiaris	2, 3, 7, 6	*1	7, 29	7	12, 13, 30			
Amara plebeja	33, 7, 1, 10, 6							7
Trechus quadristriatus	33, 3, 4, 7, 6	1	7, 21, 22	7	11, 12, 13, 30	20	31	7
Risophilus atricapillus	2, 3, 4	1					31	
Stomis pumicatus	7		7, 22	7				7
Coccinellidae:								
Coccinella 7-punctata	3, 4, 5, 6	1	7, 22, 23, 26	25		20		
Coccinella 11-punctata	3, 4, 5	1	26	24		20		
Propylea 14-punctata	3, 4, 5, 6		23					
Adalia bipunctata	1, 5		17, 23, 26	24, 25		20		
Staphylinidae:								
Tachyporus hypnorum	33, 2, 3, 4, 9, 6	28, 1	18, 29		11, 30		31	
Tachyporus obtusus	3, 4, 6, 33	27, 1	18		30		31	
Tachyporus chrysomelinus	2, 3, 4, 33	28, 1			11	20	31	
Xantholinus linearis	2, 10, *6	*1, 10, 27, 28						
Xantholinus longiventris	2, 3, 4	27			30			

Xantholinus punctulatus	1	27	29				
Othius laeviusculus	2	27					
Philonthus varius	3, *33	*1, 10, 27, 28					
Philonthus laminatus	2, 6	28					
Tachinus rufipes	2, 6, 33	1, 28				20	
Dermaptera:							
Forficula auricularia	2, 3, 5, 9	1	23, 29				31
Syrphidae:							
Episyrphus balteatus	4, 5, 6, 32	28	17, 23, 32	25, 16	11, 12, 14, 16	20	
Metasyrphus corollae	1, 6, 32		23, 32	16	16		
Metasyrphus luniger	1, 5		23	16	16		
Syrphus ribesii	32	28	23, 32	16	15, 16		
Syrphus vitripennis	32		17, 23, 32	16	16		
Melanostoma mellinum	1, 5, 6, 32	28	23, 32		16, *11, *14	20	
Melanostoma scalare	1, 32		23, 32	16	16		
Sphaerophoria scripta	1, 5, 32, 6		17, 23	16	11, 15, 16		
Platycheirus clypeatus	1, 32		32		16, *11, *14		
Platycheirus manicatus	1, 32, 6		17, 23, 32		16		
Platycheirus albimanus	1, 32		23, 32	24, 16	16		
Platycheirus peltatus	32		32	16	15, 16		
Platycheirus scutatus	32	28	32	16	15, 16		
Syritta pipiens	1	28	23				

* Denotes identification to genus only.

Authorities

1 Vickerman & Sunderland unpublished; 2 Sunderland (1975); 3 Vickerman & Sunderland (1975); 4 Potts and Vickerman (1974); 5 Dean (1974); 6 Jones (1976); 7 Davis (1968); 8 Griffiths, Raw & Lofty (1967); 9 Morris (1922); 10 Buckle (1921); 11 Pollard (1969); 12 Dempster (1967); 13 Dempster (1969); 14 Smith (1969); 15 Way, Murdie & Galley (1969); 16 Chandler (1968); 17 Dunn (1949); 18 Dicker (1944); 19 Briggs (1957); 20 Dunn (1960); 21 Dunning, Baker & Windley (1975); 22 Baker & Dunning (1975); 23 Free *et al.* (1975); 24 Banks & Macaulay (1967); 25 Banks (1962); 26 Dunn (1965); 27 Edwards (1929); 28 Morris (1920); 29 Morris (1927); 30 Coaker & Williams (1963); 31 Buxton (1974); 32 Pollard (1971); 33 Speight & Lawton (1976).

with dimethoate (Vickerman & Sunderland unpublished) could exacerbate pest problems in other crops as well as cereals. Already there is evidence that predatory arthropods can reduce aphid populations in sugar beet (Hull & Gates 1953, Lowe 1975), Brussels sprouts (Pollard 1969), potatoes (Shands, Simpson & Brunson 1972), peas (Dunn 1950) and lucerne (Smith & Hagan 1966).

THE PARTRIDGE–INSECT RELATIONSHIP

Partridges belong to the Galliformes which are predominantly herbivorous birds with chicks which eat insects and other arthropods for at least their first few days, and sometimes for several weeks (Potts 1971). The most important insect foods of grey partridge chicks in the cereal crops of the study area are the large larvae of the cereal-leaf-eating sawflies (Tenthredinidae), which occur in densities up to about 12 m^{-2} and the much smaller (about one-ninetieth the weight) cereal aphids which occur in densities (over whole fields) up to about 80,000 m^{-2}. Other important insect foods are the Heteroptera, Jassids, the larvae of the Chrysomelidae and the pupae of ants (Potts 1970a). Both within and between years the chick survival to age six weeks increases with the biomass of insect food in cereals at the time of hatching (mid-June) and the percentage of hatched young recruited to the post-breeding population increases, to an asymptote, roughly in proportion to the increases in abundance of their food (Potts 1974).

It is suggested that the above relationship is mainly one of cause and effect since the insects which contribute most to the variation in chick survival are those which contribute most to the diet. The way in which chick mortality occurs is not understood but it has been shown that the growth rate of chicks, and in particular the growth rate of their feathers was higher when insects were added to a vegetable diet (Cross 1966).

So far in these studies, insect food abundance has been measured as sawfly larvae or aphids alone (Potts 1970a), as all insects together (Potts 1973) and as a somewhat subjective index of chick food biomass (Potts 1974). The main problem with the index is in weighting the various foods; it is extremely difficult to obtain samples of insect food eaten by chicks when they are hidden in the cereal crops. The survival rate of partridge (in this case *Perdix perdix*) chicks to age six weeks varies between years, being 19 ± 2% in 1972 the lowest year and 50 ± 1% in 1970 the highest year. Each year the survival rates vary significantly between the seventeen farms in the study area, usually from about 10% to about 70%. A preliminary analysis using data collected from cereal fields shows: % chick survival = $3 \cdot 9x + 0 \cdot 73y + 0 \cdot 04z + 7 \cdot 0$ $(P < 0 \cdot 001)$ where $x = $ no. of sawfly larvae m^{-2}, y = no. of Heteroptera + Jassids m^{-2} and z = geometric mean no. aphid apterae m^{-2}. The method of calculating chick survival, a revised version of that of Potts (1973) will be published elsewhere.

The series of cold springs from about 1962–1969 reduced chick survival more than would have been the case in the pre-pesticide era, but even after average spring temperatures the chick survival rate was 15% lower than expected (Potts 1971 and unpublished). It is interesting to speculate, using the regression formula that this change could have been caused by a decline of about half in either of the first two food groups since the 1930's, about the same as the total biomass reduction caused by dicotyledon-controlling herbicides in the trials of Southwood & Cross (1969).

Some of the effects of insect shortage have, over the past three years, been overcome by a switch to eating aphids, but this source of food is threatened by aphicides, and the insecticides used, including the most specific, pirimicarb, also kill sawfly larvae at normal rates of application.

The net effect of new agricultural techniques can only be estimated when the influence of parasitism (as possibly, in sawfly larvae) or predation (as in cereal aphids) are understood, but there is no doubt that, on balance, the trends towards monoculture in cereals adversely affect partridges.

Monoculture and pest problems

Data on cereal yields from parts of the study area are incomplete but there is no evidence of a significant increase in recent years. Only 40% of farmers claim to know yields even after the crop has been sold (Britton 1969) but the Ministry of Agriculture, Fisheries and Food Statistics Branch, at Guildford assemble annual estimates of yield and these have been used to show that yields had stabilized by 1970 (Attwood 1970). Using 1966–1975 data the yield ha^{-1} of barley and wheat has shown no significant upward trend, and annual yield increments are of the order of 1·6%.

This study has shown that these stable yields are accompanied by stable incidences of weeds, of disease (Table 1) and of many arthropod groups (Table 5) though cereal aphids and mycetophagous species may be exceptions. Although there is no clear evidence of the 'marked instability of agro-ecosystems' mentioned by Murdoch (1975) it may be there as the cause of the need for the marked upward trend in the use of pesticides (Table 2 and Potts & Vickerman 1975). It certainly appears difficult to maintain stability and productivity in the modern relatively simple cereal ecosystem; the system is still complex, especially by comparison with theoretical models.

The value of ecological diversity *per se* in agroecosystems has been questioned on theoretical grounds or in particular contexts but the role of natural control systems, which are often associated with diversity is appreciated (Emden & Williams 1974), even if their practical value is as elusive as ever.

There are two major reasons why this work cannot yet shed much light on the general problem of the origins of arthropod pest problems. First there is, as we have seen, practically no knowledge of the sort of fluctuations which occurred before the recent trends towards the strict monoculture of cereals; the work is therefore pioneering in this sense. Second we do not yet understand the causes of the vast majority of the fluctuations which have been quantified over the last six seasons. We do not know how important the recent mild winters have been and therefore we cannot know for certain by how much cereal aphids are now more abundant than they were, though Heathcote (1970) indicates the increase as tenfold between 1950 and 1970. Nor is it possible to say that *Tachyporus* spp. have declined in an abnormal way. The only way to proceed from the descriptive to the prescriptive in applied biology (to follow Holling & Clark 1975) is to progressively eliminate the background noise by a sustained monitoring programme and by associated experiments and trials. Taylor (1973) suggests that a *rational* policy of pest control can be developed only by 'developing a greatly improved science of spatial, as well as temporal, population dynamics, and by keeping constant watch on the movements of actual and potential pests'. At present a fundamental change such as a decline of polyphagous predators could trigger pest outbreaks without being detected as a cause and meaningful large-scale long-term experiments do not appear to be feasible.

The original reason for much of the work described in this paper was the need to quantify the ecological effects of modern agriculture especially of pesticide use in cereals. It is of interest that after seven years one must still stress the need for more fundamental ecological research in the field, on such aspects as the effects of polyphagous predators; other workers with a similar remit, for invertebrates as well as vertebrates, have independently come to the same conclusion (Ryszkowski 1974). Longhurst *et al.* (1972) clearly pointed out that natural fluctuations in animal populations have been ascribed incorrectly to the effects of pollutants, and 'it could be easy for a serious impact on the environment to pass unnoticed through ignorance of natural population instability'. Although they were writing about oceanic plankton, which has been well monitored since 1948 (Glover, Robinson & Colebrook 1972), the same remarks apply to the animals of cereal crops today.

Acknowledgements

I am grateful to Dr Paul Vickerman and Keith Sunderland for their help in the preparation of this paper and to all the farmers, colleagues and others who have helped with the Partridge Survival Project. The arthropod work at North Farm is currently supported by a grant from the Agricultural Research Council and some of the vertebrate work by a grant from the Natural Environment Research Council.

References

ANON. (1969) *A Century of Agricultural Statistics, 1866–1966.* HMSO, London.

ATTWOOD P.J. (1970) The cereal farmer's needs. *Proc. 10th Br. Weed Control Conf.* 873–80.

BAIRD W.W. & TARRANT J.R. (1973) *Hedgerow Destruction in Norfolk (1946–1970).* Occ. Publ. School of Environmental Sciences, Univ. East Anglia, Norwich, U.K.

BAKER A.M. & DUNNING R.A. (1975) Some effects of soil type and crop density on the activity and abundance of the epigeic fauna, particularly Carabidae, in sugar beet fields. *J. appl. Ecol.* **12**, 809–18.

BANKS C.J. (1962) Effects of the ant *Lasius niger* (L.) on insects preying on small populations of *Aphis fabae* Scop. on bean plants. *Ann. appl. Biol.* **50**, 669–79.

BANKS C.J. & MACAULAY E.D.M. (1967) Effects of *Aphis fabae* Scop. and of its attendant ants and insect predators on yields of field beans (*Vicia faba* L.). *Ann. appl. Biol.* **60**, 445–53.

BARANYOVITS F. (1973) The increasing problem of aphids in agriculture and horticulture. *Outl. Agric.* **7**, 102–8.

BASEDOW T. (1975) *Predaceous arthropods in agriculture, their influence upon the insect pests, and how to spare them while using insecticides.* Semain d'étude agriculture et hygiène des plantes, 8–12 Septembre 1975, Publ. Centre de Recherches Agronomiques, Gembloux, Belgium.

BEAVEN E.S. (1947). *Barley.* Duckworth, London.

BRIGGS J.B. (1957) Some experiments on control of ground beetle damage to strawberry. *Rep. E. Malling Res. Sta.* 1956, 142–5.

BRITTON D.K. (1969) *Cereals in the United Kingdom.* Pergamon Press, Oxford.

BUCKLE P. (1921) A preliminary survey of the soil fauna of agricultural land. *Ann. appl. Biol.* **8**, 135–45.

BUXTON J.H. (1974) *The biology of the European earwig* Forficula auricularia *L. with reference to its predatory activities on the damson–hop aphid* Phorodon humuli (*Schrank*). Ph.D. thesis, University of London.

CHANDLER A.E.F. (1968) Some host-plant factors affecting oviposition by aphidophagous Syrphidae (Diptera). *Ann. appl. Biol.* **61**, 415–23.

CHURCH B.M., BOYD D.A., EVANS J.A. & SADLER J.I. (1968) A type of farming map based on agricultural census data. *Outl. Agric.* **5**, 191–6.

COAKER T.H. & WILLIAMS D.A. (1963) The importance of Carabidae and Staphylindae as predators of the cabbage root fly *Erioschia brassicae* (Bouche). *Entomologia exp. appl.* **6**, 156–64.

CROSS D.A. (1966) *Approaches toward an assessment of the role of insect food in the ecology of game-birds especially the partridge* (Perdix perdix). Ph.D. thesis, University of London.

DAVIS B.N.K. (1968) The soil macrofauna and organochlorine insecticide residues at twelve agricultural sites near Huntingdon. *Ann. appl. Biol.* **61**, 29–45.

DEAN G.J.W. (1974) Effects of parasites and predators on the cereal aphids *Metopolophium dirhodum* (Wlk.) and *Macrosiphum avenae* (F.) Hem. (Aphididae). *Bull. ent. Res.* **63**, 411–22.

DEAN G.J.W. & WILDING N. (1971) Entomopthora infecting the cereal aphids *Metapolphium dirhodum* and *Sitobion avenae*. *J. Invertebr. Pathol.* **18**, 169–76.

DEMPSTER J.P. (1967) The control of *Pieris rapae* with DDT. I. The natural mortality of the young stages of *Pieris*. *J. appl. Ecol.* **4**, 485–500.

DEMPSTER J.P. (1969) Some effects of weed control of the numbers of *Pieris rapae* on Brussels sprouts. *J. appl. Ecol.* **6**, 339–45.

DICKER G.H.L. (1944) *Tachyporus* (Coleoptera, Staphylindae) larvae preying on aphids. *Entomologist's mon. Mag.* **80**, 71.

DUNN, J.A. (1949) The parasites and predators of potato aphids. *Bull. ent. Res.* **40**, 97–122.

DUNN J.A. (1950) Pea aphid population studies in 1950. *Rep. Nat. Veg. Res. Stn. for 1950*, 21.

DUNN J.A. (1960) The natural enemies of the lettuce root aphid *Pemphigus bursarius* (L.). *Bull. ent. Res.* **51**, 271–8.

DUNN J.A. (1965) Studies on the aphid, *Cavariella aegopodii* Scop. 1. On willow and carrot. *Ann. appl. Biol.* **56**, 429–38.

DUNNING R.A., BAKER A.M. & WINDLEY R.F. (1975) Carabids in sugar beet crops and their possible role as aphid predators. *Ann. appl. Biol.* **80**, 125–8.

EDWARDS E.E. (1929) A survey of the insect and other invertebrate fauna of permanent pasture and arable land of certain soil types at Aberystwyth. *Ann. appl. Biol.* **16**, 299–323.

EMDEN H.F. VAN & WILLIAMS G.F. (1974) Insect stability and diversity in agro-ecosystems. *A. Rev. Ent.* **19**, 455–75.

FENTON F.A. & FISHER E.H. (1940) The 1939 greenbug outbreak in Oklahoma. *J. econ. Ent.* **33**, 628–34.

FREE J.B., WILLIAMS I.H., LONGDEN P.C. & JOHNSON M.G. (1975) Insect pollination of sugar-beet (*Beta vulgaris*) seedcrops. *Ann. appl. Biol.* **81**, 127–34.

FREEMAN J.A. (1945) Studies on the distribution of insects by aerial currents. The insect population of the air from ground level to 300 feet. *J. Anim. Ecol.* **14**, 128–54.

FULLER G. (1976) The agricultural aviation industry. *Farmers Weekly* **84**, (15) 11–12.

GLOVER R.S., ROBINSON G.A. & COLEBROOK J.M. (1972) Plankton in the North Atlantic—an example of the problems of analysing variability in the environment. *Marine Pollution and Sea Life* 439–45 (F.A.O. and Fishing News Books Ltd.).

GRIFFITHS D.C., RAW F. & LOFTY J.R. (1967) The effects on soil fauna of insecticides tested against wireworms (*Agriotes* spp.) in wheat. *Ann. appl. Biol.* **60**, 479–90.

HEATHCOTE G.D. (1970) The abundance of grass aphids in Eastern England as shown by sticky trap catches. *Pl. Path.* **19**, 87–90.

HOLLING C.S. & CLARK W.C. (1975) *Notes towards a Science of Ecological Management. Unifying Concepts in Ecology* (Ed. by W.H. van Dobben & R.H. Lowe-McConnel), pp. 247–51. Junk B.v., The Hague.

HULL R. & GATES L.F. (1953) Experiments on the control of beet yellows virus in sugar-beet seed crops by insecticidal sprays. *Ann. appl. Biol.* **40**, 60–78.

JONES M.G. (1972) Cereal aphids, their parasites and predators caught in cages over oat and winter wheat crops. *Ann. appl. Biol.* **72**, 13–25.

JONES M.G. (1976) The arthropod fauna of a winter wheat field. *J. appl. Ecol.* **13**, 61–85.

LIPKOW E. (1966) Biologisch-ökologische Untersuchung über *Tachyporus* Arten und *Tachinus rufipes* (Col. Staphylinidae). *Pedobiologia* **6**, 140–77.

LONGHURST A., COLEBROOK J.M., GULLAND J., LE BRASSEUR R., LORENZEN C. & SMITH P. (1972) The instability of ocean populations. *New Scientist* **1**, June, 500–2.

LOWE J.H.B. (1975) Crop resistance to pests as a component of integrated control systems. *Proc. 8th Br. Insect. Fung. Conf.* 87–92.

MORRIS H.M. (1920) Observations on the insect fauna of permanent pasture in Cheshire. *Ann. appl. Biol.* **7**, 141–55.

MORRIS H.M. (1922) The insect and other invertebrate fauna of arable land at Rothamsted. *Ann. appl. Biol.* **9**, 282–305.

MORRIS H.M. (1927) The insect and other invertebrate fauna of arable land at Rothamsted. Part II. *Ann. appl. Biol.* **14**, 442–64.

MURDOCH W.W. (1975) Diversity, complexity, stability and pest control. *J. appl. Ecol.* **12**, 795–807

POLLARD E. (1969) The effect of removal of arthropod predators on an infestation of *Brevicoryne brassicae* (Hemiptera, Aphidae) on Brussels sprouts. *Entomologia exp. appl.* **12**, 118–24.

POLLARD E. (1971) Hedges. VI. Habitat diversity and crop pests: a study of *Brevicoryne brassicae* and its syrphid predators. *J. appl. Ecol.* **8**, 751–80.

POTTS G.R. (1970a) Recent changes in the farmland fauna with special reference to the decline of the grey partridge (*Perdix perdix*). *Bird Study* **17**, 145–66.

POTTS G.R. (1970b) The effects of the use of herbicides in cereals on aphids. *Proc. 10th Br. Weed Control Conf.*, 299–302.

POTTS G.R. (1971) Factors governing the chick survival rate of the grey partridge (*Perdix perdix*). *Proc. Int. Union Game Biol.* **10**, 85–96.

POTTS G.R. (1973) Pesticides and the fertility of the grey partridge. *J. Reprod. Fert. Suppl.* **19**, 391–402.

POTTS G.R. (1974) The grey partridge, problems of quantifying the ecological effects of pesticides. *Proc. Int. Congr. Game Biol.* **11**, 405–13.

POTTS G.R. & VICKERMAN G.P. (1974) Studies on the cereal ecosystem. *Adv. Ecol. Res.* **8**, 107–97.

POTTS G.R. & VICKERMAN G.P. (1975) Arable ecosystems and the use of agrochemicals. In *The Ecology of Resource Degradation and Renewal* (Ed. by M.J. Chadwick & G.T. Goodman), pp. 17–29. Blackwell Scientific Publications, Oxford.

RYSZKOWSKI, L. (1974) *Ecological Effects of Intensive Agriculture*. Polish Scientific Publishers, Warszawa.

SHANDS W.A., SIMPSON G.W. & BRUNSON M.H. (1972) Insect predators for controlling aphids on potatoes. 1. In small plots. *J. econ. Ent.* **65**, 511–14.

SMITH J.G. (1969) Some effects of crop background on population of aphids and their natural enemies on brussels sprouts. *Ann. appl. Biol.* **63**, 326–30.

SMITH R.F. & HAGEN K.S. (1966) Natural regulation of alfalfa aphids in California. In *Ecology of Aphidophagous Insects* (Ed. by I. Hodak), pp. 297–315. Proc. Symp. Liblice, near Prague (Academia, Prague).

SOUTHWOOD T.R.E. (1972) Farm management in Britain and its effect on animal populations. *Proc. Tall Timbers Conf. on Ecological Anim. Control by Habitat Mgmt.* (1971), **3**, 29–51.

Southwood T.R.E. & Cross D.J. (1969) The ecology of the partridge. III. Breeding success and the abundance of insects in natural habitats. *J. Anim. Ecol.* **38**, 497–509.

Speight M.R. & Lawton J.H. (1976) The influence of cereal cover on the mortality imposed on artificial prey by predatory ground beetles in cereal fields. *Oecologia* **23**, 211–23.

Staniforth A.R. (1975) Cereal straw production and utilization in England and Wales. *Outl. Agric.* **8**, 194–200.

Sunderland K.D. (1975) The diet of some predatory arthropods in cereal crops. *J. appl. Ecol.* **12**, 507–15.

Taylor L.R. (1973) Monitor surveying for migrant insect pests. *Outl. Agric.* **7**, 109–16.

Terrason F. & Tendron G. (1975) Evolution and conservation of hedgerow landscapes in Europe. *Council of Europe Nature and Conservation Series* No. 8, 1–44.

Vickerman G.P. (1974) Some effects of grass–weed control on the arthropod fauna of cereals. *Proc. 12th Br. Weed Control Conf.* **3**, 929–40.

Vickerman G.P. & Sunderland K.D. (1975) Arthropods in cereal crops; nocturnal activity, vertical distribution, and aphid predation. *J. appl. Ecol.* **12**, 755–66.

Walton W.R. (1921) The green bug or spring grain-aphid; how to prevent its periodic outbreaks. *U.S. Dept. Agric. Fm. Bull.* **1217**, 3–10.

Way M.J., Murdie G. & Galley D.J. (1969) Experiments on integration of chemical and biological control of aphids on brussels sprouts. *Ann. appl. Biol.* **63**, 459–75.

Economics

The economic and social context of pest, disease and weed problems

G. A. NORTON and G. R. CONWAY
Environmental Management Unit, Department of Zoology and Applied Entomology, Imperial College, London

Introduction

Pests are one of the hazards of life. Indeed, when any organism interferes with another, either indirectly through competing for food, or directly through predation or parasitism, it tends to be regarded as a pest. But with this definition, very few creatures would not be pests. For a concept which is both meaningful and useful in terms of decision making, some form of value system has to be employed and, without entering into ethical and moral arguments, this valuation will necessarily be anthropocentric. A pest problem is, therefore, characterized not only by the state of the pest population itself but, more importantly, by the damage or illness it causes and the value placed on these consequences by human society (Fig. 1). Although this is an obvious point, experience indicates that it is one which has been disregarded in many research programmes in pest control.

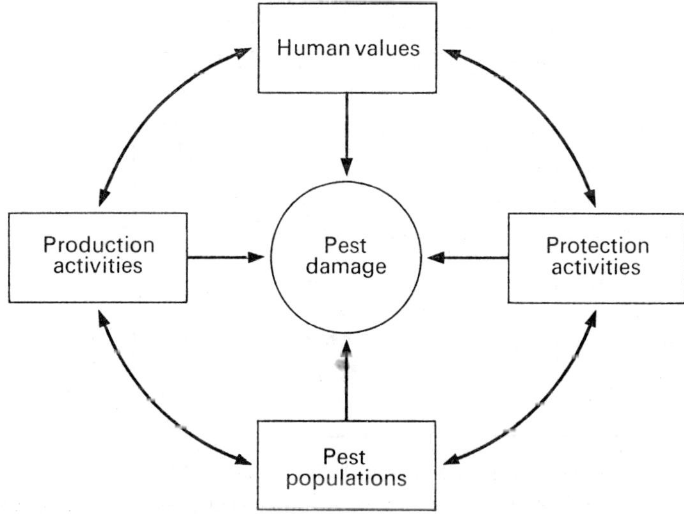

Figure 1. The pest problem.

Pests as natural hazards

Where a population periodically suffers floods, drought or earthquakes, the dimensions of the natural hazard it faces depends on the interaction of these natural events and the human use system affected (Kates 1970, White 1974). In this context, pests can also be regarded as natural hazards, locusts being

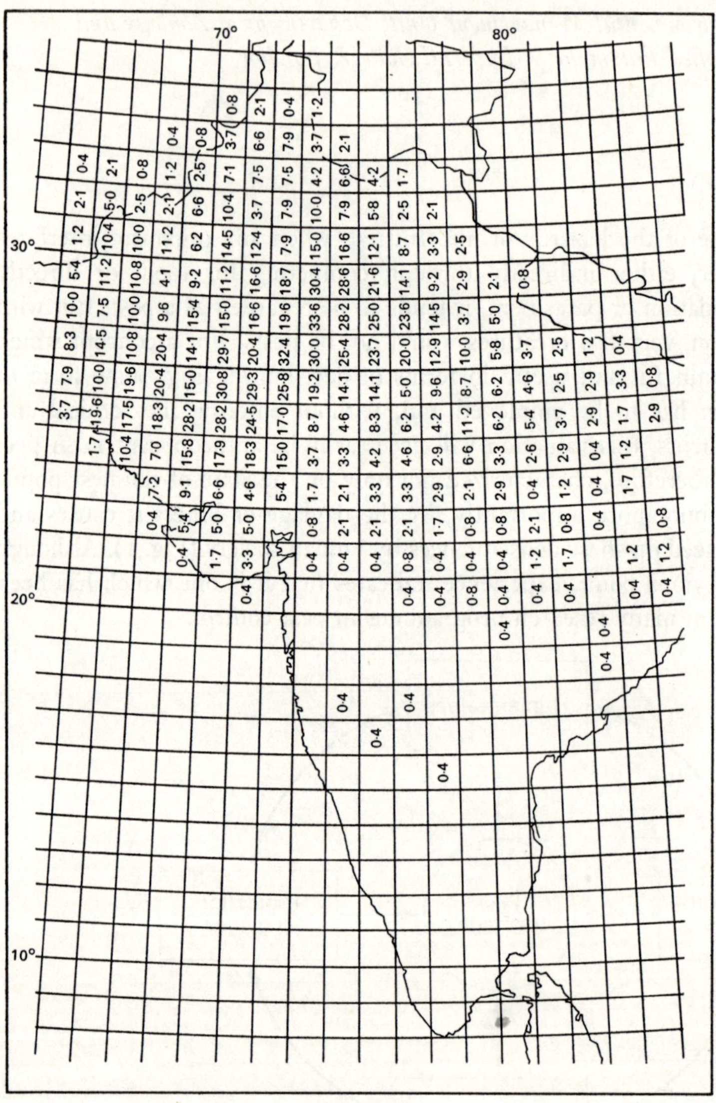

Figure 2.   India and Pakistan, 1939–1958; Desert locust swarm and hopper frequencies summed for the six months May to October, expressed as percentages of the theoretical maximum (Bullen 1966).

a particularly good example. Where a region is affected by this pest, areas having a high probability of locust attack can be identified within it (Fig. 2). Clearly, if an area that is subject to attack has no susceptible crop, there is no pest problem and no hazard. A hazard only occurs where damage can be caused and then it is a function of the probability of attack and the area of susceptible crop. Bullen (1966) combines these two measures to derive a crop vulnerability index (Fig. 3).

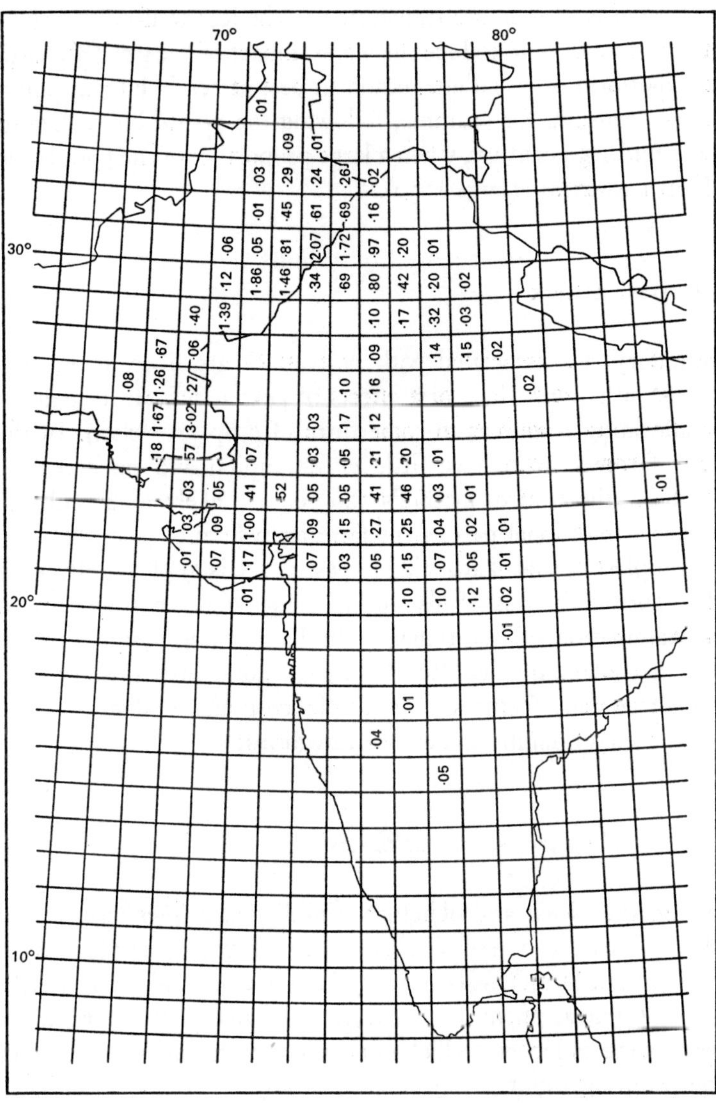

Figure 3. India and Pakistan; Crop vulnerability index values for cotton damaged by the desert locust, expressed as percentages of the theoretical maximum value (Bullen 1966).

Where there is no means of adapting to protect the crop against pest attack, the inhabitants of hazardous areas are subject to involuntary risk (Velimirovic 1975). Hunter gatherers, lacking more than the most rudimentary capacities to store grain and meat, fall within this category. Every so often, losses are suffered which are regarded as the action of supernatural forces and frequently the only form of protection is an attempt to appease these forces through religious ceremonies and offerings. In other societies, forms of social adaptation to hazards have evolved. Wedding parties and ceremonies, where gifts of food are transferred to the less well-off members of the community, can serve to reduce the worst impact of hazards (Richards 1975). An extreme form of social adaptation to pest attack is human population control, as occurred, for example, following the potato blight famine in Ireland in the mid 19th century, when a large proportion of the population migrated to North America (Large 1950).

Adaptation to pest attack

Cases of involuntary risk associated with pest attack are likely to be rare however. In most situations, the worst effects of pest attack can usually be prevented by allocating resources to some form of crop protection. In the case of peasant farmers, who have little or no capital and no access to modern technology, the control methods adopted are likely to be cultural measures that are a traditional part of the farming system.

One of the simplest means of adapting to pest attack is to plant a diversity of crops and so spread the risk of loss in yield due to any particular pest. However, an examination of traditional multiple cropping systems in the tropics reveals that adaptation is often far more subtle than this. Indeed, several pest management features of multiple cropping systems can be identified (Norton 1975), falling under two categories:

Features that reduce the attractiveness of the crop

By intercropping, the visual and olfactory stimuli that attract insects to a particular crop species can be reduced (Raros 1973), while the planting of diversionary crops can attract insects away from a vulnerable crop (Aiyer 1949). These pest management features are particularly appropriate for exogenous pests, such as locusts, which enter the crop for only a portion of their life cycle (Bunting 1972).

Features that reduce the favourability of the crop environment

Where endogenous pests are a problem, that is, pests such as nematodes or weeds that complete all or most of their life cycle within the crop, the farmer can attempt to produce an unsuitable habitat. This can be achieved by means of crop rotations, or, in the case of weeds, by intercropping to produce an unfavourable environment in terms of light and nutrients. In the Philippines, the loss in corn yield due to weeds can be reduced by hand weeding or by intercropping with mung beans (Fig. 4). Similarly, intercropping can create dispersal hazards for pests and diseases (Finlay 1974) or a more favourable environment for predators (Raros 1973).

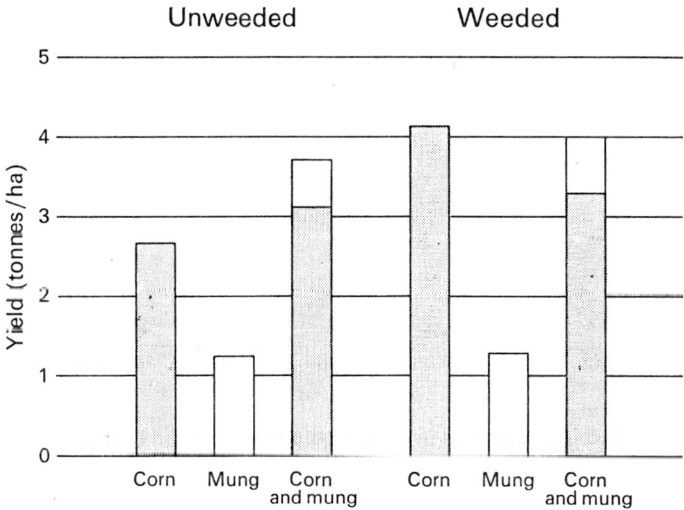

Figure 4. Effect on yield of intercropping corn and mung in weeded and unweeded situations (after I.R.R.I. 1973).

The decision problem

In allocating resources to production and protection activities, the priority of subsistence farmers will clearly be to obtain sufficient food to meet the requirements of the family. It is only when this security constraint has been met that farmers consider the production of cash crops (McLoughlin 1970, Norman 1974). The importance of this objective is described by Norman (1971), who found that certain Nigerian farmers refused to plant cotton early in the season as recommended, despite the marked increase in yield that results. The reason is quite simple. Farmers devote all their energies early in the season to planting food crops. It is only when this has been done that consideration is given to the planting of cash crops, such as cotton.

To illustrate the pest control problem facing subsistence farmers, consider a rice farmer who has the choice of planting two varieties, IR 8 and IR 20. IR 8 is a high yielding variety but one which is susceptible to tungro virus carried by the green leaf hopper (*Nephotettix* spp.). IR 20, on the other hand, is resistant to tungro virus and so when attack on IR 8 is high, that on IR 20 is much lower. In this way, IR 20 prevents high levels of attack and consequently suffers less yield loss (Fig. 5). Where the farmer

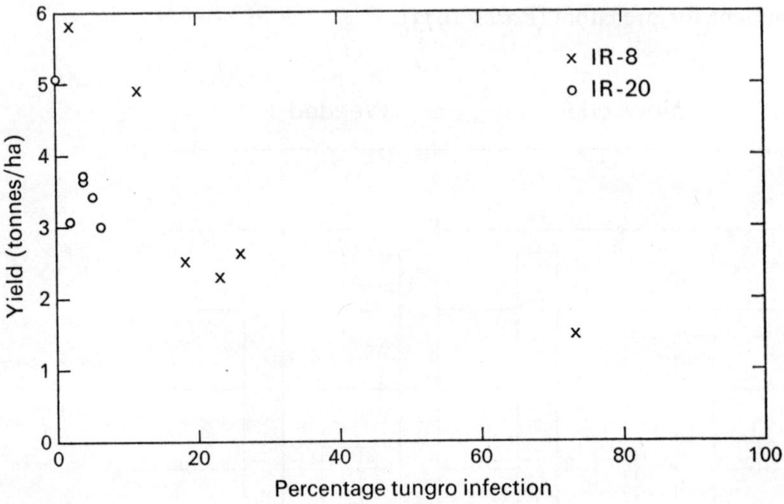

Figure 5.   Extent of virus infection and yield loss in two rice varieties: IR-8 and IR-20 (after I.R.R.I. 1973).

making the choice between these two varieties is close to subsistence level and has to produce a yield of (say) three tonnes ha^{-1} to feed his family, he will plant IR 20 provided there is at least some probability of tungro attack on IR 8 of 20% or over. On the other hand, if the farmer is entering the market economy and is something of a gambler, he may plant IR 8 and hope that the level of tungro will be less than 10%, thus giving him a higher monetary pay-off.

In practice therefore, the peasant farmer is often locked into his farming system; first, by lack of alternative actions, or money to buy them, and secondly, by the inability to withstand losses. As surplus crops are produced and markets for them become available, however, farmers are likely to become more economically motivated and more innovative (Low 1974). Hence, with the transition to cash cropping in the market economy, the techniques of agriculture will change, as will the range of pest control measures available.

Technological methods of crop protection

Where there is a shortage of capital and land but where labour is adequate, multiple cropping can often achieve both goals of risk minimization and profit maximization (Norman 1974). As resources become more available however, farmers tend to switch to those more profitable monocultures for which technology has been developed. Nevertheless, the increase in production associated with the latter must be substantial for farmers to accept the risk associated with the change (Streeter 1975, Ogunfowora & Norman 1973), and so farmers usually substitute technological means of protection, such as pesticides, for the innate, pest management properties that have been lost. They move from implicit means of adapting to pest attack to explicit forms of protection.

The increase in the number of rice farmers using insecticides in a small barrio in the Philippines indicates how the adoption of explicit pest control measures can be stimulated by agricultural innovation (Fig. 6). The first large increase in use occurred in 1956, the year in which an irrigation system was opened, while the introduction of high yielding I.R.R.I. varieties in 1966 stimulated a further increase. In other situations, the adoption of new pest control measures may be stimulated by increased attack. Gould (1965), for example, found that farmers in Kenya were reluctant to adopt new rust resistant wheat varieties. It was only after severe outbreaks of the disease, when conventional varieties were seen to be unsatisfactory, that the new varieties were adopted. This tends to agree with the findings from other natural hazards research where the adoption of new alternatives has often been triggered by crises (Slovic, Kunreuther & White 1974).

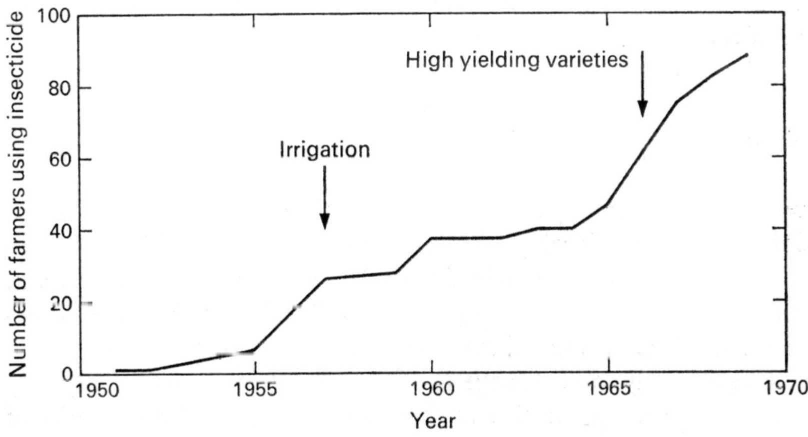

Figure 6.   Adoption of insecticides by rice farmers of San Bartolome in the Philippines, 1952–1969 (after Huke 1974).

As more effective methods of pest control become available, some farmers are able to divert even further from traditional methods of control. For instance, as effective nematicides become available, there is less need for rotations to keep eelworm populations down and so more profitable crops, such as potatoes, can be grown on a larger acreage with a shorter rotation (Jones 1973). Furthermore, as agriculture becomes more mechanized, the need to spread fixed costs, particularly of harvesting machinery, makes it more desirable to specialize in monocultural crops. To some extent, this development has been stimulated by the pressures of the competitive market and the need for greater efficiency to maintain or increase profits.

The outcome of such specialization has been a reduction in flexibility. Agroecosystems of modern, industrial agriculture have been designed more as part of a conveyor belt system than an adaptation to natural events. By concentrating on improving the agronomic components of yield, modern agricultural crop production may not only increase attack, by increasing the attractiveness of the crop and its favourability, but may also reduce its tolerance and so further increase the potential for damage.

The development of modern crop production and protection technology has therefore been an interactive and cumulative process. As more effective crop protection becomes available, more efficient production techniques can be adopted and so technological progress in one field has reinforced the technological development in the other.

The result has been a system of food production that critically depends upon the use of pesticides, a situation that appears to be in accordance with the general theory of the behaviour of the firm (Cyert & March 1963). These authors conclude that firms '. . . . achieve a reasonably manageable decision situation by avoiding planning where plans depend on predictions of uncertain future events and by emphasizing planning where the plans can be made self-confirming through some control device'.

Development of market valuation

Whereas a farmer operating at a subsistence level will make pest management decisions on the basis of ecological considerations, once he produces a surplus that can be marketed, economic factors become increasingly important. Apart from developments in crop production therefore, a change in objectives occurs, output being valued in terms of revenue rather than yield *per se*. Consequently, pest damage is often in the form of reduced quality and is defined as the proportion of crop units affected and the severity of injury to each (Wheatley 1974). Although the demands of the housewife are partly responsible for this increase in the importance of crop quality

(Bundy 1977), it is also associated with an increase in grading schemes, food processing and the packaging of produce for supermarket outlets.

Hence, the sale of crop produce to agricultural markets has produced a more extreme damage function, considerable revenue losses often being incurred with low levels of attack (Southwood & Norton 1973). Consequently, as the value of a crop increases, and particularly in horticultural crops, we might expect farmers to invest more capital in crop protection. In general, this seems to be true (Fig. 7), the exceptions—leeks and celery—being crops that are well known for their lack of serious pests.

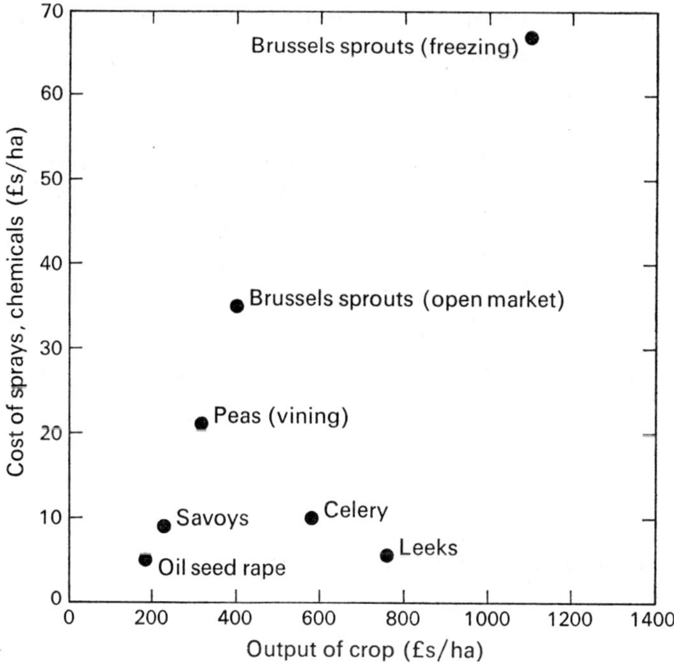

Figure 7.   The relationship between crop value and expenditure on sprays and chemicals for six crops (after Nix 1974).

The game against nature

If we regard pest control as a game against nature, it is clear that as economic and technological development occurs the game changes. In the first place, more players are involved. The technological development of pesticides not only increases the strategies that farmers can adopt against pests but also introduces the chemical industry to the game. Moreover, as well as consumers, whose influence operates through the market, considerations of environmental safety have involved environmental pressure groups and agencies. These new players have the effect of imposing new constraints on the farmer. Thus, although economic and technological innovation have reduced many of the constraints that limited the actions available to subsistence farmers, in commercial farming these are replaced by the quality standards of an affluent society, by the pressures of a competitive market, and by constraints imposed through environmental legislation.

Technological developments have also changed the time dimensions of the game. Since pest populations themselves have an objective, namely survival, which they pursue through a strategy of genetic and behavioural modification, reliance on pesticides for crop protection can lead to the development of pesticide resistance. This means that the application of a pesticide in one season can reduce its effectiveness in following seasons. The extent of the hazard that results depends upon the efficacy of pesticide use and on various biological and ecological characteristics of the pest, particularly whether it is endogenous or exogenous to the cropping system. Where the migration rate from unsprayed populations falls below a critical level, as occurs when pests are largely endogenous, Comins (1976) has shown that the rate at which resistance develops is greatly increased. To illustrate the extent this problem can reach, Table 1 shows the historical development of acaricide resistance in the Australian cattle tick.

Table 1. Development of acaricide resistance in the Australian cattle tick (courtesy of W. J. Roulston & R. H. Wharton).

Chemical	Year introduced	Year in which resistance recognized
Arsenic	1895	1936
DDT	1946	1955
BHC	1950	1952
Diazinon	1956	1963
Dioxathion	1958	1963
Carbaryl	1963	1963
Carbophenothion	1961	1963
Coumaphos	1959	1966
Ethion	1962	1966
Chlorpyrifos	1967	1970
Bromophos-ethyl	1967	1970

It would appear therefore, that the adoption of pesticides as a means of control, changes the nature of the pest hazard, replacing short-term risks by longer-term uncertainty. Through our efforts to develop a technological means of ameliorating the immediate natural hazard associated with pests, we now face the more pervasive, man-made hazards associated with widespread pesticide use.

The analysis of specific pest problems

So far in this paper we have attempted to diagnose the changing hazard associated with pests as seen in the context of technological, social and economic development. At this point we want to change direction and show that this type of approach has more than descriptive value. To analyse current pest problems and assess the best means by which they can be improved, we feel it essential that each problem is initially analysed in the overall, socio-economic context in which the decision maker operates. Thus, we need to consider the state of the three factors that most influence crop protection decisions that are made. The three factors are:

1. The nature of pest attack and damage; as determined by the features of the pest, whether exogenous or endogenous, the degree of damage caused, and the probability distribution of attack.

2. The control measures available to the farmer; depending upon the technology available, the cost of the control measures, and the farmer's ability to incorporate particular crop protection measures within his farming system.

3. The farmer's objectives which, as already seen, may be totally dominated by subsistence considerations or be geared entirely to economic motivations.

Using this framework to identify where constraints to the reduction of pest problems lie, consideration can then be given to ways of overcoming them. There are four means of doing this: by changing the structure of the agroecosystem, by increasing the pest control measures available, by increasing information on pest attack, and by improving the decision rules by which the farmer operates.

Change agroecosystem structure

Future developments in pest control will be closely linked to the agricultural development that occurs. Policies that influence the form this agricultural development takes should therefore be designed with pest hazards in mind. In subsistence agriculture, emphasis needs to be placed on maintaining the pest management properties of traditional multiple cropping systems while attempting to improve their productivity (Bradfield 1972). Similarly, there may be potential in commercial farming for the reintroduction of more adaptive protection measures. Indeed, we might suggest that multiple

H

cropping may have some lessons that temperate agriculturists could usefully learn.

Increase the control measures available

Crop protection can also be improved by increasing the range of protection measures available to farmers. By making public funds available, research and development of those methods of control thought to be desirable can be encouraged. However, even though a farmer is aware of alternative methods of control he may not be able to use them. His choice may be limited by the credit available to him, the constraints set by his farming system, or by his inability to use or obtain particular measures. Specific grants and subsidies or comprehensive extension programmes can attempt to reduce some of these constraints, whereas for others, new institutional involvement may be required. This is particularly true for exogenous pests, a regional control strategy only being possible where suitable institutional structures exist. These may be in the form of government extension agencies, food industries that closely supervise agricultural production activities, such as the British Sugar Corporation and food processing industries, or pest management consultancies (Dietrick 1973).

Improve information on pest attack

Two forms of decision making in pest control can be distinguished, depending upon the information on pest attack available (Norton 1976a). In the first case, when no information on the level of attack is available, prophylactic measures have to be adopted. These can be divided into pre-emptive measures, that are necessarily taken before pest attack occurs (e.g. cultural techniques, seed dressings), and schedule measures where control agents are applied according to a fixed programme. In the second case, control action is taken in response to specific information on pest attack. This often involves the use of an economic threshold criterion.

Apart from transforming prophylactic strategies to economic threshold strategies, an improvement in the information obtained on pest attack, as well as on the damage and control relationship, can enable more precise economic threshold strategies to be deployed. To achieve this improved information status, additional resources can be allocated to monitoring techniques, both at farm and regional levels, and more accurate forecasting devices can be developed (Way & Cammell 1973).

Improve decision rules

The final means of improving pest control involves the development of better decision rules by which farmers can operate. For this purpose, decision models in pest control are required, having two important roles to play. First, in their descriptive role, decision models can indicate the importance of their various components in determining the best strategy of control. Second, as prescriptive models, they can act as decision rules from which specific recommendations can be derived.

Decision models in pest control

Natural hazards are typically analysed using the techniques of decision theory (White 1974). Where pest problems, particularly with exogenous pests, closely resemble such hazards, this form of analysis can be usefully employed. Bullen (1966), for example, has suggested that the crop vulnerability index he devised for locusts can be used to concentrate protection in those areas of highest risk. Disease control decisions have also been analysed by Carlson (1970) using decision theory, and Gould (1965) has suggested that *Quelea* attacks could also be analysed in this way.

To illustrate the approach, Norton (1976b) has applied decision-theory to the problem of potato blight in the United Kingdom. Following Large (1958) he distinguishes four levels of attack (θ) in terms of the time at which 75% of the potato haulm is blighted. The probability of each of these levels of attack in the four main potato growing regions of the United Kingdom is given in Table 2 and the effect on yield of spraying at each level is shown by a physical pay-off matrix (Table 3). A monetary pay-off matrix can then be constructed with information on the price of potatoes and spraying costs (Table 4). Norton then showed how the long-run profitability or expected

Table 2. Probability of blight attack in the United Kingdom (Norton 1976b based on Large 1958). Level of attack expressed as the time at which 75% of the haulm is blighted.

| Region | Level of attack | | | |
	θ_1 (mid-August)	θ_2 (end of August)	θ_3 (mid-September)	θ_4 (end of September)
South-West	0·5	0·5	0	0
Fens	0·4	0·2	0·1	0·3
Southern	0	0·5	0·2	0·3
Northern	0	0	0·5	0·5

monetary outcome can be calculated for each strategy (Table 5) by weighting pay-off according to the probabilities of attack. On the basis of this expected monetary criterion, Norton concluded that it is rational to spray in the South-West and the Fens but not in the South or the North.

Pest control decisions are not made solely on the basis of expected outcomes however, but are modified according to the farmer's attitude to risk.

Table 3. Physical pay-off matrix for blight control strategies (tonnes ha^{-1}) (Norton 1976b based on Large 1958)

Level of attack	Strategy	
	Unsprayed	Sprayed
θ_1	22	26
θ_2	26	28
θ_3	29	29*
θ_4	30	29*

* 3% loss by wheel damage.

Table 4. Monetary pay-off matrix for blight control strategies. Values are £s ha^{-1}. Price of potatoes £40 tonne^{-1}, cost of spraying £20 ha^{-1} (Norton 1976b).

Level of attack	Strategy	
	Unsprayed	Sprayed
θ_1	880	1,010
θ_2	1,040	1,090
θ_3	1,160	1,130
θ_4	1,200	1,130

Table 5. Expected monetary value of blight control strategies. Values in £s ha^{-1} (Norton 1976b).

Region	Strategy	
	Unsprayed	Sprayed
South-West	960[a]	1,050[b]
Fens	1,036	1,074[b]
Southern	1,112[b]	1,110
Northern	1,180[b]	1,110

[a] Sum of the products of the monetary pay-offs (Table 4) and probabilities (Table 2) for each level of attack, i.e. (880 × 0·5) + (1,040 × 0·5) = £960.
[b] Best strategy.

The extent to which risk minimization conflicts with the goal of profit maximization will depend on the farmer's ability to assess the level of attack and to adopt appropriate measures. Hence, a more important conclusion from the potato blight analysis is that improved information, in the form of a forecasting scheme, would only be of value in the Southern region and in the Fens. This type of analysis has also been used by Cammell & Way (1977) to assess the costs and benefits of the forecasting scheme for black bean aphid, *Aphis fabae*.

When information on pest attack is available at a time when appropriate decisions can be made, the decision problem is more concerned with the efficiency of crop protection measures. This is particularly so for endogenous pests, such as soil nematodes, where populations build up over time. An illustration of the type of analysis required has been given by Norton (1976b), concerning the control of potato cyst eelworm by a soil fumigant (Fig. 8).

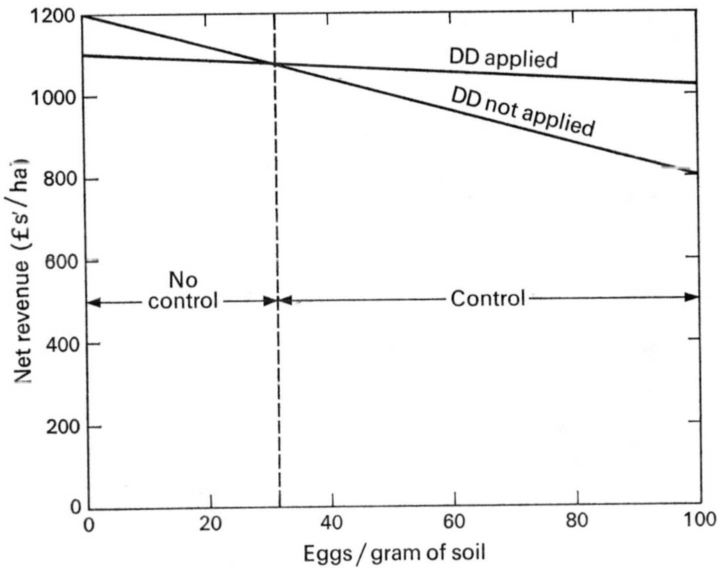

Figure 8. The economic threshold (31 eggs/gm of soil) for potato cyst eelworm using the soil fumigant DD. Loss of yield = 0·1 tonne ha^{-1} egg^{-1} gm^{-1} of soil; price of potatoes = £40 tonne^{-1}; cost of control = £100/ha; efficiency of control = 80%. Norton (1976b).

Clearly, the level of the threshold will depend upon the cost of the insecticide, the price of the crop, the damage coefficient, the effectiveness of control and the time horizon considered. Moreover, the threshold may be modified by individual farmers according to their belief in the accuracy of the information incorporated in the model, as well as the structure of the model itself.

Table– 6.   Optimal application times for ten spraying strategies directed against a single froghopper brood (Conway *et al.* 1975).

| Strategy | Time of application (days from first adult emergence) | | | | | | | | | | | | | | | | | | % of adult-days removed |
	1	2	3	4	5	6	7	8	9	10	11	12	13	14	15	16	17	18	
1																			0·0
2											NR*								19·3
3									NR					NR					32·7
4										R*									37·6
5							NR				NR					NR			42·8
6								NR				R							49·0
7							NR			NR			R						57·2
8							R							R					60·9
9						NR		R							R				67·4
10				R						R						R			74·9

* R = residual spray; NR = non-residual spray.

The problem of determining economic thresholds is considerably more complicated where control measures can be applied at various times against a multi-voltine pest. This is illustrated by the work of Conway *et al.* (1975) on the sugarcane froghopper in Trinidad. The froghopper has four broods a year and the loss of sugarcane appears to be a linear function of the number of froghopper adult-days in the year. For the sake of analysis, Conway *et al.* assume that a non-residual and a more expensive residual insecticide are available and can be sprayed up to three times within a brood. The optimal spraying times for a single brood are shown in Table 6 and the net revenue lines are given in Fig. 9. The economic threshold here is approximately

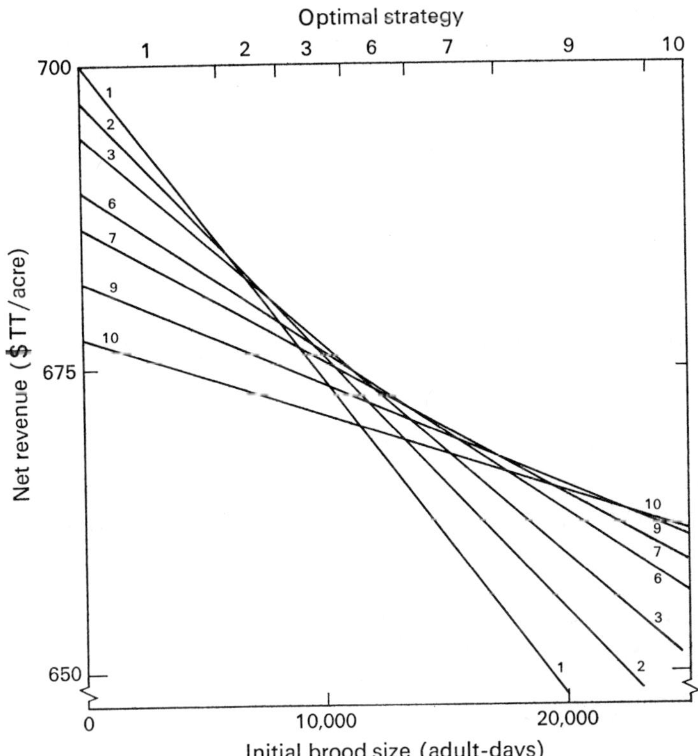

Figure 9. Optimal spraying strategies (see Table 6) against a single froghopper brood (Conway *et al.* 1975).

6,000 adult-days, since above this level at least one spraying strategy is profitable. The most profitable strategy for a level of attack above this threshold is given by the highest net revenue line.

When the analysis is extended to all four broods, the economic threshold, and hence the optimal strategy, becomes critically dependent on the density dependent relationship which exists between the broods. By assuming that this relationship can be expressed as:

$$A_{t+1} = \lambda A_t^{1-b}$$

where A_t, A_{t+1} are the number of adult-days in broods t and $t+1$, λ is the net rate of increase, and b is the measure of density dependence (see May *et al.* 1974) and by using the technique of dynamic programming (Watt 1963, Shoemaker 1973), Conway *et al.* (1975) determine the optimal strategies for different values of b (Table 7).

Table 7. Effect of density dependence on optimal spraying strategies (after Conway *et al.* 1975).

Initial size of 1st brood (adult-days)	b = 0 (none) Brood				b = 0·5 (undercompensating) Brood				b = 1·25 (overcompensating) Brood			
	1st	2nd	3rd	4th	1st	2nd	3rd	4th	1st	2nd	3rd	4th
2,000	7*	3*	1*	1*	1	10	3	1	1	10	10	1
5,000	10	5	1	1	6	10	3	1	1	9	10	1
10,000	10	10	1	1	10	10	3	1	2	9	10	1

* Control strategies against individual broods. Numbers refer to strategies listed in Table 6.

As well as considering the dynamic effect of spraying within the season, decision makers may also wish the economic threshold to take into account the effect of a particular spraying strategy on the efficacy of the chemical in subsequent seasons. In developing an analytical method of managing pesti-

Table 8. Effect of resistance on the cost of optimal spraying strategies (Comins 1977). Compare with Table 7.

Density dependence	Initial size of 1st brood (adult-days)	Total cost ($TT)	
		No resistance	Resistance
b = 0·5	2,000	59	59
	5,000	73	77
	10,000	84	99
b = 1·25	2,000	90	98
	5,000	95	100
	10,000	104	109
b = 0	2,000	35	35
	5,000	49	53
	10,000	60	75

cide resistance, Comins (1977) uses the sugarcane froghopper problem to illustrate how the cost of optimal strategies changes in the presence of resistance (Table 8). He does this by incorporating the cost of future losses due to resistance in the present cost of the insecticide. For undercompensating density dependence ($b = 0.5$, Table 8), the resistance cost, as might be expected, is greatly increased where populations are large. For overcompensating density dependence ($b = 1.25$) however, the impact is more complicated, the net effect being a constant resistance cost for all population sizes.

The value of decision models

Since they provide a means of assessing the importance of biological parameters in the decision making process, decision models, in their descriptive role, can indicate where research effort is best directed. The fact that the objectives and constraints of decision makers can be incorporated in such models means that they can also be used to test the practical feasibility of proposed methods and strategies of control.

In their prescriptive role, decision models can illustrate the principles involved in designing a good control strategy. One approach is to derive optimal solutions under a variety of assumptions (cf. Conway *et al.* 1975). Holling, Jones & Clark (1976) investigating the spruce budworm problem, have adopted a different approach to prescription. They allow real life decision makers the opportunity of testing various management strategies on a simulation model, and thus enable them to gain valuable experience and insight before employing particular strategies in the field.

Decision models also have a role to play in the formulation of extension recommendations. Good extension is clearly a critical factor in the improvement of pest control and the need to understand the social and economic context in which recommendations are made has been emphasized by Norman, Hayward & Hallam (1974). In making pest control recommendations, the following questions need to be asked:

Will the recommendations enable farmers to meet their objectives more effectively?

To what extent will farmers be able to meet these recommendations?

How sensitive is output to variations about the recommended practice?

If the answers to any of these questions indicate that the recommendations are not feasible, further consideration should be given to the means of relaxing farm constraints or modifying the recommendations to meet them.

Conclusions

Over the past twenty years, pest control workers have become increasingly aware of the ecological problems associated with pest control. As a consequence, ecological techniques, such as integrated and biological control, have gained a central place in pest control theory and practice. In other areas of resource management however, a number of authors (e.g. Chapman 1974, Slovic *et al.* 1974, Firey 1960) have emphasized the need to view management problems as much in their social and economic environment as in their physical and ecological context if realistic solutions are to be found. We would conclude by suggesting that the same is equally true of pest management problems, and that economics has its place alongside ecology at the centre of pest control theory and practice.

Acknowledgment

Tables 2, 3, 4 and 5 and Fig. 8 are reproduced by kind permission of Elsevier Scientific Publishing Co., Amsterdam.

References

AIYER A.K.Y.N. (1949) Mixed cropping in India. *Indian J. agric. Sci.* 19, 439–543.

BRADFIELD R. (1972) Maximising food production through multiple cropping systems centred on rice. *Rice, Science and Man*, pp. 143–63. International Rice Research Institute, Los Banos, Philippines.

BULLEN F.T. (1966) Locusts and grasshoppers as pests of crops and pasture—a preliminary economic approach. *J. appl. Ecol.* 3, 147–68.

BUNDY J.W. (1977) Consumer expectation and innovation in processed food production. In *Origins of Pest, Parasite, Disease and Weed Problems* (Ed. by J.M. Cherrett & G.R. Sagar), pp. 227–35. Blackwell Scientific Publications, Oxford.

BUNTING A.H. (1972) Pests, population and poverty in the developing world. *J. Roy. Soc. Arts.* 120, 227–39.

CAMMELL M.E. & WAY M.J. (1977) Economics of forecasting for chemical control of the black bean aphid, *Aphis fabae* Scop., on the field bean, *Vicia faba* L. *Ann. appl. Biol.* (in press).

CARLSON G.A. (1970) A decision theoretic approach to crop disease prediction and control. *Am. J. Agric. Econ.* 52, 216–33.

CHAPMAN G.P. (1974) Perception and regulation: a case study of farmers in Bihar. *Trans. Inst. Br. Geog.* 62, 71–93.

COMINS H.N. (1976) The development of insecticide resistance in the presence of migration. *J. Theor. Biol.* (in press).

COMINS H.N. (1977) The management of pesticide resistance. *J. Theor. Biol.* (in press).

CONWAY G.R., NORTON G.A., SMALL N.J. & KING A.B.S. (1975) A systems approach to the control of the sugar cane froghopper. In *Study of Agricultural Systems* (Ed. by G.E. Dalton), pp. 193–229. Applied Science Publishers, London.

CYERT R.M. & MARCH J.G. (1963) *A Behavioral Theory of the Firm*. Prentice-Hall, Englewood Cliffs, New Jersey.

DIETRICK E.J. (1973) Private enterprise pest management based on biological controls. *Proc. Tall Timbers Conf. Ecol. Anim. Cont. by Habitat Manag.* 4, 7–20.

FINLAY R.C. (1974) Intercropping soybeans with cereals. Paper presented to the Regional Soybean Conference, Addis Ababa, October 1974.

FIREY W. (1960) *Man, Mind and Land: a Theory of Resource Use*. Free Press of Glencoe, Illinois.

GOULD P.R. (1965) Wheat on Kilimanjaro: the perception of choice within game and learning model frameworks. In *General Systems Yearbook* (Ed. by L. von Bertalanffy), 10, 157–66.

HOLLING C.S., JONES D.D. & CLARK W.C. (1976) *Ecological policy design: lessons from a study of forest/pest management*. Institute of Resource Ecology, University of British Columbia, Vancouver. R–6–B.

HUKE R.E. (1974) San Bartolome and the green revolution. *Econ. Geog.* 50, 47–58.

I.R.R.I. (1973) *Annual Report for 1972*. International Rice Research Institute, Los Banos, Philippines.

JONES F.G.W. (1973) Management of nematode populations in Great Britain. *Proc. Tall Timbers Conf. Ecol. Anim. Cont. by Habitat Manag.* 4, 81–107.

KATES R.W. (1970) Natural hazards in human ecological perspective: hypotheses and models. Working paper No. 14, Natural Hazard Research, University of Toronto.

LARGE E.C. (1950) *The Advance of the Fungi*. Jonathan Cape, London.

LARGE E.C. (1958) Losses caused by potato blight in England and Wales. *Pl. Path.* 7, 39–48.

LOW A.R.C. (1974) Decision taking under uncertainty: a linear programming model of peasant farmer behaviour. *J. Agric. Econ.* 25, 311–20.

MAY R.M., CONWAY G.R., HASSELL M.P. & SOUTHWOOD T.R.E. (1974) Time delays, density-dependence and single-species oscillations. *J. Anim. Ecol.* 43, 747–70.

McLOUGHLIN P.F.M. (1970) Introduction. In *African food production systems: cases and theory* (Ed. by P.F.M. McLoughlin), pp. 3–39. Johns Hopkins Press, Baltimore.

NIX J. (1974) *Farm Management Pocketbook* (6th Edition). Wye College, Ashford, Kent.

NORMAN D.W. (1971) Economic and non-economic variables in village surveys. *Samaru Agricultural Newsletter* 13, 61–4.

NORMAN D.W. (1974) Rationalising mixed cropping under indigenous conditions: the example of Northern Nigeria. *J. Devel. Studs.* 11, 3–21.

NORMAN D.W., HAYWARD J.A. & HALLAM H.R. (1974) An assessment of cotton growing recommendations as applied by Nigerian farmers. *Cotton Grow. Rev.* 51, 266–80.

NORTON G.A. (1975) Multiple cropping and pest control: an economic perspective. *Meded. Fac. Landbouww. Rijks. Univ. Gent.* 40, 219–28.

NORTON G.A. (1976a) Pest control decision making: an overview. *Ann. appl. Biol.* 84, 444–7.

NORTON G.A. (1976b) Analysis of decision making in crop protection. *Agro-ecosystems* (in press).

OGUNFOWORA O. & NORMAN D.W. (1973) An optimization model for evaluating the stability of sole cropping and mixed cropping systems under changing resource and technology levels. *Bull. Rural Econ. Sociol.* (Ibadan, Nigeria) 8, 77–96.

RAROS R.S. (1973) *Prospects and Problems of Integrated Pest Control in Multiple Cropping*. Saturday Seminar. International Rice Research Institute, Los Banos, Philippines.

RICHARDS P. (1975) 'Alternative' strategies for the African environment. In *African environment: problems and perspectives* (Ed. by P. Richards) pp. 102–17. Special Report No. 1, International African Institute, London.

SHOEMAKER C. (1973) Optimisation of agricultural pest management. II. Formulation of a control model. *Math. Bio. Sci.* **17**, 357–65.

SLOVIC P., KUNREUTHER H. & WHITE G.F. (1974) Decision processes, rationality, and adjustment to natural hazards. In *Natural hazards: local, national, global* (Ed. by G.F. White) pp. 187–205. Oxford University Press, London.

SOUTHWOOD T.R.E. & NORTON G.A. (1973) Economic aspects of pest management strategies and decisions. In *Insects: Studies in Population Management* (Ed. by P.W. Geier, L.R. Clark, D.J. Anderson & H.A. Nix), pp. 168–84. Ecol. Soc. Aust. (Memoirs 1), Canberra.

STREETER C.P. (1975) *Reaching the developing world's small farmers*. Special report from the Rockefeller Foundation, New York.

VELIMIROVIC H. (1975) An anthropological view of risk phenomena. Research memorandum, International Institute for Applied Systems Analysis, Schloss Laxenburg, Austria.

WATT K.E.F. (1963) Dynamic programming, 'look-ahead programming', and the strategy of insect pest control. *Can. Entomol.* **95**, 525–36.

WAY M.J. & CAMMELL M.E. (1973) The problem of pest and disease forecasting—possibilities and limitations as exemplified by work on the bean aphid, *Aphis fabae. Proc. 7th Br. Insectic. Fungic. Conf. 1973*, 933–54.

WHEATLEY G.A. (1974) Insect problems in crops for processing. *Pestic. Sci.* **5**, 101–12.

WHITE G.F. (1974) Natural hazards research: concepts, methods, and policy implications. In *Natural Hazards: Local, National, Global* (Ed. by G.F. White) pp. 3–16. Oxford University Press, London.

Consumer expectation and innovation in processed food production

J. W. BUNDY *Birds Eye Foods Ltd, Walton-on-Thames, Surrey*

Food production has increasingly become an industrial process especially in developed countries. Industrialization brings problems, some quickly and others later. The purpose of this paper is to examine some of the ways the processed food industry makes demands on agriculture and agricultural systems. Standards set by the consumer have important implications for those interested in pest, disease and weed problems. In particular, as will appear, crop deficiencies previously accepted become serious problems when the quality demand is increased.

The industry and the consumer

The vegetable processing industry involves nationally probably less than 5% of the gross value of farm products but it is specialized with high margin crops; the industry is concentrated particularly in the eastern counties of England and the units of production are relatively large. It is also a growth industry. Currently about 50% of the national outdoor vegetable acreage is used for processed vegetables. Vegetable consumption is more or less static in terms of quantity with a national consumption of vegetables of 2·5 million tonnes plus 3 million edible tonnes of potatoes, but the way in which vegetables are presented and bought is changing rapidly.

Vegetables are distinct from some other farm products in that they are normally consumed in their natural form and, as such, contamination in the field is often obvious and should be absent. Control of defects must therefore be high and not just to the point where crop yields are not adversely affected. Control must be at a level where processing factories can handle bulk intakes which do not incur excessive factory cost but still allow consumer standards to be met.

The evidence from research carried out on the shopping habits of younger women indicates that they seek greater self-fulfilment in making creative use of both work and leisure. Seventy-two per cent shop only once a week, and a small but significant number shop only once a month. One-third shop at freezer centres whether they have a freezer or not, and nearly half of them buy frozen food in bulk and three-quarters buy most of their food in large supermarkets. Over 90% of them own refrigerators and over one-third own freezers (national average 18%). Forty-two per cent of housewives go out to work; at the same time job security is becoming less important than money. Most consumers correctly estimate the current rate of inflation and believe that food prices have outstripped it in the past year; 30% say their house-keeping costs have increased even faster. They are price conscious, but are prepared to pay premium prices for convenience foods in order to buy time. The younger woman likes to try new food stores and specialist shops. She believes that natural foods are healthier, but that modern foods are more convenient and have better keeping quality. Few consumers want information about nutrition on the packs and most are vague about the language of metrication (data from recent surveys).

Questioned on damage to the environment, younger women do not consider that traffic noise or pesticides in the soil are damaging, rather that they are simple facts of life. All women reckon that factory waste in rivers, rubbish and litter and sewerage in the sea are the worst offenders. Some commentators have claimed that the growth of convenience foods ranks among the most perniciously erosive social forces of all time, but the consumer says she will be using more frozen meat, fish and vegetables, more tea bags, more ready deserts, more instant coffee, more prepared dough, more instant potato tomorrow than she does today.

The consumer has a distinct and definite expectancy in food, in the way in which it is bought, and this expectancy has to be met in terms of availability, quality, cost and variety. We shall deal with these requirements separately, and shall emphasize where and why the pest, disease and weed problems originate and, in many cases, remain.

Availability of supplies

The changes in consumer habits in vegetable buying have affected vegetable production generally and the biggest change is in the type of vegetable grown. Traditional vegetable crops have been replaced by those most suited to freezing and these are grown in the Eastern Counties of England in proximity to the factories which process them. Vegetable production in other areas, e.g. sprouts in the West Midlands, market peas and broad beans in the South-West and carrots in the Northern Region, has declined in consequence. Similarly, field scale vegetables have replaced fodder crops, barley, sugar beet and even grass in the Eastern Counties. The acreage shift is determined by both the market and the ability to produce extensively in a mechanized system. These changes have led to crop specialization by growers, partly through increasing acreage requirement, but mostly from harvest mechanization which has allowed full exploitation of group machinery operations in the field. Our own experience in the last fifteen years indicates changes in contracted hectares per farmer of from 10 to over 40 in peas and from 2 to 20 in green beans. Thus there is a change in the type of vegetable grown, in the area and the size of the farm holdings where it is produced, and in the methods used to harvest them.

It has to be concluded that soils and climate are less relevant than business management and mechanization as controlling factors of geographic distribution of vegetable crops. The changes are as important and far reaching as those of the 18th century and are as significant geographically as they are agriculturally.

Mechanization itself has caused dramatic changes in crop production methods. Fields have been merged and hedgerows removed, new crop cultivars have been introduced especially those which are determinate in growth where the pods or edible parts mature at one time on the plant to optimize usable yield from a single harvest. Plant densities have changed to suit mechanical and varietal systems and the growing seasons of crops are extended to lengthen harvest times.

To be successful, crops must fulfil their yield predictions in usable quality. Varietal yield stability remains a faint hope and rare reality and expensive shortfalls or embarrassing peaks are all too common. Crop losses occur from soil-borne causes, such as eelworm in peas, sclerotinia in green beans, or fusarium in peas and beans, from virus diseases carried by air-borne vectors as in top yellows in peas, or soil-borne vectors as in pea early virus browning, from seed-borne diseases such as halo blight or ascochyta in peas, or airborne diseases such as mildew or botrytis. Pest infestation remains unpredictable and leads to crop losses through ineffective control procedures often applied inadequately in high density crops, or through effective procedures applied too late due to lack of forecasting data. Some success has been achieved in the control of pea midge and pea moth, but serious losses continue annually from other pests such as migrant diamond black moth on sprouts, silver Y moth on green beans, or cabbage root fly larvae on sprouts. Weeds, through their effects on growing crops and their interference with harvesting create major problems in meeting raw material requirements. Small amounts of weed vegetation can affect the efficiency of harvesting and processing even though they need not adversely affect crop yield. Their presence affects harvesting rate when speed of harvest at selected maturity is important. Weeds also directly affect cutting efficiency. Knotgrass and black bindweed wrap themselves around the cutting blades, causing stoppages and loss of valuable time. The presence of tough, branching weeds such as fat hen can easily cause damage to the picking reel of bean harvesters, reducing efficiency, increasing harvest time and causing pod damage, making the crops unsuitable for the quality market. The main weeds in spinach crops, chickweed and annual nettle, often require hand removal at harvest to avoid crop contamination. Higher levels of weed infestation in these crops often lead to slower processing rates at the factory or even crop rejection simply because it cannot be accepted for processing. The presence of weeds, and the side effects of herbicides, can considerably affect the rate of maturity of the crop, throwing the harvesting sequence off-plan and leading to by-passed crops.

The challenge of meeting consumer demand in terms of available suitable product has therefore introduced changes in type and geographic distribution of cropping, has increased the scale and specialization in farming rotations and has led to large-scale mechanization. These changes have themselves brought problems of availability of supplies where varietal performance remains too unstable to produce planned yields, where intensive cropping encourages concentration of pests, and where mechanical efficiency in high density crops is impaired by weed population. The low incidence at which some pests and weeds need to be present brings a control level requirement which is well beyond the point at which yields need be adversely affected.

Quality attainment

Probably the most defiant and contentious character sought by processors on behalf of the consumer is quality, the definition of which varies with the processor and the market he seeks to supply. Vegetable quality may be represented by size, maturity, texture, colour or appearance, and many other varietal or production criteria may be willingly forsaken if a processor believes a product has the quality he seeks.

Extensive research is carried out to select the best cultivar and the conditions most suited to meet quality specifications; but inherent varietal quality shortcomings arise through lack of intra- and inter-plant uniformity at the time of maturity. Diseases which were considered economically unimportant when crops were handpicked, have now assumed new significance. These include particularly, pod or leaf spotting blemishes which affect the visual appearance of the harvested product. Infected produce, previously left on the plant by handpickers, present a major problem when the crop is machine harvested, for even at low incidence its presence in factory intakes can incur expensive inspection and seriously affect through-put costs. Physiological defects such as internal browning in sprouts, cavity spot in carrots and curd discoloration in cauliflower can be equally damaging to quality. Such defects may appear less dramatic than the damage seen in crops in the field from pest and disease attacks, but they are far more insidious and costly to the processor and ultimately to the grower since they are totally unpredictable and uncontrollable and if present may leave no alternative but crop rejection. Product quality may be depressed by con-tamination of the harvested crops by buds, flowers and berries of such weeds as mayweed, creeping thistle and poppy. These contaminants are roughly the same size, specific gravity and colour as the pea, making detection and removal difficult. There are obnoxious weeds whose presence in crops, if not removed, incurs crop rejection. The berries of nightshade, a species generally found on light land and which is late germinating and affects late-sown crops, are dark green and the same size, shape and density as a pea. Perhaps worst of all is the potato berry, produced by volunteer potato plants (the most prevalent and damaging of all arable weeds!). A national survey of the presence of potato volunteers in processed vegetable crops during 1975 showed 15% pea crops, 25% green bean crops, and 5% broad bean crops to be infested to varying degrees. All required treatment and removal before acceptance of the crop for processing.

Development work on chemical control procedures is extensive and processors normally work to a code of practice which incorporates the recommendations of the Pesticide Safety Precautions Scheme and Agri-cultural Chemicals Approval Scheme to ensure freedom from taint and residues. But in a branded product, every consumer complaint is of the greatest concern to the processor; indeed, product failure is an offence against the Food Hygiene Regulations. Thus any procedure which eliminates only say 70% or 80% of the damaged product provides nothing like the level of control required to make any impression on the problems of the processor whose aim is to satisfy consumer expectancy.

Mechanical harvesting, itself a solution to the problem of adequate supplies, creates its own problems in product quality by the damage it causes during harvest and by the contamination levels it still allows in the form of extraneous vegetable matter present in the harvested product. The absence of discrimination of the machines themselves has brought a new significance to certain diseases. Despite the improvements in recent machines, they need to be still more effective in the field in minimizing damage and separating waste effectively, and this is even more important than improving capacity performance. Equally, the crop cultivars which are grown need to be bred and produced to ease the task of the machine designers; the machines must have a weed-free environment in which to perform.

Cost achievement

The consumer profile showed that women believe that food prices have outstripped inflation in the past year. Forty per cent of families consider themselves financially worse off than two years ago and a further 30% certainly no better off. Processed vegetables have to remain cost competitive. Raw material, even if it is available and of acceptable quality, contributes to product cost in at least two ways, directly through its own inherent cost of purchase and indirectly, through its effect on factory costs. Here again, we can see how pest, disease and weed problems contribute to increasing cost at the expected consumer level.

Quality in vegetables is related to maturity at harvest and optimum maturity in terms of consumer quality is usually immature in terms of potential crop yield. Price to the producer has to respect lower yields through a higher price for better quality. Existing cultivars, even those bred for the purpose, are extremely inefficient in producing higher yields at optimum consumer maturity. They remain too indeterminate, both within a plant and between plants within a crop, at the time of harvest. This inefficiency is avoided to some extent by using high plant densities but this in turn creates greater seed costs which have to be repaid. The relationship between seed rate and optimum plant stand is commercially important and significant to the ultimate cost of the finished product, particularly in crops such as sprout hybrids and silverskin onions, where the seed cost per hectare varies from £250 to £2,000. Soil-borne diseases, pest attacks and weed competition are some of the major causes leading to seedling mortality and hence higher seeding rates to ensure adequate plant stands and yields.

Extensive and often expensive control measures have to be taken to keep pest infestation to acceptable levels and the cost of chemicals and their application have to be recompensed to the grower. Extra cost may arise to cover the risk of crop losses. Crop rejections which occur because of relatively high pest contamination have to be paid for in a higher price for the acceptable product. Improved forecasting is badly needed for accurate prediction of the imminence of pest, parasite, disease and weed problems and would certainly allow some reduction in the rates and frequencies of chemical application while retaining adequate control levels.

In addition to high raw material costs, further costs arise at the factory due indirectly to raw material problems. Factory production programmes may have to be significantly altered because pest infestations remain too high to allow production to continue. The natural Brussels sprout season extends from September to March, but processing is delayed until mid-October due to inadequate control of cabbage root fly larvae in the sprouts themselves, and ends usually in mid-January because later harvest carries too high a risk of internal browning and external blemishes arising from previous attacks of insects and diseases such as powdery mildew. Inadequate use of factory investment increases overhead costs, again contributing to higher product cost. Growers often choose crops and cultivars which fit their labour and harvest programme and accept lower potential yields to get the spread. Equally the processor who must aim to optimize his factory investment and hence minimize his production cost, must also extend the harvest seasons. Thus lower yielding, early or late maturing cultivars can take priority over higher yielding standards and there are many examples of this. This again has to be paid for in the raw material through early or late planting bonuses. An extreme example is where crops cannot thrive in the locality of the factory, as in green beans which do not develop in an environment north of the Wash. Northern processing factories do not have a bean season. As investment gets more difficult to find and justify, the search for extending harvest seasons through breeding or disease and pest prevention, will increase in importance.

Product variety

The consumer, defined earlier in this paper, has an increasing strength of identity. She needs to be more creative, wishes to try new things but does not necessarily have more confidence in a product because she has seen it advertised. So there is an increasing interest in 'other' vegetables, baby carrots, pearl onions, haricot beans, green sprouting broccoli and so on. Such product development has required field work to establish suitable cultivars, plant densities, pest and disease control procedures, growing and harvesting techniques and production experience. Success is limited and problems still occur, mainly from the need to contain weeds, pest and diseases at an acceptable consumer level. The problems are largely agricultural, e.g. cavity spot in carrots, low crop productivity and poor head conformity in sprouting broccoli, eelworm and discoloration in silverskin onions, and remain the main causes for lack of confidence in expanding process production.

New consumer constraints

Perhaps the modern consumer's expectation is too demanding and unrealistic but we are coming into an era where social responsibility is expected to increase, where concern about pollution and resource depletion is openly expressed, where consumer pressure is greater through informed labelling, product composition, open date marking, unit pricing and nutritional and health information, where advertising persuasion is being resisted and where communication is essential. Legislation will increasingly impose restraints on the use of various valuable chemicals, and mandatory tolerance levels of chemical residues on crops will bring tighter control on application practices. So we must look to the new opportunities from R & D which will continue to provide the changing consumer needs.

Future research developments

The increasing intensity of cropping, crop specialization, harvesting methods and control procedures whilst meeting more immediate needs may in themselves create new problems, many yet to emerge. Is specialization and large-scale production encouraging land sickness even if crop rotation is carefully practised? Are sporadic migrant pest infestations becoming indigenous annual problems? Are routine spray programmes accelerating genetic resistance in pests, dominance of resistant weeds, or new opportunities for disease? Will genetic improvement to eliminate physiological problems in existing varieties break down existing resistance to established diseases or desirable agronomic characters? If research provides solutions to our needs, are we even then prepared to grasp them if it means reducing, in some other way, our ability to satisfy consumer expectancy?

In the meantime work is being done which may resolve some of the problems in modern practices. To those who seek greater yield stability of cropping and avoidance of yield losses, there is encouragement in the integrated approach to pest and disease control being developed at Wellesbourne, Invergowrie, John Innes, Rothamsted and elsewhere. Plant resistance to insects already exists in many crops, but as yet there are only a few examples in vegetables (e.g. lettuce resistance to root aphid). Work in progress includes brassica resistance to cabbage root fly, resistance of roots to carrot fly and sprout resistance to cabbage aphis. Extreme resistance or immunity is rare, but the work may allow moderate resistance to be bred into these crops which, combined with lower rates and levels of insecticide, will allow dramatic practical advances. Work at Rothamsted on chemical control of nematodes with aldecarb and oxamyl can contribute to yield stability of crops such as peas and onions. Improvement in herbicide formulations, mixtures and spray programmes, the use of plant growth regulators which influence growth, but also affect plant resistance to pests (Long Ashton work on chlormequat chloride on blackcurrant aphis) are other examples.

Quality improvement may come from new work at Wellesbourne on methods of breeding for improved uniformity in carrots and reduction of genetic variability in onions, or from work at Invergowrie on the factors influencing yields and quality in calabrese.

Costs may be reduced if successful results from pest forecasting systems at Rothamsted can be introduced into commercial programmes thereby allowing anticipation of infestation and therefore precautionary, and hopefully, reduced sprays to be applied in good time. New type machines will find their place and reduce crop damage, being aided by breeding programmes such as leafless peas at John Innes or morphological and physiological improvements in sprout varieties at Invergowrie. It must be recognized that in seeking solutions to such extensive problems, the chemical manufacturers, plant breeders or machine engineers must themselves obtain satisfactory returns on their research investment. They may feel that the processed vegetable acreage is not enough incentive for such an investment, or that farms cannot justify the cost of their research products if available. However, it is the factory end-product cost which really matters and not just the raw material price. It may be more economic for the processor to pay a higher price for clean usable raw material from the field if inspection, labour cost and investment can be reduced at the factory and consumer complaints avoided. It is also true that the processors' needs will increasingly apply to all forms of vegetable presentation as consumers become more insistent on higher standards generally.

This spring, 100,000 ha of outdoor vegetables will be planted and used for processing. The farm value of the crops will be £50,000,000 and the ancillary costs of machinery, seed, fertilizer and chemicals used will be £25,000,000 on present values. These are estimated figures but are of the right order. The operation will be reasonably effective but not very efficient. Too much seed will be used, too many hectares will be lost or rejected, yields will be lower than they should be and harvesting machines will operate below par and cause too much crop damage. Growers will sustain too many losses, processors will incur excess costs and consumers will pay more than they could have done, or accept less than they may have wanted. Much of this inefficiency or loss will have been caused, directly or indirectly, by pest, disease and weed infestation which need not have happened, or which could have been better anticipated or controlled.

I welcome the opportunity to state, and hopefully clarify, the problem and I look to the ecologist to recognize and accept the consumer challenge.

The economics of pesticide research

CADWALADR J. LEWIS *I.C.I. Plant Protection Division,
Haslemere, Surrey*

When an industry is enjoying a growth rate of about 20% per annum for a decade or so, it often ignores many of the elements that exist in the business environment within which it operates and often fails to identify the changes that are generated by the rapid growth. But when growth rates decelerate to 5%–6% the business environment becomes a prime determinant of corporate policy. If, as is the case in the pesticide industry, research represents a significant cost centre and growth is research generated, a fall in the growth of sales is tantamount to a reduction in the productivity of research. This paper analyses the inter-relationship between research and the business environment of the pesticide industry over the last thirty years.

The pesticide industry as we know it today, began thirty years ago with the development of BHC, DDT and the phenoxyacetic acid herbicides. Its growth to the current level of world sales of nearly $5,000m was the result of a happy coincidence of two developments, namely, an intensified awareness of the need to increase food production, and the realization by chemical companies that organic chemistry could well include a reservoir of compounds that had pesticidal effects.

The first and most persistent influence that drew attention to the world's food needs was the FAO. Its pessimistic predictions of hunger and malnutrition among a high proportion of mankind, although challenged severely over the years, contributed significantly to the urgency with which governments and foundations such as Ford and Kellogg applied themselves to devising methods to increase food production. Large investments were made in irrigation schemes and fertilizer plants; ambitious plant breeding and seed multiplication programmes were initiated and aid schemes were introduced to alleviate food shortages in developing countries. The political and economic climate generated by this activity was conducive to the growth of agriculture and the growth of industries, including the pesticide industry, that served agriculture. It also made for confidence in the future.

Countries that enjoyed sophisticated and well developed agricultural systems had to achieve increased food production with a falling number of workers in agriculture. For example, between 1949 and 1959, agricultural employment in the U.S.A. and U.K. (regular workers) declined by nearly 30% and by 1965, there was a further 25% reduction in both countries. Investment in mechanization increased and with it, a ready adoption of new herbicides to control weeds. Adequate returns from the swollen fixed capital became a criterion of success. Risks of depredation by insects and diseases had therefore to be reduced to the minimum. Herbicides in addition to controlling weeds also provided a means of increasing crop density to levels that precluded the use of mechanical weed control.

The significance of the discovery of BHC, DDT and the phenoxyacetic acid herbicides was quickly appreciated by a number of chemical companies. Between 1945 and 1950, twenty companies had introduced to the market twenty-eight new pesticides, the majority being insecticides. By 1960, about fifty companies had entered the pesticide industry and between them had introduced about 145 compounds of varying importance to the farmer. Success rates varied widely. One company introduced ten products by 1960 while thirty-eight companies had each launched less than four products in the period 1945 to 1960. By 1974, the number of chemical pesticides introduced had increased to over 500 and the number of pesticide companies pursuing R & D had grown to over seventy-five. Only about 250 of these pesticides are now considered viable in the market but the majority of the seventy-five companies remain active in research.

Research costs

Precise information on the Research and Development (R & D) costs incurred by pesticide companies is not available but it is generally considered that they account for between 5% and 10% of the value of sales. This compares with the 2·5%–3% estimated for chemical industry as a whole. A number of attempts have been made to estimate the R & D costs up to the launching of a pesticide on to the market. Table 1 summarizes these attempts.

Table 1. Estimates of total R & D cost per product marketed (Lever & Strong 1973).

	Year of estimate					
	1956	1964	1967	1969	1970	1972
Total cost in $ million	1·2	2·9	3·4	4·1	5·5	10+
'Present value' of cost in $ million at time of discovery*	1·0	2·1	2·6	3·6	3·9	7+

* The total cost discounted back to the year of discovery at 10% per annum.

Another way of indicating the effort required to launch one pesticide on to the market is to show the number of chemicals screened per successful product (Table 2); the screening of chemicals for biological activity and all the research associated with the exploratory process of research represent a significant and inescapable proportion of the total R & D cost.

Table 2. Number of compounds passing through each R & D stage per product marketed (Lever & Strong 1973).

Activity	Year					
	1956	1964	1967	1969	1970	1972
Synthesis and initial bio-screen	1,800	3,600	5,500	5,040	8,000	10,000
Advanced screening	60	36	*	126	80	*
Field evaluation	6	4	*	9	4	*
Development	2	2	*	2	2	*
Sales	1	1	1	1	1	1

* Not available.

Up to the mid 1960's the performance of a chemical in the biological screen was the main determinant of its being eventually launched in the market place. Although pesticide companies had increasingly subjected their products to more stringent health and environmental tests as they progressed through the R & D stages, the 'environmental screen' became a much more significant obstacle during the late 1960's. To compensate for the increase in failure rate, the number of chemicals subjected to the initial screening was increased. This added to the cost, as indeed did the more intensive toxicological and environmental testing that became an essential pre-requisite to the registration of products for use in the major countries.

By the 1970's a third screen was added, which can be best described as an economic screen. Much of this paper describes the way in which this economic screen developed over the years and the problems that it poses for the pesticide industry.

Growth patterns of pesticide use

Figure 1 describes a best guess growth curve of world sales of pesticides from 1954 to 1974 by which time sales had reached about $5,000m. It is based on a combination of data derived from surveys carried out by the journal *Farm Chemicals* (Anon. 1975), FAO statistics, and data from the better documented markets such as the United States of America, Japan and some west European countries. An attempt has been made to eliminate the effect of the post-1971 world inflation on pesticide prices. This rough and ready correction using various indices that reflect overall inflation brings down the 1974 sales to about \$3,000m in 1970 U.S. dollar value. Following a steady average annual growth of 7% compound in the period 1954 to 1961, the value of world pesticide sales increased markedly to an average annual rate of 15% for the subsequent eight years. This reflects the higher prices paid for more sophisticated and efficient products as well as the growing consumption of pesticides. The heterogeneity of pesticides creates difficulties in measuring volume growth, but, inasmuch as such a measure is relevant, the growth was 5% to 7% per annum during the 1960's.

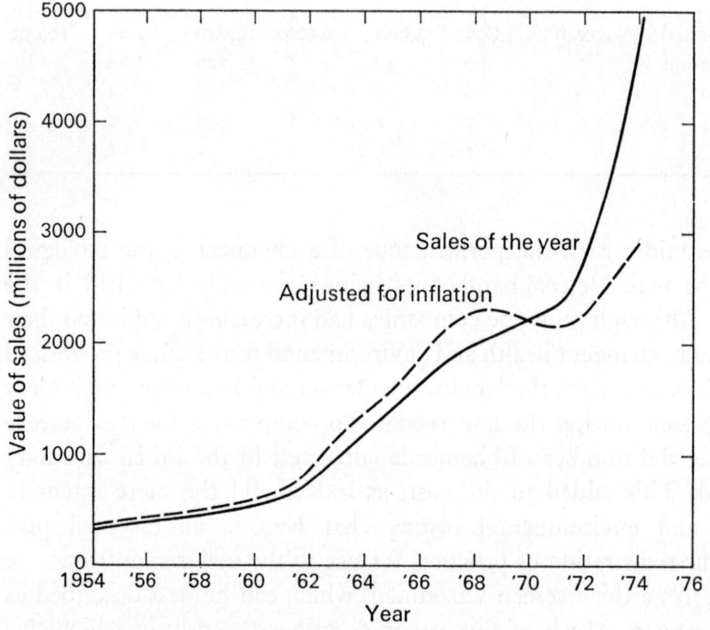

Figure 1.   Growth in world pesticide sales 1954–1976.

Herbicides

Herbicides featured as a major component of growth in terms of value; their sales increasing at an average rate of over 20% per annum during the 1960's, and by 1971, herbicides accounted for 47% of world pesticide sales. Growth in the sales of insecticides and fungicides was considerably lower—of the order of 6%–8% per annum in value and possibly 2%–4% in volume. This lower growth rate was due to the early establishment of the use of organo–chlorine and organo–phosphate insecticides prior to 1961 and the persistence with which farmers continued to use these early introductions.

In the case of fungicides, inorganic chemicals, the thiocarbamates and captan (all introduced prior to 1950), continued to dominate the market although they have recently been supplemented by the newer systemics. In addition a large proportion of insecticide and fungicide introductions were replacements for the older chemicals when resistance developed or when older products did not meet the more stringent toxicological and environmental demands. The introduction of a new insecticide or fungicide, therefore, did not increase the total treated area to anything like the extent experienced with herbicides.

The dominance of herbicides in the growth pattern was largely due to the developments that took place in the United States of America where high labour costs and a high proportion of arable area devoted to row crops formed an acceptable environment for herbicide use. By the early 1970's the value of herbicides sold in the United States was over 50% of the total world sales of herbicides and by 1974 nearly 90% of the area of corn, soya and cotton were treated, compared with levels of 5%–20% in 1960. West Europe also followed a similar trend. In the United Kingdom, for example, over 90% of cereals and sugar beet and over 75% of the potato area is now treated with herbicides.

Geographical

Growth of pesticide use was concentrated in the United States of America, West Europe and Japan, where by 1974 over 70% of world pesticide sales took place. The United States alone accounted for 40%. All these countries participated in the growth of herbicide use, but in Japan, rice fungicides were equally important, particularly in recent years. Developing countries account for only 7%–10% of world sales, and outside the advanced plantation sector these sales are by and large made up of early generation insecticides and fungicides.

Crops

Pesticide use has been concentrated on a small selection of crops. About 60% of the world pesticide sales in 1974 were in five crops: cotton, corn, rice, soya and wheat in that order. If to these is added the group 'fruit and vegetables' the proportion increases to 85%. The relative importance of herbicides, insecticides and fungicides in these selected crops is shown in Table 3.

Mixtures

A wide choice of pesticides and mixtures of pesticides is now available for the control of most weeds, pests and diseases for the majority of crops. An example of this development is the use of herbicides in maize in the United States of America; many other analogous situations could also be quoted.

Table 3. Pesticide use in selected major crops (% of value of total pesticide for each crop; USDA 1975).

Pesticide	Crop					Fruit & vegetables	All crops cited
	Cotton	Corn	Rice	Soya	Wheat		
Herbicide	26	78	35	92	75	10	43
Insecticide	71	20	47	6	10	38	37
Fungicide	3	2	18	2	15	52	20

In 1965 atrazine held a virtually monopolistic position as a herbicide but in 1965, propachlor was introduced, in 1960 alachlor, in 1968 butylate and in 1970 cyanazine. The first three are complementary to atrazine in that they control grass weeds. By 1974, the corn farmer had a choice of at least ten products or product mixtures, the mixtures normally containing atrazine. The share of the market held by atrazine used alone fell to about 30% of its 1965 market share although the expenditure on atrazine used alone increased marginally during the period, most of the increase being due to inflation. Of greater significance, is that not one of the newer herbicides when used alone exceeds a 7% market share, but one, alachlor, when used as a mixture with atrazine, can now claim about 20% market share. Both products have very similar dose response curves. In a market as big as the United States maize market (about \$350m at user prices), a product with as little as 5% market share generates a fair sized income, although its position may be vulnerable. But in many outlets across the world a similar market share would yield less than \$5m; a significant number yield less than \$1m.

Recent trends

A description of the post-1970 trend in pesticide sales is bedevilled by the effects of the oil crisis, world wide inflation, the elusive Peruvian anchovy, the withdrawal of millions of acres from the land bank in the United States of America and more recently the build up of inventory levels in 1974/5. The increase in crop acres, mainly corn and soya in the United States, increased pesticide consumption in that country by an amount equal to 7%

annual world sales in 1973/4. In other words, much of the post-1973 increase in the world pesticide sales arose, not from an increase in the intensity of pesticide use nor the introduction of novel pesticides but from an increase in one country of acres available for treatment. It is difficult to picture what the growth would have been without these step changes in external factors; one can only speculate that the post-1970 average annual growth might have been in the region of 5%–8%.

The future

The growth in the sales of pesticides in the decade ending in 1970 was primarily the result of a rapid adoption of novel herbicides in arable crops in the developed sectors of the world's agriculture. Today there are strong indications that the farmers' need for herbicides is largely satisfied and that the future overall growth in the herbicide, insecticide and fungicide field will revert to the level associated with insecticides and fungicides, until the developing countries begin to exert a stronger demand for pesticides. New introductions of herbicides, insecticides and fungicides will tend more and more to replace existing products and contribute less to the overall growth in sales; increased prices of new introductions will become a more significant element in the value of sales. This prognosis of course relates to the industry as a whole. Individual companies within the industry will each have to form a judgement as to the probability of their research being successful in discovering, developing and being able to introduce the right superior product at the right time, and at a competitive price that fairly reflects enhanced activity.

Developments in the manufacturing sector of the pesticide industry have added to the severity of the economic screen. A significant proportion of the important pesticide introductions that are in current use have lost, or are about to lose, their patent protection. Their success in the market makes them attractive for other companies to manufacture. This more often than not results in falling prices. Innovator companies therefore face a loss in income from both a fall in market share and increases in selling costs; the financing of R & D becomes more difficult as a result. There is the added problem that new discoveries, being more sophisticated in their chemistry, are generally more costly to manufacture than older generations of products.

Pesticide introduction trends

Given that the economic screen is as severe as described above, it is interesting to examine in greater detail the pattern of new introductions of pesticides over the period 1945 to 1974. The data in Fig. 2 are based on information provided in the *Pesticide Manual* (BCPC 1974) on the date of introduction for the majority of the compounds described. This date can be misleading in some cases but using a three-year moving average, a number of possible discrepancies can be significantly reduced.

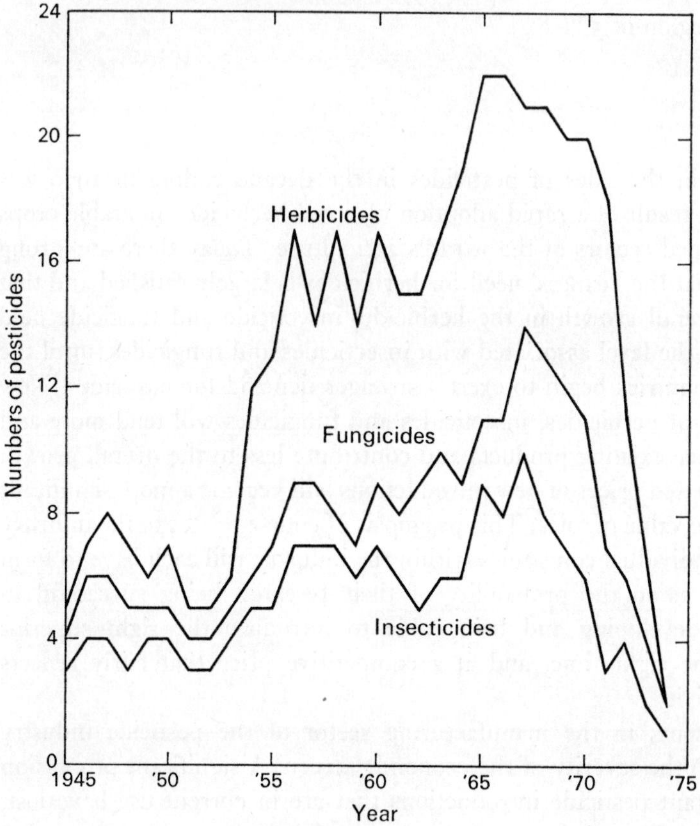

Figure 2.   Annual introductions of pesticides 1945–1975 (three-year moving average).

Following a period of nine years when the introduction rate was about six products per annum for the whole industry, 1954 saw the beginning of a rapid growth in the productivity of R & D. Fig. 2 shows that the main component of this growth was herbicides right up to 1970. A study of 260 pesticides which are representative of the total and introduced between 1950 and 1970 shows that about 50% failed to reach an annual sales level of $2·5m by 1970 (1970 dollar) and only about fifty had reached individual sales levels of over $5m. Of these fifty, twenty-five pesticides managed to reach a sales level in excess of $20m per annum. When these rewards are compared with the escalating costs shown in Table 1 it becomes clear that, even when allowance is made for inflation since 1970, only a small proportion of introductions generated sufficient income to finance the R & D costs incurred in their discovery and development, and to finance the escalating costs of ongoing R & D and the search for future discoveries. It should also be noted that the growth in sales (in real terms) in the early 1970's of 5%–8% is roughly equal to the estimate of the industry's annual expenditure in R&D.

The post-1970 precipitous fall in the number of introductions raises an interesting and important question of whether the cause lay in the increasing severity of the environmental screen and later, the economic screen. The description of the developments that have taken place in the market place and in the manufacturing sector suggests that the downturn in the number of introductions may well represent a decay in the herbicide, insecticide and fungicide cycle of research-generated business activity which has hitherto been the main concern of the pesticide industry. The pattern of introduction in the late 1970's and 1980's will most probably be in line with that of the first decade described in Fig. 2.

It would be inappropriate to detail the responses that pesticide companies are likely to make to the situation described in this analysis. But it is clear that future pesticide research wherever it is carried out, must pay far greater attention to the developments taking place in the business and agricultural environment into which the product of research will be introduced. In the herbicide, insecticide and fungicide sector of research, economic criteria must impinge upon earlier stages of the research process than has been hitherto the case and it is being increasingly acknowledged that the experience and expertise gained in discovering agents that reduce the effects of competitive elements in the crop environment will in future be progressively applied to discovering agents to promote and engineer plant growth. Such agents may well be the components of the next cycle of discoveries of what is now a crop protection industry.

References

ANON. (1975) Farm Chemicals.

BCPC (1974) *Pesticide Manual* (Ed. by J.T. Martin). British Crop Protection Council, London.

LEVER B.G. & STRONG W.M. (1973) The economic and technical evaluation of new pesticides. OEPP/EPPO Bull. 3 (3).

USDA (1975) *Pesticide Review* 1967–1974.

Problems past, present and future

Prediction of new weed problems, especially in the developing world

C. PARKER* *Agricultural Research Council Weed Research Organization, Begbroke Hill, Yarnton, Oxford*

As a first step in predicting new weed problems in developing countries, the origins of existing weed problems will be surveyed under four main headings—sudden genetic change, distribution, chemical selection and cultural selection. Some of the likely future changes in weed problems will then be considered under these same headings and finally some possible methods by which such changes can be monitored will be discussed.

Origins of weed problems

SUDDEN GENETIC CHANGE

The simplest way in which a sudden genetic change could occur would be by mutation, when a previously harmless wild plant is suddenly converted to a troublesome weed. Baker (1974) listed a number of characteristics of the ideal weed including discontinuous germination, rapid growth and plasticity. Sudden mutation in one of these characters could greatly alter the weediness of a species. There is no definite evidence for such mutation having occurred recently but there are examples of important weed types arising by another type of sudden change, namely, hybridization, with or without subsequent polyploidy.

The best-known examples are perhaps *Galeopsis tetrahit*, which almost certainly arose by hybridization of *G. speciosa* and *G. pubescens* (Müntzing 1930) and *Spartina × townsendii* originating from hybridization of the native *S. maritima* and the introduced *S. alterniflora* (Salisbury 1964). Another species which is now thought to have originated in a similar way is *Salvinia molesta* one of the two most serious of all floating aquatic weeds. Mitchell (1972) in his paper naming *S. molesta* as a new species notes that

* H.M. Ministry of Overseas Development Tropical Weeds Specialist.

this aggressive type previously known as *S. auriculata* is quite distinct from the latter species in its strict sense, and does not produce viable spores. One of the earliest herbarium specimens of *S. molesta* is from the Rio de Janeiro botanic gardens in 1941. At that time *S. biloba* and *S. auriculata* were known to occur in the gardens and it is possible that *S. herzogii* had also occurred previously. *S. molesta* is in some respects intermediate between *S. biloba* and *S. herzogii*, 'raising the possibility that *S. molesta* is a hybrid of horticultural origin'. No other specimen has been recorded from South America, but before disappearing from Rio de Janeiro it was apparently distributed.

Less dramatic changes in genotype can result from recombination within or hybridization between closely related species. Many of the major weed problems must certainly have evolved quite naturally in this way and such evolution will inevitably continue especially where different ecotypes of a weed previously separated geographically, have the opportunity to come together. A particularly serious risk arises where there are wild relatives of crop species growing where seed crops are grown. Introgression of a weedy characteristic into the crop can occur resulting in a proportion of the crop having weedy characters such as premature seed dispersal, dormancy (sorghum, red rice) or premature flowering (sugar beet). The 'weed beet' problem is now particularly acute in Belgium and several other European countries. Production of triploid 'monogerm' seed involves a male–sterile diploid and a tetraploid pollinator. Unfortunately in southern Europe where much of the seed is produced there are wild annual beets (*Beta vulgaris* and *B. maritima*) which release their pollen earlier in the day than the tetraploid pollinator. Consequently some commercial seed has become contaminated with diploid strains carrying the dominant annual character, resulting in premature flowering (bolting) and no useful root production. Seeds from these plants have produced seedlings in subsequent crops, multiplying and becoming a regular component of the weed flora, a particularly serious situation when the rotation comes back once again to sugar beet (Longden 1974).

DISTRIBUTION

In the past few centuries, by far the greatest source of new weed problems has been by accidental introduction from other areas. A substantial proportion of the more important weeds of the United States of America, Australia and South Africa are not indigenous to those countries but have been introduced as contaminants of crop seed, fodder or packing materials. Simmonds & Greathead (1977) refer to examples. Wild (1968) concluded that only slightly more than half the 'important' weeds of Rhodesia are even African in origin. Thirty-two per cent have come from America and 13% from Europe and Asia.

Thanks to increased awareness and improved phytosanitary laws, such accidental introduction is probably much less common now on the international level than it was before but it certainly still occurs. Where the importation of soil is not effectively prohibited there is a particular danger, and a number of important weeds including *Sorghum halepense* and *Oxalis latifolia* have recently been introduced to Oman in this way (Parker 1973). Other species apparently introduced and causing important new problems within the past twenty to twenty-five years include *Parthenium hysterophorus* in India (Jayachandra 1971) and *Eupatorium odoratum* in West Africa (Ivens 1974).

Apart from accidental introduction there have also been many celebrated cases of deliberate introduction of serious weed problems under the misguided idea that the species in question would be useful. Johnson grass (*S. halepense*) is generally understood to have been introduced to the United States of America as a forage just as the prickly pear (*Opuntia stricta*) was to Australia. *Lantana camara* and water hyacinth (*Eichhornia crassipes*) have been spread widely as ornamentals. It is not certain how the climber *Mikania micrantha* was introduced to South East Asia from tropical America but it was distributed within Malaysia as a ground cover until it was realized that it was much more seriously competitive than the legumes (Wong Phui Weng 1964). The introduced Kudzu vine (*Phaseolus lobatus*) in the southern parts of the United States of America is still harmless and beneficial as an anti-erosion species in some situations but is causing increasing problems of maintenance in others owing to its extremely rapid growth (Dickens 1974).

There are other examples where there was no particular use envisaged, but the plant may have been introduced as a scientific curiosity. The spread of *Salvinia molesta* is not fully documented but it appears that specimens were transferred in the 1930's from Rio de Janeiro to one or more botanic gardens in Europe. From these it found its way to the University of Sri Lanka, where it was allowed to escape and become a severe field problem which still persists. Introduction to India and South East Asia followed. The very date of introduction is known for Indonesia, Nguyen-van-Vuong (1974) recording that it was introduced to the Bogor Botanic Garden on 12 December 1950 from the Botanical Gardens, Montreal, Canada. It has since become a considerable problem in rice growing in both Java and Sumatra. It is an interesting quirk of fate that having almost certainly originated in South America it apparently became extinct there after some years (probably due to flood action), and would no longer exist at all had it not been artificially distributed by man.

CHEMICAL SELECTION

The acquisition of resistance to pesticides by insects and disease organisms is well recognized and documented but the demonstrations of corresponding intra-specific selection within weeds has been relatively rare.

Harper (1956) discussed the question in some detail and explained why the development of such resistance in weeds would be slow, owing to their relatively long life cycle, overlapping of generations due to dormant seed and slow spread. In the body of his paper he could quote only one suspected instance, of increased resistance of *Cirsium arvense* to 2,4-D (Abel 1955). In an addendum he quoted one further example, the resistance of *Erechtites hieracifolia* to 2,4-D in Hawaii (Hanson 1956).

Twelve years later, Hammerton (1968) surveyed the subject again and was able to quote two further examples from Hawaii involving *Cynodon dactylon* (resistant to TCA and dalapon) and *Commelina diffusa* (resistant to 2,4-D) (Hanson 1962). After another six years, Baker (1974) quoted a number of further instances including the particularly dramatic one of *Senecio vulgaris*. Ryan (1970) had reported that on a nursery site where simazine and atrazine had been used regularly over a ten-year period, the surviving population of *S. vulgaris* was able to withstand either herbicide at 18 kg ha^{-1} whereas unselected populations were killed by about 1 kg ha^{-1}. Since then a comparable strain of *Amaranthus retroflexus* has been reported from Washington State, United States of America (Peabody 1973). This strain of *A. retroflexus* only suffers 60% kill by 4 kg atrazine ha^{-1} and 20% kill by 16 kg simazine ha^{-1} whereas unselected populations are completely killed by 0·5 kg ha^{-1} (Thompson *et al.* 1974). More recently still a strain of *Chenopodium album* has been detected in Canada which has a comparable degree of resistance to these triazine herbicides (Bandeen & McLaren 1976).

No such dramatic changes have been reported from the United Kingdom but Holliday & Putwain (1974) have demonstrated a gradual tendency to resistance in populations of three weed species subjected to regular treatments of simazine in apple orchards. *Senecio vulgaris*, *Chenopodium album* and *Capsella bursa-pastoris* from sites with nil or only one year of simazine spraying were killed to the extent of 85%–95% by 0·7 kg simazine ha^{-1}, whereas after nine years of simazine treatments only 72%, 25% and 25% respectively were killed by this same dose. After twelve years, mortality of *S. vulgaris* was reduced to 44%. Ellis & Kay (1975a) showed how regular use of MCPA had resulted in selection of a population of *Tripleurospermum maritimum* at least twice as resistant as an unselected population. Such selection of moderately resistant types is perhaps an inevitable outcome of the use of the same class of herbicides over a period of years, on a population which has a natural range of variability in response. Many species have been reported to show variation in susceptibility particularly different clones of perennial species, for example *Agropyron repens* (Buchholtz 1958) and *Cynodon dactylon* (Rochecouste 1962), even before any selection by herbicide has occurred. Ellis & Kay (1975b) have shown how selection can occur even after a single generation, in *T. maritimum*. The build-up of the resistant type in the population will be accelerated by outbreeding and Holliday & Putwain (1974) suggest this is why *Chenopodium album* and *Capsella bursa-pastoris* showed increased resistance more rapidly than the predominantly self-pollinated *S. vulgaris*. Holliday & Putwain (1974) suggest that this moderate type of resistance is probably controlled by polygenic inheritance whereas the much higher degree of resistance in *S. vulgaris* reported by Ryan (1970) may be due to a single major gene.

Inter-specific selection by herbicides is now by far the most important factor resulting in change of weed problems. The changes in weed population associated with the use of herbicides in Germany have been studied in detail by Bachthaler (1967, 1969) and these together with much other evidence have been usefully reviewed by Fryer & Chancellor (1970a, b). The whole history of herbicide development since 1950, particularly for small-grain cereals in the United Kingdom has been one of adjustment to changing weed flora (e.g. Woodford 1964). After 2,4-D and MCPA had been in use for some years, mecoprop and 2,3,6-TBA were developed to control *Stellaria media* and *Galium aparine* and then dichlorprop and dicamba for control of *Polygonum* spp. Later still ioxynil was introduced to help control mayweeds (*Matricaria* and *Tripleurospermum* spp.). Meanwhile there was a continuous increase in the importance of grass weeds particularly *Avena fatua* necessitating the development initially of barban and tri-allate and later of benzoylprop-ethyl, chlorfenprop, flamprop-methyl and difenzo-quat. However, not all these compounds control other grass weeds such as *Alopecurus myosuroides* and other substituted triazine and urea herbicides have been introduced especially for these.

In maize in the United States of America there have been corresponding changes resulting in increases again in annual grass weeds, particularly *Digitaria sanguinalis*, *Setaria* spp. and *Panicum dichotomiflorum*. The extent to which any or all of these species have become more resistant to the triazine herbicides over the years is not altogether clear but they are fortunately now being controlled by other herbicides such as alachlor. In other parts of the world *Rottboellia exaltata* is now a more serious problem, proving resistant to alachlor as well as the triazines. In soya beans, it is the broad-leaved weeds that have benefited from repeated use of grass-killing herbicides such as trifluralin and alachlor and major problems now include *Abutilon theophrasti*, *Sida spinosa* and *Cassia obtusifolia*. In rice there have been less serious changes in the annual weed flora thanks to the early development of propanil and other herbicides selective against annual grasses. Possible exceptions now include localized problems from red or wild rice, especially *Oryza punctata* in Swaziland (Armstrong 1968) and *Ischaemum rugosum* in West Africa and parts of Asia. There has, however, been a more widespread tendency to change in favour of perennial weeds in rice especially *Scirpus maritimus* in Philippines and Italy, *Eleocharis kuroguwai* in Korea and a number of sedges and other monocots in Japan. In upland crops generally in the tropics, herbicide use has tended to favour various perennials, especially *Cyperus* spp. and this tendency has been particularly pronounced in plantation crops where regular use of paraquat has led to a preponderance of tolerant perennials including *Cyperus* spp., *Borreria latifolia* and *Imperata cylindrica*.

CULTURAL SELECTION

Problems under this heading include those occurring in response to changes in crop husbandry in the widest sense. Improved seed cleaning has for instance resulted in the loss of *Agrostemma githago* as a weed in Europe over the past fifty years. By contrast, the increase in *Avena fatua* can be attributed to various changes in cultural practice such as combine harvesting, cereal monoculture and greater inter-farm trade in straw as well as to herbicide use. Various factors have probably contributed to the increase in the problem of the potato as a weed (Kadijk 1972) but the most important has probably been mechanization of harvesting.

Hammerton (1968) predicted that the increased use of minimum cultivation techniques would tend to result in greater problems from perennial weeds and Cussans (1975) showed that tine cultivation had not resulted in any serious increase of *Agropyron repens* (compared with mould-board ploughing) but that direct drilling (zero tillage) certainly leads to problems not only from *A. repens* and certain perennial broad-leaved weeds but also from some small-seeded annual grasses (*Poa* spp. and *A. myosuroides*) and volunteer cereals. The use of the new short-statured wheat and rice varieties has not often been reported to favour any particular weed species but there have certainly been indications of more serious weed problems generally owing to the reduced competitiveness of the crop. Economic selection is perhaps a form of cultural selection in the broadest sense. For instance *Avena* spp. have suddenly become much more serious in herbage seed crops in the United Kingdom not because of increased occurrence but because these crops are now subject to stringent EEC regulations which result in rejection of all but the cleanest crop samples.

Future changes

At this point I have to suggest that the best predictions are the ones that eventually prove false. Prediction should ideally be self-defeating in that it should be used as the first step in prevention, so I would like to think that we can predict trends and then take action to prevent the new problems arising. The process might therefore be better described as anticipation.

SUDDEN GENETIC CHANGE

The origins of these new problems must be very largely unpredictable depending as they do on random mutation and hybridization. The only type that might be anticipated is the inadvertent hybridization of crops with wild relatives. There is therefore a need to beware of a repetition of the weed beet incident wherever specialized breeding or seed production techniques are used involving open pollination in an area where wild relatives occur. Sorghum, millets and rice could all perhaps be affected.

DISTRIBUTION

Accidental distribution is again almost completely unpredictable but one may usefully consider which species would create the greatest problems if they were to be further distributed. *Echinochloa crus-galli* for instance has so far surprisingly not reached West Africa. *Eichhornia crassipes* does not yet occur in the Volta or Niger river systems in Africa. The Philippines and a number of other countries and many individual islands in South East Asia are still free of *Salvinia molesta* and so are the whole of tropical America and the Caribbean. *Striga hermonthica*, almost certainly the most damaging of this genus of root parasites, does not yet occur in India and no species of this genus yet occurs in the New World except for the localized infestation by *S. asiatica* in the Carolinas in the United States of America. This latter infestation was discovered in 1955 and has been the subject of an intensive research and eradication campaign ever since. Unfortunately it was not noticed until 200,000 ha were infested and eradication is proving an extremely lengthy and expensive operation. Hopefully, with adequate publicity and education the introduction of these and many other major weed species might be detected more rapidly in future, so that eradication is a more practical possibility; but experience in two particular situations is not encouraging. In Eastern Ethiopia there is only one small infestation of *Eichhornia crassipes* in the Awash River which has been known for at least ten years and should not be difficult to eradicate, yet it is still there in spite of strong recommendations for its destruction by many senior advisers both within the country and from outside. Similarly *Eupatorium odoratum* persists in the Botanic Garden of Legon University in Ghana in spite of recommendations for its removal.

CHEMICAL SELECTION

Intra-specific selection of resistant weed strains will tend to occur where there is monoculture involving the use of one class of herbicide repeatedly over a number of years; but with the wider range of herbicides now available the likelihood is rather less than in the past.

So far as inter-specific selection is concerned, the increased use of herbicides in developing countries must inevitably be accompanied by many of the same shifts in weed flora that have occurred in the developed world and we may safely predict a steady increase in the type of problems which *Avena fatua* and *Cyperus rotundus* typify. There are however, two good reasons why the shifts should not be so serious or at least not the same. Firstly there is now a wide range of compounds already available to control most of the known annual weed problems so that combinations and rotations of herbicide can be introduced from the start, which will prevent the changes. Secondly there should be a greater availability of hand labour in most developing countries to ensure some follow-up hand weeding after herbicide use, to remove or reduce the resistant component of the weed flora. Working against these two potential advantages however, are firstly the high cost of the more specialized herbicides, forcing the tropical user to start with the older, cheaper herbicides which will start him off on the same sequence of events as has occurred elsewhere; and secondly, over-enthusiasm for chemicals and a reluctance to use hand labour following investment in herbicide.

Particular problems may result from species which do not occur in developed agriculture and for which none of the available herbicides is selective. Fortunately by chance, herbicides developed for grass weeds of North America or Europe are also effective on species such as *Rottboellia*, *Phalaris* spp. and *Snowdenia polystachya* (a problem peculiar to the Ethiopian highlands) and many others for which no deliberate screening has been attempted. At present I am not aware of individual species which completely defy all available compounds—and even the seemingly impossible problem of wild and red rice in rice crops may be overcome with the aid of herbicide antidotes (Smith 1971, Parker & Dean 1976).

In general, annual weed problems in developing countries are likely to follow a similar pattern of change in response to herbicide use as in Europe and the United States of America but hopefully to a less serious degree. For several reasons the increased seriousness of perennial weeds is likely to be paralleled or even exaggerated in developing countries. Firstly the means for chemical control of perennials are still not so well developed and the more successful treatments are even more costly than those for annual weeds. Furthermore, perennial weeds are not so readily taken care of by hand-weeding and the average tropical farmer lacks mechanical tractor power which the more sophisticated farmer can use as a means of holding perennials in check. In the past I believe the status of *C. rotundus* as the 'world's worst weed' has been exaggerated but it may yet come to deserve that title more fully. In plantation crops selective control of annual weeds by chemicals is easier to achieve and although there have been changes in weed flora resulting from repeated use of the same herbicide (e.g. an increase in *Borreria latifolia* in tea with use of paraquat), such changes have not resulted in acute problems. Perennial weeds do however, present a great threat and the use of chemicals, associated with less intensive manual weeding, has tended to encourage *C. rotundus* and other grass and sedge species, and will probably continue to do so. A particularly serious threat arises from certain vines which climb up the stems of the crop and are not readily killed by herbicide without danger of crop phytotoxicity. These are kept down by regular hand weeding but rapidly escape from herbicide control and tend to smother the crop. The 'strangler vine' (*Morrenia odorata*) is a major pest of citrus in Florida (Tucker & Philips 1974) and *Mikania micrantha* is an outstanding problem in tea and oil palm in Southern Asia. In forestry in Malaysia and Bangladesh, *M. micrantha* is a problem even where herbicides are not yet being used, and it seems likely to increase considerably in importance until improved (biological?) control measures are developed. *Merremia* spp. are apparently causing comparable problems in forestry in the Solomon Islands and New Hebrides (Whitmore personal communication) and a particularly virulent scrambling mistletoe *Struthanthus orbicularis* is serious in Belize (Waller personal communication). These problems seem likely to spread and become even more important wherever there is a tendency for hand labour to be replaced by chemicals.

CULTURAL SELECTION

Prediction or anticipation under this heading depends firstly on prophesying the likely changes in cropping systems in developing countries which in itself is a daunting task. To some extent the changes may parallel those that have occurred in the developed world such as increased farm size and greater mechanization. Collectivization of farms results in such changes and the corresponding changes in weed flora may then be anticipated by analogy with what has already happened in other parts of the world.

The higher price of fuel however, will impose serious limits on the extent and intensity of mechanization with the main consequence that whatever the scale of farming, there must be a tendency to minimize cultivations. In the tropics this is in any case desirable from the point of view of conserving soil structure and organic matter and preventing erosion. Where tropical agriculture is not already improved, cultivations are probably no better than minimal, the traditional animal-drawn plough being equivalent to little more than a superficial tine cultivation. No abrupt change is therefore anticipated in cultivation practices which would lead to greatly increased perennial weed problems. The more likely and most rapid changes are to be expected in crop variety, crop species, use of fertilizer and irrigation. The use of shorter-stemmed crop varieties has already been mentioned as a source of generally increased weed problems but there is now a realization of this among plant breeders who are tending to favour intermediate straw length in the newest varieties. Changes in crop species will occur in response either to market demand or perhaps to ease of cultivation. There is increasing concern about the need for increased supplies of protein-rich grain legumes. This may in due course lead to higher prices for such produce and cereals might then give way somewhat to legumes. This is not however, likely to affect weed problems directly—only indirectly if herbicides are used.

Changes in cropping pattern may happen more rapidly in response to changes in ease of cultivation. In particular it is possible that the ease of controlling weeds chemically in maize will result in an increase in maize growing at the expense of cotton, for example. There will be the danger then of changes towards tolerant weeds, especially perennials as noted in the previous section, associated with the more intensive use of herbicides. The increased use of fertilizer will encourage those weed species best able to benefit from higher fertility. One such species is *Cyperus rotundus* and Okafor & De Datta (1976) have shown how rice yields are more seriously affected by this weed with high levels of nitrogen fertilization. Irrigation can lead to a pronounced change in weed flora and once again *C. rotundus* is a likely beneficiary. On the other hand, where irrigation is used to grow flooded rice crops there will be the opportunity of killing *C. rotundus* and many other weed species by drowning.

Other more complex changes in cropping systems could also lead to changes in weed flora but thanks to the natural conservatism of the tropical farmer, such changes are likely to be slow and are in any case most likely to be introduced as part of elaborate 'packages' of techniques which hopefully will be designed to reduce the risk of new troublesome weed problems. The international research institutes such as I.I.T.A. (International Institute of Tropical Agriculture), I.R.R.I. (International Rice Research Institute) and I.C.R.I.S.A.T. (International Crops Research Institute for the Semi-Arid Tropics) have weed agronomists on their staff, who are devoting particular attention to the problem of weeds in new multiple cropping systems. The possibilities for suppressing weeds in ways other than by herbicides, such as by dense planting and high levels of crop competition, are being seriously studied.

One situation where a package of treatments will be especially necessary is in the transition from shifting cultivation to more continuous cropping. In the past, weeds have been a dominant factor in the traditional system, very often forcing the abandonment of land after a few seasons of cropping. Where more continuous cropping is introduced as a means of increasing food production or because of increased pressure on the land, weed problems will be correspondingly more acute and one of the more important species especially in West Africa, will be *Imperata cylindrica*. Another probable change in cropping system will be from transplanted to direct sown rice, partly as an economy in labour but also as a means of greater productivity. By starting with a dry-sown crop early in the rainy season (before there is enough water for transplanting) two crops may be grown instead of the usual one. Dry-sown rice however, is notoriously weedy and poses serious weed control difficulties whether traditional or chemical control methods are used. The particular danger in dry-sown rice, especially where it is subsequently flooded is from the various types of wild and red rice. Their germination is prevented under wet-sowing or transplanting conditions but they germinate freely under a dry-sowing system and then flourish with subsequent irrigation or flooding. They are already a severe problem in parts of North and South America, Africa and Asia and are likely to benefit from any change from transplanted to dry-sown rice.

New weed species may be discovered simply through the use of new land, not previously cultivated. In general however, it is rare for serious weed problems to be so localized. Clearance of bush or forest may be followed by the troublesome regrowth of particular local species, not found elsewhere, but species with seriously weedy characteristics have usually succeeded in spreading over a considerable area already and are not likely to have been completely missed by the weed inventory. A change of crop can however, lead to the exaggeration of an existing weed problem. In Sudan for instance, sugarcane planted on old sorghum land has been seriously affected by the parasite *Striga hermonthica*. The parasite was present and damaged the sorghum to some extent but it became economically more important on the cane.

Biological control methods are receiving increased attention and will no doubt prove successful for a limited number of weeds in the future. They should not result in direct encouragement of other weeds but will sometimes allow replacement by secondary species. Fears of such replacement have hindered plans for biological control of *Eupatorium odoratum* in Nigeria. Forestry authorities felt that successful removal of *E. odoratum* might lead to increase of *Imperata cylindrica* which could be more competitive and more difficult to remove from young forestry plantings. These objections were eventually overruled and the control programme is now going ahead but it is too early to judge whether *I. cylindrica* will be favoured. Another situation in which the selectivity of a biological control method could have the same disadvantages as incomplete chemical control is in the use of the grass carp (*Ctenopharyngodon idella*) for aquatic weed control. *C. idella* is somewhat selective and is likely to leave certain *Myriophyllum* and *Vallisneria* spp. which could therefore increase and become more important than at present.

Methods for prediction

Locally it should not be difficult to monitor changes in weed population on an annual basis using quadrats or some other sampling procedure. It will not be feasible to use the same, fixed quadrat each year however, as it may be necessary to leave weeds undisturbed (for identification) beyond the time they would normally be controlled, and this interference with normal procedure could influence the long-term pattern. Useful indications should be obtained where the land is cultivated and observed at the same time each year, but if the cropping pattern changes from year to year, results will be less helpful. In the United Kingdom for instance, there are enormous variations in the number of emerged individuals of certain species from year to year depending on whether the land is used for spring cereals or autumn cereals. The exact timing of the rains or irrigation in the tropics could cause corresponding, though smaller variations. The only way to avoid this problem would be to use a soil sampling technique and then force the germination of the seeds in the soil by repeated soil stirring in pots under glasshouse or laboratory conditions (Roberts & Dawkins 1967). At the more practical extreme one might not bother to monitor the potential weed problem in this way but only assess the weeds surviving the normal weed control treatments. The changes in *surviving* weed flora after treatment may be very unrepresentative of the potential weed flora as indicated by buried viable seeds or even seedling emergence, but would give much quicker warning of new problems than either of the other methods.

The ultimate aim could be a full study of the population dynamics and the response of individual weed species to anticipated changes in chemical and cultural practices, involving the monitoring of all parameters of seed production, seed loss and the buried seed reserves. Upchurch (1974) in considering weed control in the 1980's suggests that '. . . computerized modelling programmes could be useful in making projections on rates of spread and the conditions conducive to spread . . .' Such an approach is already being attempted for *Avena fatua* (Cussans 1976) but it is clear from the volume of data and effort that are required that this is hardly a practicable method for deciding which out of fifty or more common weed species is going to be the major problem of the future. It may be justified for a few selected weeds that are already major problems or clearly threatening to become so. It could not be justified for a large number of weeds of marginal importance, least of all in a developing country.

On a more modest level, simple biological studies of individual species which might be suspected of increasing in importance are justified and provide ideal academic exercise for university students. Such studies on life cycles and reproductive patterns may reveal ways in which a minor weed species can be prevented from becoming a major one.

In the all-important sphere of chemical selection the main predictive tools must be hindsight and analogy with situations in the developing world. Where herbicides are introduced on an extensive scale in crops with a mixed weed flora the same changes are likely to occur as have occurred in the developed world. But there is also perhaps scope for more systematic attention to the full weed spectrum of the available herbicides. Susceptibility tables published by chemical companies should always include the minor weed species as well as the major, especially if they are resistant. If the information is then pooled for all the herbicides being used in a particular region, the potentially dangerous minor weeds could then be detected in good time. In developing countries unfortunately the weed flora may not be adequately known and there will be a lack of information on the herbicide susceptibility of all the species. Harper (1965) emphasized 'it is of the greatest importance that occasional failures of previously successful measures for weed control should be critically examined—so that—if or when resistance is demonstrated the pockets of resistance may be isolated and eliminated'. In that paper he was concerned with the occurrence of intra-specific resistance but one could equally well heed his warning in respect of tolerant species.

At present the failure of a herbicide might be brought to the attention of a representative or agent of the company selling the product and possibly of a local extension officer. It may result in compensation of the farmer, who might then switch to another company or product. A different product may then control the problem but only after it has begun to spread. By the time it comes to the notice of a weed scientist, or other official who can recommend methods for effective control, eradication may be very difficult. There is perhaps a case for the appointment of an officer who would be automatically informed of such instances and could immediately appraise whether the incident indicated a potentially dangerous new problem.

Meanwhile greater attention should be paid to general procedures for preventing the changes in weed flora which otherwise may occur with the use of herbicide. Such procedures include the rotation of herbicides with different modes of action (with or without rotation of crops) and the use of treatments which provide complete weed control. Inadequate doses of herbicide providing only partial kill will increase the risk of resistant types surviving and increasing. Most important and most practical especially in the developing countries is the use of supplementary hand or hoe weeding following herbicide use to ensure that most surviving weeds are prevented from seeding. Such additional hand weeding may be too late to result in any economic benefit in the current crop and might be discouraged if the economics of the weeding practices were being looked at on too short term a basis. Extra small expense of this sort however, is likely to be more economic in the long term than a change of weed flora to tolerant or resistant species which require more expensive herbicides and/or more supplementary hand weeding.

Deliberate distribution of weeds on the international level should be prevented by phytosanitary regulations and the movement of soil should be prohibited; nevertheless the speed and volume of modern trade and travel provide ample opportunities for the introduction and dispersal of new weed problems, and occasional inadvertent introductions will occur. By far the most important barrier to the build-up of new problems from such introductions must be the alertness of weed scientists, agronomists and other agricultural officials who should be educated in the weed species of countries other than their own and so be able to quickly recognize important newcomers.

References

ABEL A.L. (1955) The rotation of weedkillers. *Proc. Br. Weed Control Conf. 1954* 1, 249–55.

ARMSTRONG K.B. (1968) Weed control on a Swaziland rice and sugar cane estate. *Proc. 9th Br. Weed Control Conf.* 2, 687–92.

BACHTHALER G. (1967) [Changes in arable weed infestation with modern crop husbandry techniques.] *Abstra. 6th int. Congr. Pl. Prot. Vienna* 167–8.

BACHTHALER G. (1969) [Development of the weed flora in Germany in relation to changes in method of cultivation.] *Angew. Bot.* 43, 59–69.

BAKER H.G. (1974) The evolution of weeds. *A. Rev. Ecol. Syst.* 5, 1–24.

BANDEEN J.D. & McLAREN R.D. (1976) Resistance of *Chenopodium album* to triazine herbicides. *Canad. J. Pl. Sci.* 56, 411–12.

BUCHHOLTZ K.P. (1958) Variations in the sensitivity of clones of quackgrass to dalapon. *Proc. 15th N. cent. Weed Control Conf.* 18–19.

CUSSANS G.W. (1975) Weed control in reduced cultivation and direct drilling systems. *Outl. Agric.* 8, 240–2.

CUSSANS G.W. (1976) Population dynamics of wild oats in relation to systematic control. *Rep. Weed Res. Org.* 1974–75, 47–56.

DICKENS R. (1974) Kudzu: Friend or foe? *Weeds Today* 5, 9.

ELLIS M. & KAY Q.O.N. (1975a) Genetic variation in herbicide resistance in scentless mayweed (*Tripleurospermum inodorum* (L.) Schultz Bip.). I. Differences between populations in response to MCPA. *Weed Res.* 15, 307–15.

ELLIS M. & KAY Q.O.N. (1975b) Genetic variation in herbicide resistance in scentless mayweed (*Tripleurospermum inodorum* (L.) Schultz Bip.). III. Selection for increased resistance to ioxynil, MCPA and simazine. *Weed Res.* 15, 327–33.

FRYER J.D. & CHANCELLOR R.J. (1970a) Herbicides and our changing weeds. In *The Flora of a Changing Britain* (Ed. by F.H. Perring), pp. 105–18. Classey, Hampton.

FRYER J.D. & CHANCELLOR R.J. (1970b) Evidence of changing weed populations in arable land. *Proc. 10th Br. Weed Control Conf.* 3, 958–64.

HAMMERTON J.L. (1968) Past and future changes in weed species and weed floras. *Proc. 9th Br. Weed Control Conf.* 3, 1136–46.

HANSON N.S. (1956) Dalapon for control of grasses on Hawaiian sugar cane lands. *Down to Earth* 12, 2–5.

HANSON N.S. (1962) Weed control practices and research for sugar cane in Hawaii. *Weeds* 10, 192–200.

HARPER J.L. (1956) The evolution of weeds in relation to resistance to herbicides. *Proc. 3rd Br. Weed Control Conf.* 1, 179–86.

HOLLIDAY R.J. & PUTWAIN P.D. (1974) Variation in the susceptibility to simazine in three species of annual weeds. *Proc. 12th Br. Weed Control Conf.* 2, 649–54.

IVENS G.W. (1974) The problem of *Eupatorium odoratum* L. in Nigeria. *PANS* 20, 76–82.

JAYACHANDRA (1971) *Parthenium* weed in Mysore State and its control. *Current Sci.* 40, 568–9.

KADIJK E.J. (1972) The volunteer potato problem in the Netherlands. *Proc. 2nd int. Meet. Selective Weed Control in Beet Crops, Rotterdam 1970* 208–9.

LONGDEN P.C. (1974) Sugar beet as a weed. *Proc. 12th Br. Weed Control Conf.* 1, 301–8.

MITCHELL D.S. (1972) The Kariba weed: *Salvinia molesta. Brit. Fern Gaz.* 10, 251–2.

MUNTZING A. (1930) Über Chromosomenvermehrung in *Galeopsis*-Kreuzungen und ihre phytogenetische Bedeutung. *Hereditas, Lund.* 17, 131–54.

NGUYEN VAN VUONG (1974) Some aspects of the autecology of *Salvinia* spp. Paper presented at *Sth East Asian Workshop on Aquatic Weeds*, Indonesia, July 1974, p. 33.

OKAFOR L.I. & DE DATTA S.K. (1976) Chemical control of perennial nutsedge (*Cyperus rotundus* L.) in tropical upland rice. *Weed Res.* 16, 1–5.

PARKER C. (1973) Weeds in Arabia. *PANS* 19, 345–52.

PARKER C. & DEAN M.L. (1976) Control of wild rice in rice. *Pestic. Sci.* (in press).

PEABODY D. (1973) Aatrex tolerant pigweed found in Washington. *Weeds Today* 4, 17.

ROBERTS H.A. & DAWKINS P.A. (1967) Effect of cultivation on the numbers of viable weed seeds in soil. *Weed Res.* 7, 290–301.

ROCHECOUSTE E. (1962) Studies on the biotypes of *Cynodon dactylon* (L.) Pers. *Weed Res.* 2, 136–45.

RYAN G.F. (1970) Resistance of common groundsel to simazine and atrazine. *Weed Sci.* 18, 614–16.

SALISBURY E. J. (1964) *Weeds and Aliens.* Collins, London.

SIMMONDS F.J. & GREATHEAD D.J. (1977) Introductions and pest and weed problems. In *Origins of Pest, Disease, Parasite and Weed Problems* (Ed. by J.M. Cherrett & G.R. Sagar), pp. 109–24. Blackwell Scientific Publications, Oxford.

SMITH R.J. (1971) Red rice control in rice. *Proc. 24th a. Meet. sth. Weed Sci. Soc.* 163.

THOMPSON L. Jnr., SHUMACHER R.W. & RIECK C.E. (1974) An atrazine resistant strain of redroot pigweed. *Weed Sci. Soc. Amer. Abstr.* 196.

TUCKER D. & PHILLIPS R. (1974) The strangler vine: A major weed pest in Florida citrus groves. *Weeds Today* 5, 6–8.

UPCHURCH R.P. (1974) Weed control in the 1980s. *Proc. 31st N. cent. Weed Control Conf.* 29–31.

WILD H. (1968) Weeds and aliens in Africa: The American immigrant. *Publ. Univ. Coll. Rhodesia*, p. 30.

WONG PHUI WENG (1964) Evidence for the presence of growth inhibitory substances in *Mikania cordata. J. Rubb. Res. Inst. Malaya* 18, 231–41.

WOODFORD E.K. (1964) Weed control in arable crops. *Proc. 7th Br. Weed Control Conf.* 3, 944–64.

Prediction of new storage pest problems

J. A. FREEMAN *Ministry of Agriculture, Fisheries and Food,*
Pest Infestation Control Laboratory, Slough, Berkshire

Prediction of the status of stored product and domestic insects and mites
can be long or short term.

Long-term prediction seeks to estimate how changes in agricultural,
commercial and industrial processes and in the domestic scene are likely to
affect the introduction and establishment of new pests and the status of
existing ones. Short-term prediction is concerned with the way in which
the populations present in a product, store, processing establishment or
domestic dwelling are likely to develop.

The relation between laboratory experiment and prediction

Most stored product insects and mites are relatively easy to culture in the
laboratory under controlled conditions of temperature and relative humidity,
diet, light regime etc. By rearing them at a number of combinations of temperature and relative humidity, it is possible to obtain, for each set of conditions, information regarding, for example, survival, period of development
of each stage, oviposition, mortality and sex ratio. By using ideal diets the
maximum potential of any species can be determined and the results compared with development on products normally attacked (Cox 1975, Le Cato
1976). The basic food requirements may be studied by using diets of carefully
controlled composition (Fraenkel & Blewett 1943, Woodroffe 1965). The
ability of species to resist adverse conditions outside those suitable for complete development must also be determined, e.g. winter survival in Britain
in unheated warehouses (Solomon & Adamson 1955). The results can be
plotted on a lattice diagram to show the limits within which (a) survival is
possible; (b) complete development can occur; and (c) the life cycle is
completed most rapidly or population increase is most rapid; the last two
are not necessarily identical (Fig. 1).

Potential for population increase may be estimated by using the parameter r, the intrinsic rate of increase (the parameter $\pm$ is the logarithm of the self-multiplicative rate of increase (Howe 1953a, b, 1963, 1971)) which takes into account rates of development, oviposition, mortality and sex ratio. It has a disadvantage that it is based on the concept of populations of stable age at each set of conditions which seldom occur in practice.

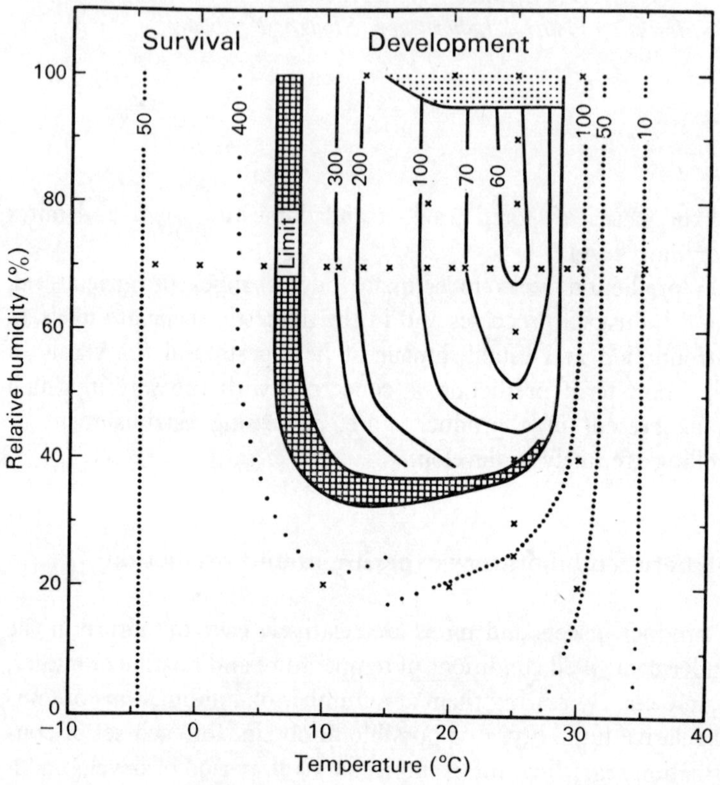

Figure 1. Developmental time (days) and physical limits for survival of *Ptinus tectus* (Howe & Burges 1953).

It is possible to make predictions based on comparisons of physical limit data with conditions in stores and Howe (1963) states: 'If a (developmental) diagram has been constructed for a particular food for a species and we have information about the conditions in a warehouse, it is reasonably easy to predict whether or not the species would flourish if introduced into that warehouse. On such a diagram an optimal zone for development can be marked, although the limits must be chosen arbitrarily; because, in stored product species, the mortality varies little between conditions except in marginal ones. At the same time the limits outside which it will multiply too slowly to have pest potential can be shown. This will probably be 3–5 °C above the ill-defined minimum developmental temperature. Armed with information about warehouses or less surely with summaries of the outdoor meteorological conditions in an area, one can now make intelligent first guesses as to the potential distribution and importance of this pest on this food in various parts of the world.'

A technique which can be used is to superimpose on a lattice diagram a climatograph representative of the conditions likely to be experienced, and see the extent to which these match. By this and other methods the distribution in the world of a number of stored products insects and mites has been explained and predictions made regarding their potential for spread. This has been done for the flour mite,* *Acarus siro* L. (Cunnington 1965); for beetle and moth pests in flour mills (Freeman 1962); for the cigarette beetle, *Lasioderma serricorne* (F.) (Howe 1957), and for the Australian spider beetle, *Ptinus tectus* Boield. (Howe & Burges 1953).

Ptinus tectus, which does best under temperate climatic conditions, was first recorded in Great Britain in 1892 (Chitty 1904), probably reaching this country from Tasmania at a time when increasing speed of ships enabled it to survive the adverse conditions of the voyage through the tropics. It is now one of the major warehouse pests in this country. The application of laboratory data enabled Howe & Burges (1953) to explain its existing world distribution and to predict that it should be found in certain mountainous areas in the tropics where its presence had not already been recorded (Figs. 2 and 3). In 1962 the species was found fairly widespread in stores in the Highlands of Kenya (Coombs unpublished).

A stored product species with a potential for wider distribution is the nut moth, *Paralipsa gularis* (Zell.), which in this country is particularly associated with the processing of almonds. It is distributed in the northern hemisphere and occurs on imports of rice from Italy, hazel nuts from Turkey and walnut kernels from France. Smith (1956, 1960, 1961, 1965) has shown that the insect can develop between 15 °C and 35 °C with an optimum near 31 °C. There is a diapause in the final larval stage which is resistant to cold. The species is therefore similar in biology to the warehouse moth, *Ephestia elutella* (Hübn.) and it is reasonable to suppose that eventually it will extend its range from a solely northern hemisphere distribution to suitable areas in the southern hemisphere.

Another stored product insect which has a potential for spread is the khapra beetle, *Trogoderma granarium* Everts., which breeds most rapidly under hot, dry conditions. It became a serious pest of groundnuts in Northern Nigeria in 1949 (Howe 1952), and got a foothold in western North America in the 1950's, spreading from its original Middle-Eastern home. It also

* Common names follow according to Thomas, Janson & Aitken (1968).

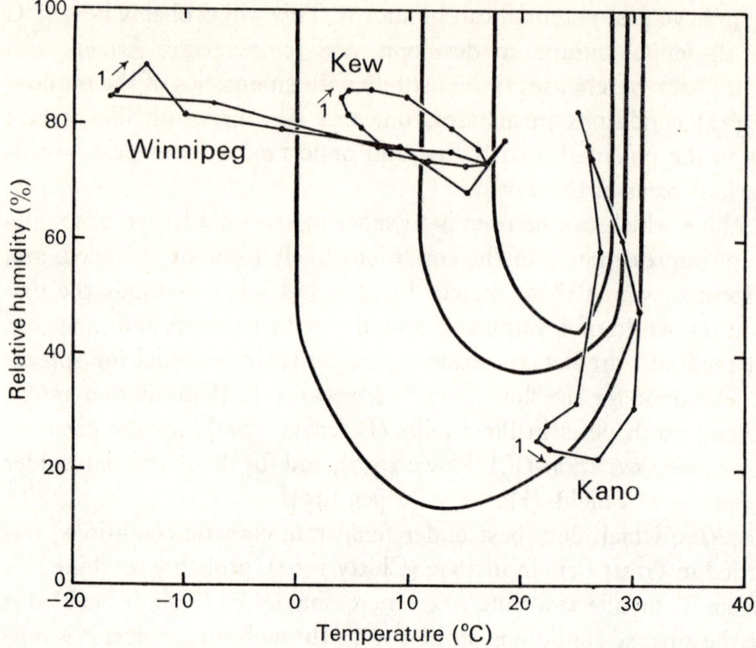

Figure 2. The relation of climate to the survival and development of *Ptinus tectus*. Heavy lines enclose three zones, an inner of rapid development (100 days), a middle line for 300 days and outer conditions lethal in 100 days. The three climatographs are for mean daily temperature and relative humidity for each month of the year (Howe & Burges 1953).

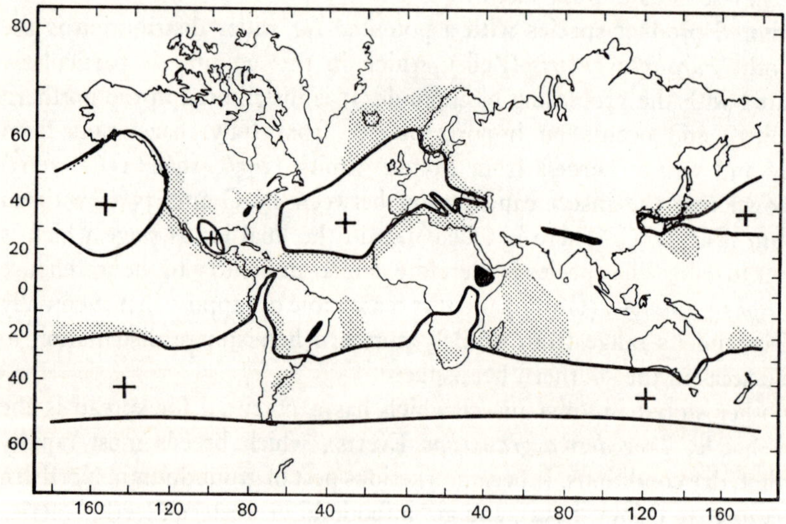

Figure 3. The theoretical range of *Ptinus tectus* (marked + or shaded black) as estimated from laboratory data. Doubtful areas are stippled. (Howe & Burges 1953.)

invaded East and South Africa somewhat later. Howe & Lindgren (1957) and Howe (1958) discussed its distribution in relation to its potential spread in North America in the 1950's, and Howe (1963) its world-wide distribution. It was eradicated from the western United States of America and Mexico in the late 1950's and early 1960's and no other established infestations in the Americas are known at the time of writing in 1976, nor is it present in Australia. That it is still spread about the world in international trade is shown by its regular interception on oilcakes and other products imported into Great Britain from Burma, Sudan, Senegal, India and Nigeria (Freeman 1974).

Whereas this distribution can be explained in relation to laboratory data on limiting and optimum physical conditions, the ability of *Trogoderma granarium* to succeed in competition with other species may be assessed by plotting diagrams showing the conditions, based on the parameter r, for optimum rates of population increase for various competing species. By this means one can predict which species is likely to be dominant under particular microclimates.

The grain weevil, *Sitophilus granarius* (L.), the rice weevil, *Sitophilus oryzae* (L.), the lesser grain borer, *Rhyzopertha dominica* (Fab.) and the khapra beetle, *Trogoderma granarium*, respectively, become dominant under conditions of competition as one passes from cool, damp conditions, as in temperate countries, to those which are hot and dry (Fig. 4) (Howe 1963). It does not follow however, that single species dominance necessarily occurs in practice. Provided the conditions are not extreme, i.e. only possible for *Trogoderma granarium*, the occurrence of a particular pest population will depend on the species which, by chance, happen to be present at the time the grain is put into store, on what happens subsequently (e.g. spontaneous heating or not), or even on the orientation in relation to the sun, of the grain store. Thus in Iran, wheat which had been stored on the ground floor of a silo was infested heavily by *Sitophilus granarius, Rhyzopertha dominica, S. oryzae* and *Trogoderma granarium*, each species being dominant in a different part of the bulk. *T. granarium* was most numerous in the southeast corner, which received most heat from the sun (Freeman 1958).

Although the climatic conditions in Rangoon would appear to favour the development of *Sitophilus oryzae* rather than *Trogoderma granarium* on stored rice, it was the latter species which caused most damage during long periods of storage in the 1950's. The microclimate in stores may have been drier and hotter than indicated by standard meteorological observations.

These examples indicate that broad conclusions derived from comparisons of limiting conditions and potential rates of increase with standard meteorological observations of those areas of the world in which insects and mites may establish themselves, do not necessarily apply in particular cases.

It is the microclimatic conditions in the places where the insects and mites live which determine their ability to survive and build up to pest proportions. Our knowledge of such microclimates in stores and commodities is fragmentary.

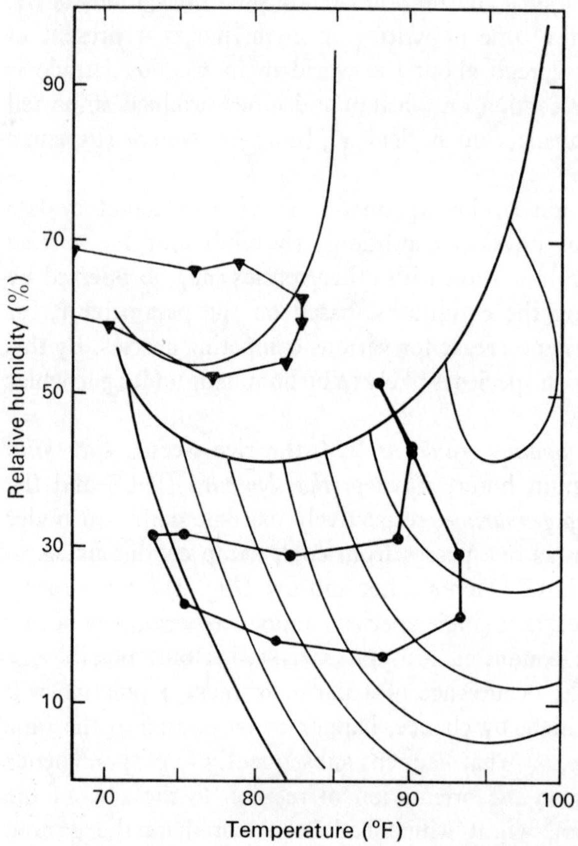

Figure 4. Climatographs (mean daily temperature and relative humidity for each month of the year) for Cairo (▲) and Khartoum (●), superimposed on a lattice chart showing the most favourable temperature humidity zones for the breeding of (from top left to bottom right) *Sitophilus oryzae*, *Rhyzopertha dominica* and *Trogoderma granarium*. (Howe 1958.)

Conditions tend to be more variable in buildings (being more directly affected by external changes such as the effects of wind, sun etc.) than in bulk grain or stacks of bagged goods where changes occur very slowly. The properties of grain are such that no measurable change in temperature from summer to winter is likely to occur in the centre of a bulk of grain 18 ft (5·5 m) square. With increasing depth the advancing temperature wave is so slowed by bulk wheat that at 5·5 ft (1·7 m) it will be three months and at 11 ft (3·3 m) six months behind external conditions (Oxley 1948).

In heated buildings differences from outside conditions may be even more marked, especially when any processing such as flour milling produces heat. Dyte (1965) showed that temperatures inside certain machines in flour mills fluctuated some 10–17 °C above the monthly mean minima of the standard readings taken at the nearest meteorological station. Often such standard readings represent the only available source of information. Information relating the outside atmosphere, the climate within the store, and the type and condition of produce, particularly in the tropics, has been summarized by Ward & Calverley (1972).

Spontaneous heating caused by the activity of microorganisms in damp pockets of grain or by insect metabolism may invalidate predictions by creating conditions suitable for the development of insect populations whose minimum developmental temperature would otherwise be too high for successful population increase in unheated stores (Howe 1962, 1965). By a combination of heating and protection inside stacks of bagged grain, species such as *R. dominica* and *S. oryzae* have survived the winter in Britain; the rice moth, *Corcyra cephalonica* (Staint.), the rust red flour beetle, *Tribolium castaneum* (Herbst.) and the copra beetle, *Necrobia rufipes* (Deg.), have been observed to breed in midwinter in Liverpool in groundnuts heating by insect infestation (Freeman 1950). Since temperatures within stacks or bulks of produce stored in unheated warehouses in this country rarely exceed 15 °C for very long, these increases of temperature due to spontaneous heating play an important part in encouraging insect development and in making prediction difficult.

A difficult question to determine in practice is the minimum population necessary to establish an infestation within the likely period of storage. Surtees (1965) showed that individual male and female saw-toothed grain beetles, *Oryzaephilus surinamensis* (L.) could find one another in a bag holding about 20 kg English wheat kept at 25°C and 70% relative humidity, the time to fertilization of the female being from four to twelve weeks. In large bulks a concentrating effect was observed since, moving more or less at random, any beetles present would tend to concentrate in the warmest and dampest parts of the bulks, with the consequent risk of numbers becoming sufficient to induce insect heating.

In general the smaller the initial population the longer will be the period before the insects are noticed or give rise to appreciable populations. Thus it took some three years for populations of the warehouse moth, *Ephestia elutella* (Hübn.), to build up to epidemic numbers in reserve stocks of wheat put into store in warehouses in 1938. Spectacular migrations of larvae producing sheet webbing on the surface of bulks and the walls and ceilings of buildings, took place in 1941 (Freeman 1948), causing considerable alarm to the storekeepers who had not experienced such infestation in the past in commercial stocks normally stored for short periods. Such infestations were experienced during World War I (Dendy & Elkington 1919), although similar occurrences were familiar to those storing cocoa, dried fruit and tobacco during the 1920's and 1930's (Munro 1966). These outbreaks and that of *Oryzaephilus surinamensis* in imported dried fruit in 1941 (Freeman 1948) emphasize the fact that stored product infestation problems often occur when produce is left in store for much longer periods than has previously been normal commercial practice.

Studies of the development of *Ephestia elutella* populations in a London warehouse gave useful guidance in estimating the risks of outbreaks and in the timing of control measures (Richards & Waloff 1946, Waloff & Richards 1946). The use of trap bags (Pinniger 1975), trap boxes and trap spears (Loschiavo 1975a, b, Loschiavo & Atkinson 1973) can give better estimates of initial populations than visual inspection, but it must be remembered that birds' nests may act as reservoirs of infestation by mites such as *Glycyphagus domesticus* (Deg.), and insects such as *Ptinus tectus* and *Hofmannophila pseudospretella* (Staint.), serious pests of flour (Woodroffe & Southgate 1951, Woodroffe 1953).

There are therefore two main difficulties in making short-term predictions. The first is ascertaining the initial pest population, the second particularly in unheated buildings, is knowing the physical conditions likely to occur during the period ahead, including the risk of spontaneous heating. These are perhaps why many of those responsible for storage prefer to play for safety by using chemical or physical protective measures as routine, even if no pest species can be detected.

Predicting changes in the degree and type of infestation of imported products

Many of the stored products infestations in this country occur because infested commodities are imported from abroad, mainly by sea but also by air. The laboratory work already described enables us to forecast the likely development of imported infestations and to advise action appropriate to the time of year. For example, the fumigation of cocoa beans infested by *Ephestia cautella*, *Lasioderma serricorne* and *Tribolium castaneum* and imported in the autumn is not recommended because winter conditions in unheated warehouses will normally check the increase in populations and eventually kill the insects; similar lots imported in early summer must be treated for safe storage.

For many years there has been virtually unrestricted entry of food, feeding-stuffs and raw materials for industrial processing, bringing with them large numbers of many species of storage pests. During and since World War II this entry has been monitored by the regional staff of the Pest Infestation Control Laboratory of the Ministry of Agriculture, Fisheries and Food and the corresponding organization of the Department of Agriculture and Fisheries for Scotland. On the basis of the information obtained, pressure has been brought to bear on countries overseas to control infestation at source and information has been provided, either directly or through overseas technical aid, to assist those without their own control facilities. The number of cargoes seen is about 7,500 annually and includes cereals, pulses, cereal and pulse products, oilseeds, oilseed products, spices and drugs, cocoa beans and coffee, confectionery nuts, dried fruit and various materials for technical processing.

Many species of insects are intercepted each year, but only some seventeen species (fourteen beetles and three moths) occur regularly (Table 1).

Table 1. Interception of stored product insect pests on commodities imported into Great Britain between 1956 and 1974.*

	1956–60	1961–65	1966–70	1971–74
Average no. of cargoes seen per annum	7,773	7,855	7,183	7,150
% infested	48	39	31	22
Rate of occurrence per 100 cargoes inspected (Rank order of magnitude in parenthesis)				
Tribolium castaneum	291 (1)	221 (1)	172 (1)	102 (1)
Ephestia cautella	251 (2)	193 (2)	140 (2)	98 (2)
Dermestes maculatus	15 (13)	18 (11)	31 (4)	23 (3)
Necrobia rufipes	64 (5)	49 (3)	35 (3)	22 (4)
Lasioderma serricorne	51 (6)	33 (7)	26 (6)	19 (5)
Oryzaephilus mercator	65 (4)	46 (4)	23 (7)	18 (6)
Corcyra cephalonica	47 (8)	36 (5)	30 (5)	16 (7)
Alphitobius diaperinus	26 (11)	22 (8)	13 (9)	11 (8)
†*Oryzaephilus surinamensis*	67 (3)	35 (6)	17 (8)	9·5 (12)
Dermestes frischii	7 (17)	6 (15)	9 (12)	8·5 (10)
†*Plodia interpunctella*	36 (9)	17 (12)	11 (10)	8 (11)
†*Sitophilus granarius*	21 (12)	6 (15)	5 (16)	7·5 (12)
Tenebrioides mauritanicus	36 (9)	15 (13)	8 (14)	7 (13)
Sitophilus oryzae	51 (6)	20 (10)	9 (12)	7 (13)
Trogoderma granarium	15 (13)	21 (9)	11 (10)	7 (13)
†*Ahasverus advena*	15 (13)	7 (14)	8 (14)	6·5 (16)
†*Cryptolestes ferrugineus*	11 (16)	6 (15)	3 (17)	4 (17)

* Based on inspections carried out by Regional Infestation Control Advisers of M.A.F.F. and corresponding staff in D.A.F.S.

† May be troublesome in unheated stores, all insects listed being potential pests in heated premises, or where processing produces heat.

Only five of the most commonly introduced species (four beetles and one moth) are cold-hardy and capable of breeding in unheated stores, but there is always the risk of introducing less common species with pest potential. Thus, between 1957 and 1969 some 440 species of beetles were recorded on imports, amongst them a number of potential pests (Aitken 1975).

The success of the policy outlined above is evident because since 1956, the earliest year for which the observations have been analysed in detail, the incidence of infestation (as measured by the percentage of cargoes in which living insects or mites were found) has declined fairly steadily from 51% in 1957 to 22% in 1974 (Figs. 5a and b).

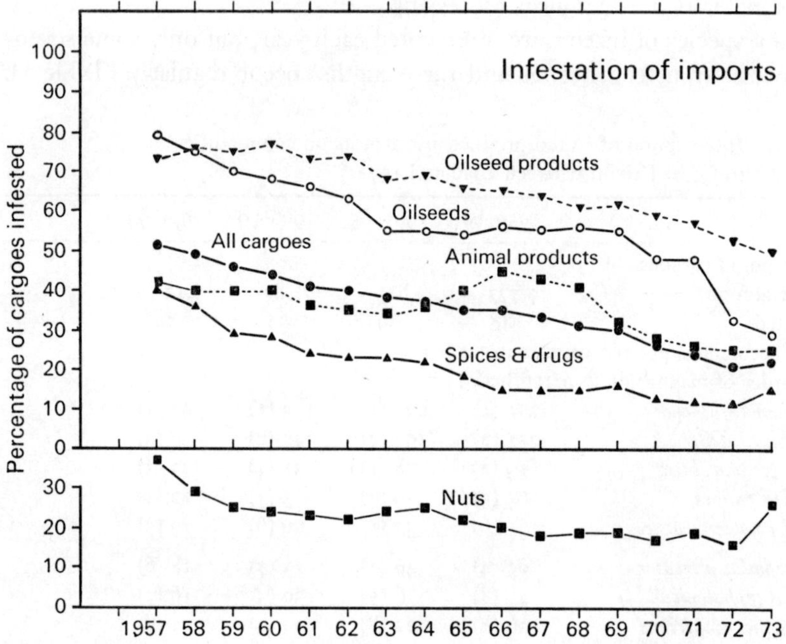

Figure 5(a).   Infestation of imports into Great Britain 1956–1974. Three-year running average of the percentage of cargoes in which living insects or mites were found. (Modified from Freeman 1974.)

This overall reduction is the result of integrating the rates for different commodities and different countries of origin. There are some where the rate of reduction is low, e.g. oilseed products, which may not be free from infestation for many years, whereas it should not be more than a few years before cereals are virtually pest free. Again, some individual commodities still have a very high incidence, for example, groundnut oilcake and cocoa beans from Nigeria and bones from India (Table 2), whereas others are virtually free, e.g. wheat from Canada and Australia.

It is possible to predict for particular groups of commodities the period needed to reach any given percentage incidence of infestation, on the assumption that the rates of decline of recent years will continue. However, complete freedom of commodities from infestations is unlikely to be achieved for commercial reasons as well as for technical ones such as lack of treatment in the country of origin, cross and residual infestation in transport, and failures of control methods. Something around 5% may be the eventual practical level, with most of the infestations that occur being in the 'light' category.

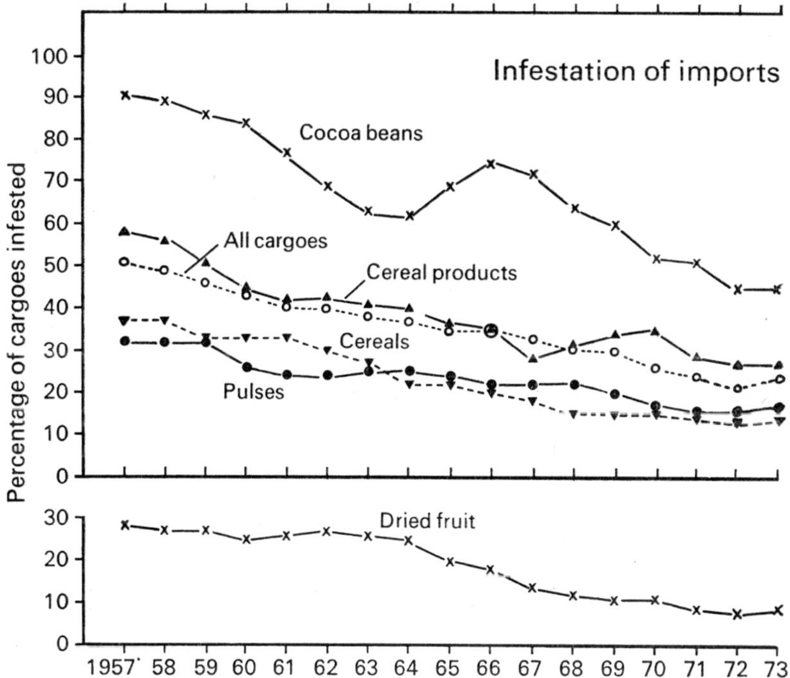

Figure 5(b). Infestation of imports into Great Britain 1956–1974. Three-year running averages of the percentage of total cargoes and of those of oilseed products, animal products, spices and drugs and confectionery nuts in which insects or mites were found. (Modified from Freeman 1974.)

The reductions in incidence of infestation are the result of a number of factors, mainly the quality requirements of importers influenced by rising consumer standards, and public health requirements. There are also some countries which apply plant quarantine regulations against storage pests in agricultural products for consumption or industrial processing, as well as to materials for sowing or propagation. The khapra beetle (*Trogoderma granarium*) into the United States of America, East Africa and Australia and virtually any stored grain pest into the Peoples' Republic of China are

Table 2.   Infestation of cargoes imported into Great Britain.*

	Per cent infestation						Annual rate of reduction of % infestation[a]	Rate of infestation when x = 0 (notional in year 1955)	Predicted year of nil infestation[b] (y = 0)
	1956	1956–60 (mean)	1961–65 (mean)	1966–70 (mean)	1971–74 (mean)	1974			
All commodities	51	48	39	31	22	22	1·77	53	1985
Cereals	36	36	27	17	13	14	1·6	40	1980
Dried fruit	26	27	24	13	9	13	1·3	32	1980
Oilseeds	83	74	60	55	31	25	2·7	83	1986
Cereal products	58	53	41	34	22	25	1·7	56·5	1988
Rice	49	45	43	33	22	15	1·6	53	1988
Nuts	47	32	24	19	23	12[d]	0·7[c]	26[e]	1993
Cocoa beans	91	88	66	67	43	39	2·7	94	1990
Pulses	30	30	24	20	17	18	0·9	32	1991
Animal products	45	41	37	35	25	26	1·0	45	2000
Oilseed products	66	73	70	64	52	49	1·4	80	2012

[a] Slope of line of best fit by least squares method. Base line for x = 0 is notional year 1955.

[b] Year of no infestation assuming the trend 1956–74 continues unchanged.

[c] Period 1959–71.

[d] 1971.

[e] 1958.

* Based on inspections carried out by Regional Infestation Advisers of M.A.F.F. and corresponding staff in D.A.F.S.

examples. It was the refusal of the Chinese to accept wheat from Australia infested with grain insects which induced the authorities there to carry out an extensive clean-up of stores, to arrange for all grain to be treated with malathion on arrival into store, and to apply nil tolerance of living insects in export grain. The effect of this on imports of Australian wheat into Great Britain was to reduce average incidence of infestation from 85% of cargoes inspected to less than 10% over a period of six years (1960–66) and to keep it at that level since (Freeman 1974).

A similar effect on the incidence of infestation in Greek currants resulted from the adoption of compulsory fumigation before export, following complaints from the British dried-fruit importers (Freeman 1974) (Fig. 6).

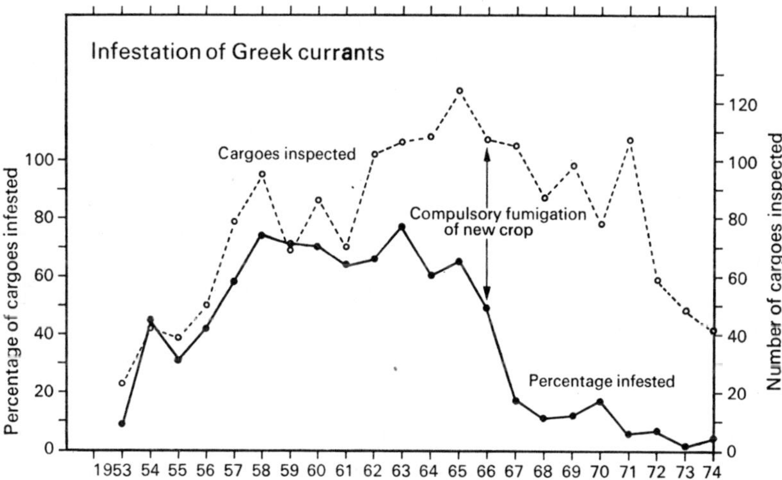

Figure 6. Infestation of Greek currants imported into Great Britain 1953–74. Percentage of cargoes of dried fruit in which insects or mites were found. (Modified from Freeman 1974.)

The observations up to 1966 relate almost entirely to cargoes brought more or less directly from ports overseas in conventional ships. These were discharged at a relatively small number of large ports (e.g. London, Avonmouth, Liverpool, Manchester, Hull, Glasgow, Leith), and the cargoes could be economically and effectively monitored there and action taken to deal with any severely infested lots. Since that time there have been four developments which have made the introduction of infested commodities easier and the consequent risk of established new pests greater:

1. Trans-shipment of cargoes in continental ports into conventional coasters or dry-cargo containers. 2. Long distance roll-on, roll-off lorries bringing susceptible commodities rapidly from as far away as Iran. 3. The use of dry-cargo containers carried in specialized ships. (Container inspection has

increased from nil in 1967 to about 3,000 in 1975, 54% of all import inspections.) 4. Carriage of barges by sea from one country to another. These barges, loaded inland, are floated into special ocean-going ships and at the importing country are floated off again and move inland along the canal systems.

With all these developments there is a tendency for a much larger number of smaller ports to be used, and because of the greater rate at which goods pass through the system, detailed inspection similar to that done for a conventional ship is impossible unless it is done at the place where the container is eventually unloaded. The advantages and disadvantages of the container have been summarized by Freeman (1967) and Smith (1974). The main advantage is that a container can be fumigated and the commodity is then protected from reinfestation. The disadvantage is that if the commodity is infested the infestation instead of being detected and dealt with in the port may be carried into the food factory or warehouse inland. The container, with residual infestation, may then carry some infestation elsewhere.

Wider use of the container provides an opportunity to control infestation in primary products in the country of origin. The roll-on, roll-off lorry presents greater opportunities for rapid diffusion of pests. Another favourable effect of containerization has been the replacement of many old warehouses, full of crevices harbouring residual insect and mite populations, by modern depots, the container acting as a movable warehouse. There is so far, no particular change in incidence of infestation which can be linked with the increased use of containers.

A development which has led to an appreciable increase in imports of spices and of unusual African and Asian foods into this country, has been the need to provide supplies for Chinese and Asian restaurants and for immigrant populations from various countries. Since these commodities are often infested, and are imported in small lots which do not pass through the hands of traditional wholesalers, there exists a risk of widespread introduction and establishment of potential domestic pests such as the Dermestid beetle *Attagenus fasciatus* (Thunberg) (= *Attagenus gloriosae* (F.)), which is especially common on oilcakes and myrobalans from India. It is known as a household pest in Sweden (Aitken 1975).

Although there is no evidence that air traffic has produced serious spread of stored products insects and mites, it is undoubtedly implicated in the spread of cockroaches. One major British airline finds it necessary to treat all aircraft on a routine basis every six weeks (Bailey 1972). The risks of introduction of cockroaches through airports are shown by the following examples:

1. The first difficulties in controlling dieldrin-resistant cockroaches (*Blattella germanica* (L.)) in Britain occurred in 1960 in buildings at an inter-

national airport near London (Green *et al.* 1961). As dieldrin resistance was common at that time in the United States of America, it is possible that the cockroaches had been introduced, although the resistance could have developed in this country since dieldrin had already been used at the airport. 2. Two established infestations of the cockroach *Periplaneta brunnea* (Burm.) were found at the same airport, the first in 1964 (Bills 1965), the second in 1966 (Bills 1967). This cockroach is common in the southern United States.

Changes in industrial, agricultural and domestic practices affecting infestation

Changes in industrial, agricultural and domestic practice ashore similar to those which have occurred in international transport normally take place in response to technical or economic factors and are seldom taken deliberately to prevent infestation. All too often they encourage it, as in a flour mill whose bare joists were enclosed with hardboard for decorative reasons. It all had to be torn down a few months later because the dust collecting in the spaces inaccessible for cleaning proved excellent breeding places for the confused flour beetle, *Tribolium confusum* J. du V.

Whilst the information may be available from past experience or current research, unless the flour miller, the silo constructor or farmer seeks advice in advance of making technological changes, the first news of the change to the entomologist is when the new problem comes to light. Since food storage and processing are carried out indoors, the existence of trouble can be kept secret, unlike outbreaks of pests in a farmer's fields. For these reasons forecasting the effects of technological changes is difficult and their nature is best illustrated by some examples from the past.

FLOUR MILLING AND FLOUR STORAGE

The system of reduction milling using steel or porcelain rollers for grinding wheat, which replaced grindstone milling, was perfected in Austria by about 1875, and since it produced higher yields of whiter flour, it spread rapidly through the world during the following fifteen years. No serious pest problems appear to have affected the grindstone milling process, possibly because of the high temperatures reached during grinding and the cooling to which the meal was subjected between grindings.

In 1877 *Ephestia kuehniella* Zeller, which is almost certainly of Mediterranean origin, was recorded as a pest in a mill in Germany, and as roller milling spread, so infestations by this species occurred—in England in 1887

K

in imported flour (Klein 1887), and in a mill in 1889 (Ormerod 1890); in Canada in mills in Ontario in 1888 (Bryce 1889) and in the United States (California) in the same year (Johnson 1895).

The damage done is principally the webbing together by the caterpillars of the floury stocks passing through the machines, reducing their efficiency and eventually blocking the chutes. The webbed material also acts as a breeding place for many species of beetles.

Subsequent changes in milling technology which have tended to reduce infestation have been the use of synthetic fibres instead of susceptible natural silk in screens; substitution by metal or plastics of wood in machines and chutes; the introduction of conveying by pneumatic power instead of bucket conveyors, thereby eliminating one reservoir of *E. kuehniella*; and the storage and delivery of flour in bulk, or in multiwall paper sacks which replaced the returnable flour sack which was an important means of reintroduction of pests into mills. Sacks also transferred pests between mills, via the baker's loft which contains flour from several sources. During recent years *Tribolium castaneum* has tended to replace *T. confusum* as one of the principal beetle pests in the milling machinery of some British mills. *T. confusum* is cold-hardy and has a lower minimum temperature for complete development (about 17 °C) and a lower optimum range (30–33 °C) than *T. castaneum* which is cold-susceptible, with a minimum of 20 °C and an optimum of 32–35 °C (Howe 1965). It is thought that changes in milling practice, including longer periods of continuous milling, installation of central heating and milling at a lower grain moisture content, have raised temperatures inside the machinery and so favoured the development of *T. castaneum* (Kennedy 1973). This change is in line with observations made by Freeman (1962) who showed that in mills in temperate countries *T. confusum* and *Ephestia kuehniella* occurred most frequently, whereas in warmer countries *T. castaneum* became more common, and in hot dry climates only beetles occurred. In hot damp climates *E. kuehniella* was replaced by the rice moth *Corcyra cephalonica* (Staint.).

A slight change of mean temperature in a mill could result in the complete elimination of one species of *Tribolium* in favour of the other (Park 1954). Similarly, the maintenance of a temperature above the developmental or survival minimum may enable a species to survive and multiply, which it could not do so previously.

Infestation by *Tribolium confusum* of imported flour stored for long periods in Great Britain during World War II occurred only when it was stored in cotton mills heated (to preserve the machinery) to temperatures between 15 and 21 °C, and not in similar flour kept in unheated stores (Freeman 1976). Flour is not normally stored commercially for long periods, passing nowadays mainly in bulk from mills to bakeries where it is used within a few days. Security stocks are however held by the government and infestation by mites (mainly *Acarus siro* L.) has been prevented by requiring it to be milled to a moisture content not exceeding 13% and then maintaining this low moisture by enclosing the jute bags in polythene outers. This decision was based on laboratory work which showed that whereas *Acarus siro* can breed at temperatures between 0 and 5 °C, it cannot do so at a relative humidity less than 62·5%, the equivalent of 13% moisture content of flour (Cunnington 1965). This moisture content however is not low enough to prevent development of *Ephestia kuehniella* brought with the flour from the mill, but such infestations, being contained within the polythene outers, are readily controlled by localized fumigation (Smith 1969a, b). *E. kuehniella* can breed in flour which is virtually free from moisture, apparently obtaining water by breakdown of carbohydrate (Fraenkel & Blewett 1943, 1944).

MALTINGS

The traditional technique of malting barley has involved drying the germinated barley over a kiln to a moisture content of less than 3% and keeping the malt dry until required. This has been achieved in the past by storing it in chambers surrounding the kiln. In such chambers temperatures range from 40 °C near the kiln to less than 20 °C against the outside walls.

Such conditions provided good protection against most existing pests especially *Sitophilus granarius*, but when the khapra beetle, *Trogoderma granarium*, was introduced into maltings, probably on barley from the Indian sub-continent between 1908 and 1918 (Walker 1917, Mason 1921, Miles 1928), it found ideal conditions for rapid reproduction (temperatures between 30 and 40 °C). The larvae, which penetrate deep into cracks in buildings especially in diapause, are cold-hardy and resistant to starvation, so that once established *Trogoderma granarium* is very difficult to eradicate (Burges 1959). It has become much less important recently in this country because malt is now more commonly stored in silos or in other stores away from the hot kiln (Hunter, Tulloch & Lambourne 1973). Since the malt cools rapidly to temperatures less than 21 °C there is no risk of development of *Trogoderma* and its low moisture content protects it from attack by other species.

GRAIN STORAGE

Before 1939, most grain was harvested by reaper and binder, dried in the stook in the field and stored in the corn-stack. Thrashing took place as required during autumn and winter, the sound grain being delivered off the farm in sacks, the tailings being stored in the farm granary and used for feeding livestock.

After 1945 there was a rapid change to combine harvesters, use of grain driers and storage of grain in bulk on the farm mainly in silos, but also in bulks on floors.

Grain stored in the stack suffered considerable damage from attack by rats and mice, but none of the grain storage insects could live there (Munro 1940), infestation being normally confined to tailings, and principally caused by the granary weevil, *Sitophilus granarius*.

The saw-toothed grain beetle *Oryzaephilus surinamensis*, now the major insect pest of stored grain in Britain, has been established here at least since Roman times, having been found, with other stored product species including *Sitophilus granarius* and *Tribolium castaneum*, in excavations of Roman villas (Osborne 1971, 1973). *O. surinamensis* is recorded as a 'constant inhabitant' of farms by John Curtis (1883), writing between 1841 and 1857. It is however not mentioned in any of Miss Ormerod's reports (1877–97). Munro (1940) records its occurrence on farms, mainly in association with imported feeding-stuffs.

Both *O. surinamensis* and *S. granarius* are cold-hardy (Solomon & Adamson 1955), the former requiring a minimum of 18 °C for complete development and 21 °C for at least annual doubling of numbers (Howe 1956, 1965), whilst *S. granarius* needs 12 °C and 15 °C respectively.

In considering the problem of storing home-produced grain in Britain, the following points should be kept in mind.

1. Grain is normally harvested at such a high moisture content (upwards of 20%) that artificial drying is needed for safe storage.

2. Although grain is harvested at a time of year when daytime temperatures are high, provided it is properly dried and is not infested, it should cool naturally because the trend of mean temperature is downward and so it should remain below the developmental minimum of most storage insects for many months.

3. Most of the crop is disposed of before the following harvest, so that only a small proportion remains in store until the following summer when rising air temperatures provide conditions suitable for insect attack (Table 3).

Grain dried by hot air and not properly cooled, may go into store at temperatures of 25–35 °C (Armstrong & Howe 1963). Grain which does not need drying may come off the field at ambient temperatures of 20 °C (Wayman 1969). Large bulks of grain cool slowly, but isolated bins of 50 t capacity and 20 ft (6 m) deep may cool from 26–32 °C to 20 °C in about 10–13 weeks and reach a safe temperature of say 15 °C by early winter (Armstrong & Howe 1963). Grain harvested by combine, dried and stored without being screened or aspirated may contain broken and damaged grains and weed seeds and these are readily attacked by the larvae of *O. surinamensis* and rust red grain beetle, *Cryptolestes ferrugineus* Steph.

Apart from heat from the drying process, there are at least three other sources; the effect of sunlight on the storage buildings; spontaneous heating due to development of fungi and bacteria in pockets of moist grain (16% moisture content and over) and spontaneous heating due to insect metabolism (Howe 1962).

Table 3. Monthly stocks of grain on British farms 1974–1975.*

	Wheat (000 tons)		Barley (000 tons)	
Grain produced by 1974 harvest	5,950		8,950	
Quantity stored on farms at end of:		% of harvest		% of harvest
September	4,890	82	5,870	66
October	4,500	76	5,530	62
November	3,910	66	5,080	57
December	3,526	59	3,526	39
January	2,840	48	3,390	38
February	2,378	40	2,871	32
March	2,000	34	1,799	20
April	1,430	24	1,150	13
May	1,036	17	753	8
June	570	10	490	5

* Source: Grain Bulletin Vols XXI and XXII, Commonwealth Secretariat, London.

A bin of warm grain in which insects are present at the time of storage thus becomes the arena for a race with time. To establish itself *O. surinamensis* must build up a population sufficiently large to cause spontaneous heating before natural cooling of the bulk brings the temperature sufficiently low to arrest development. Armstrong & Howe (1963) estimated that if the initial temperature were 35 °C and the temperature remained about 25 °C for three months, an initial population of 500 *O. surinamensis* could multiply to 5×10^6, prevent cooling and initiate spontaneous heating. This situation is different from that described earlier regarding outbreaks of *O. surinamensis* on dried fruit, where storage during summer months enabled populations to increase.

The demonstration of the need for some source of heat to enable *O. surinamensis* populations to establish themselves led to a further technological change in techniques of grain storage. In-bin drying with warm air was already an established practice and the same principle was used in the development of systems of aeration with unheated ambient air aimed at cooling grain to temperatures at which insect development could be prevented or, at least, reduced to levels at which damaging rates of population increase could not occur. Burges & Burrell (1964) showed that if the temperatures of grain bulks of moisture content not exceeding 14·5% can be reduced to not more than 17 °C (or a temperature at which any species takes more than 100 days to develop) it should be safe against insect attack. A reduction to 12 °C is even safer. They set out in graphical form (Fig. 7) the conditions of temperature and moisture content necessary for safe storage.

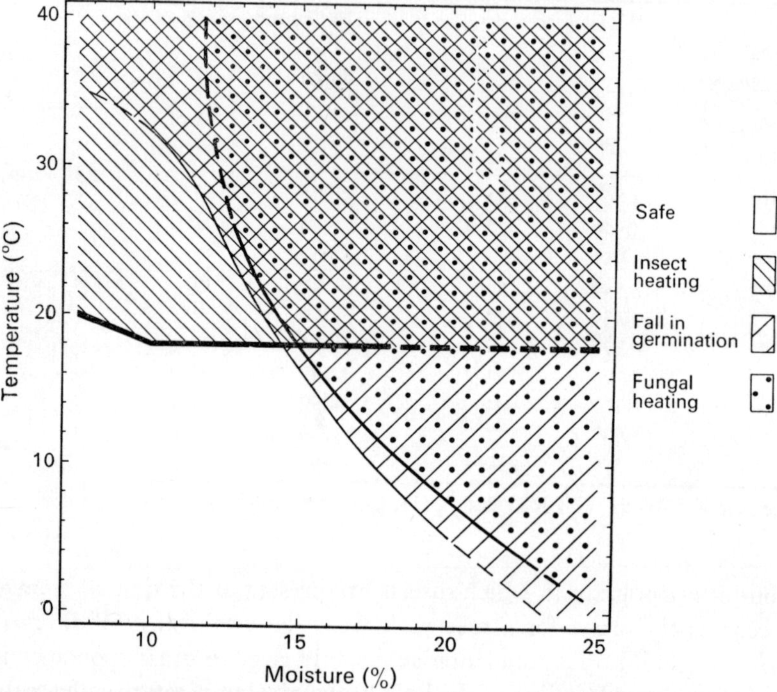

Figure 7. Relationship of storage temperature and grain moisture content to insect heating, fall in germination and damp grain heating. (Modified from Burges & Burrell 1964.)

In practice grain cooled to such a temperature in autumn continues to get colder during the winter, the rate depending on the size of the bulk. This method of grain storage, now widely used in the United Kingdom and elsewhere in the world, not only prevents insect infestation development, but by reduction of temperature below 17°C enables grain of high moisture content to be stored. In some countries refrigerated air is used.

The technique does not however solve all the problems, since some species of mites which attack stored grain can complete their life cycles at low temperatures. For example *Acarus siro*, the flour mite, has a developmental minimum of between 0 and 5 °C (Cunnington 1965). Cooling does however, reduce the risk of damp grain heating and of moisture migration due to insect heating, both of which produce conditions favourable for the development of mite populations (Burrell & Havers 1976). Even so, a combination of cooling and chemical control is necessary to prevent the development of mite infestation.

The control of mites in grain stored on British farms (Griffiths *et al.* 1976) and elsewhere, is of considerable potential economic importance, not only because of the present attitude of some buyers but also because the regulations of the E.E.C. require that all grain taken into intervention storage must be free from living insects and mites and be kept so until delivery ex store. It is proposed to use cooling by aeration in such stores for large bulks of grain but it is by no means certain that techniques which have worked on small bulks will necessarily apply to large and there may well be unforeseen problems if grain remains in store for several years. One may even find it more practicable to use the predator *Cheyletus eruditus* to keep down infestation of *Acarus siro*, *Glycyphagus destructor* and *Tyrophagus longior* than to use acaricides. Much depends on employing the predator under the conditions of temperature which favour its increase more than that of the prey (Solomon 1962, 1967, Norris 1958, Pulpan & Verner 1965).

PLANT BREEDING

Changes in susceptibility of plant products to attack by storage pests may be brought about incidentally by plant breeders seeking to change such characteristics as yield, length of growing season, plant height, or resistance to pathogens. No attempt appears to have been made to breed for resistance to storage pests, although this may be due until recently to lack of satisfactory criteria for measuring resistance.

Giles & Ashman (1971) pointed out that good husk cover of maize cobs gave maximum protection from field infestation of standing maize by maize weevils (*Sitophilus zeamais* Motsch.) flying out from nearby stores. Good husk cover is however associated with short ears, whilst selective breeding in favour of increased yields has produced varieties with longer cobs, less well enclosed and hence more liable to infestation. Dobie (1974, 1975), who has developed the idea of an 'index of susceptibility', found that local varieties of maize grown in Malawi which had short ears, tight fitting husks, and a white semi-flint grain, were much less susceptible to attack than improved imported varieties which had poor husk cover and soft floury kernels. Because of storage losses a number of the African farmers reverted to their traditional varieties for their own consumption, selling the hybrid grain to the official buying agency.

Although the increased yield from the improved variety, even with the greater insect damage, may be more than from the less susceptible traditional varieties, this incident indicates the need for plant breeders to take account of storage pest susceptibility in breeding programmes.

NEW CROPS IN BRITAIN

The problems in Britain described so far have their origins in new ways of harvesting and storing traditional cereals like wheat and barley. Some problems have already occurred and others may do so in future with those crops which are attracting more interest at present such as maize grain, beans, oilseed rape and linseed. The main problem with maize grain is spontaneous heating due to storage at too high moisture; at the normal time of harvest in late autumn, temperatures are too low for there to be much risk of insect attack, although this could occur if grain is held over until the following summer. Storage of oilseed rape at too high moisture content (over 8%), has resulted in heating due to attack by fungi, and there has also been attack by mites (mainly *Acarus siro* and *Glycyphagus destructor*), which tunnel into and destroy the contents of the seeds. Similar trouble has been experienced in France, where *Tyrophagus putrescentiae* Schr. is the major pest, associated particularly with attack by moulds at an oilseed rape moisture content exceeding 10% (Fleurat-Lessard & Anglade 1973).

Linseed is liable to attack by the same pests as oilseed rape, including surface webbing by the white shouldered house moth, *Endrosis sarcitrella* (L.).

POULTRY HOUSES

New problems have arisen from the mass housing of chickens in various types of controlled environment buildings such as deep-litter houses, those with droppings pits and in battery houses used for egg production.

The first infestations of the lesser mealworm beetle *Alphitobius diaperinus* (Panz.) were found in broiler and deep-litter houses in Great Britain in 1949. The insect, which occurs frequently (Table 1) on imported raw materials such as crushed bones, wheat offals, groundnut and other oilcakes and rice bran, especially from the Middle and Far East (Gradidge *et al.* 1969, Brett & Kennedy 1973), and used in the manufacture of feeding-stuffs, must have been carried into the poultry houses in the food.

This species has found the warm, moist and mouldy conditions in deep litter and droppings pits favourable for development as these reproduce the conditions found in the floors of poor quality warehouses in damp, tropical countries. The exact status of *A. diaperinus* in the economy of the poultry houses is not clear as the beetles do not appear to harm the poultry.

Recently there have been reports of damage to the woodwork of enclosed air conditioned battery houses by larvae of the leather beetle, *Dermestes maculatus*, another tropical insect especially common on imported bones, hides and skins and as a scavenger on oilcakes. The larvae tunnel into woodwork for pupation thereby weakening it. Similar damage used to be caused in fishmeal factories by this species and by the bacon beetle, *Dermestes lardarius* L., which has lower temperature requirements and which is also known to occur in poultry houses and to cause similar damage. In fishmeal factories the trouble was eventually solved by using steel framed rather than timber supported buildings. It is known that the poultry house complex where the damage by *D. maculatus* has occurred manufactures its own feeding-stuffs from imported raw materials including bones, thereby ensuring a high risk of introduction of the pest.

In common with other producers, minimum temperatures in the battery houses have been raised from about 18 °C to about 21 °C. Since *D. maculatus* requires a minimum of about 20 °C for complete development (Oxley 1949), at which temperature population increase would be slow, it may be that the slight increase from 18 to 21 °C has been the prime cause of the outbreak.

Replacement of polystyrene insulation panels in some poultry houses has been necessary because of damage done by the tunnelling of larvae of the moth *Niditinea fuscipunctella* Haw, which like the Dermestids, breeds in the droppings and when mature migrates upwards.

In poultry houses with droppings pits the nature and extent of the infestation depends very much on the rate of drying of the droppings, those which are wet being liable to attack by flies and those which are dry, by beetles.

A new problem in battery egg production houses, threatening continued use of this method of production, is severe biting of egg collectors by a species of itch mite of the genus *Pyemotes*. This mite is predacious on insect larvae and the development of a satisfactory method of controlling the mites and so preventing attack is held up until more is known of their biology and ecology, especially the primary host or hosts, since control of these may provide the solution of the problem.

All these problems have arisen out of new techniques in the management of poultry by intensive methods.

CHANGES IN THE DOMESTIC SCENE

The extension of central heating, with houses generally being kept warmer and drier, together with wider use of synthetic fibres and moth-proofed carpets, has led to a considerable reduction in incidence of clothes moths, especially the common clothes moth *Tineola bisselliella* (Humm.). However species such as the varied carpet beetle (*Anthrenus verbasci* (L.)) and the fur beetle (*Attagenus pellio* (L.)) appear to have replaced them as major causes of damage, breeding in birds' nests in roof spaces and penetrating thence into inhabited parts of houses. Halstead (1975) has drawn attention to a number of potential pests in this family (Table 4).

It may take many years for potential domestic pests to establish themselves after initial introduction. Thus *Anthrenocerus australis* (Hope) first

Table 4.   Some domestic pests recently established or having a potential for establishment in Britain (Based on Halstead (1975) and Woodroffe (1967)).

	Recorded in imports into Britain	Established in		Notes
		Mainland Europe	Eastern U.S.A.	
Attagenus fasciatus (Thun.) (= *A. gloriosae* (F.)	Yes	Yes		Indoor pest in Sweden: not cold hardy. Commonly intercepted in oilcakes and myrobalans especially from India (Aitken 1975).
Attagenus smirnovi Zhantiev	No	Yes		Indigenous to Kenya: established in U.S.S.R. and Denmark.
Anthrenus sarnicus Mrocz.	No	Not known		First discovered Guernsey 1961 and in London 1966. Has similar requirements to *A. verbasci.*
Anthrenus coloratus Reitt.	Yes	No	Yes	Mediterranean, Asia and Africa. Intercepted on bones from Iraq (Aitken 1975).
Trogoderma angustum Solier	No	Yes	No	Originally S. American: now in Germany and Denmark.
Novelsis aequalis (Sharp)	No	No	Yes	N. America.
Reesa vespulae (Milliron)	No	Yes	Yes	N. America: now also in Scandinavia, Germany and U.S.S.R.

recorded in Great Britain in 1933 does not seem to have become widespread as a household pest since the recording of a number of occurrences during 1956–1957 (Bezant 1957a, b, Woodroffe & Parkin 1957). It has been intercepted from time to time in imports from Australia, its original home (Aitken 1975).

Another potential domestic pest is the Tenebrionid beetle *Tribolium destructor* Uytt., recorded first in this country in 1939 (Potter 1939), and mainly associated here with rodent baits based on oatmeal, one infestation being in a factory making pet foods (Brett 1973). It does not appear to have established itself permanently, although an infestation in a flat in London occurred in 1971 (Halstead, Richardson & Walker 1975). It is a major pest in Denmark where its distribution was associated with infested bird-seed, and its success there appears to be due to the maintenance of temperatures not less than 20 °C in dwellings in winter. The beetle is not resistant to cold, has a minimum for development of 15–17·5 °C and an optimum of around 30 °C (Halstead *et al.* 1975, Halstead 1975).

A change in housing practice has led to two species of *Dermestes* becoming domestic pests. These are *Dermestes haemorrhoidalis* Kuster, known in Britain since 1953, and *D. peruvianus* Castelnau, since 1905, which until recently were mainly confined to markets and poorly heated dwelling houses. They have been recorded from time to time on imports of animal products mainly from South America (Aitken 1975). The new problem is the invasion of high-rise flats by these beetles, which then establish themselves in kitchens, canteens and restaurants, as high as the 27th floor. Colonization takes place by flying beetles attracted to lights, and once in the building they spread through service ducts. Central heating maintains conditions above developmental minima (Bezant 1963, 1965, 1973).

Changes in biology affecting pest potential

Most if not all of the changes in pest status so far described have been explicable and might have been forecast on the basis of information provided by laboratory experiments on biology, especially physical limits, egg production, mortality, and nutritional requirements and by field observation of habits. They did not require any change to have taken place in the requirements or limitations of any of the species studied.

There are however two types of biological change in the pest which may affect its pest status, one possible change is development of tolerance of low temperatures of which there is so far no evidence, the other which has occurred, is resistance to pesticides.

TOLERANCE OF LOW TEMPERATURES

Both cold-tolerant and cold-susceptible insects must have been introduced regularly into Britain for many years. The occurrence of *S. granarius* and *O. surinamensis* (cold tolerant) and *T. castaneum* (cold susceptible) in Roman remains has already been mentioned. The development of international trade, especially during the past hundred years or so has greatly increased the opportunity for such regular introductions (Table 1), but there is no evidence that any species not cold-hardy have been able to establish themselves other than in heated buildings (e.g. *Trogoderma granarium* in maltings; *Tribolium castaneum* in flour mills and plant bakeries; *Lasioderma serricorne* in tobacco factories, especially in sample rooms; and *Ephestia cautella* in chocolate factories).

Every population introduced on an imported commodity which is stored in an unheated building for an extended period experiences a selection pressure in favour of those individuals which can develop, survive and multiply at lower temperatures than the norm. There is no evidence yet that such selection has taken place, though this could be one factor in the replacement of *T. confusum* by *T. castaneum* in flour mills already mentioned. Smith (1975) has remarked on the spread of *Tribolium castaneum* in grain stores in the prairie provinces in Canada and has ascribed this to higher minimum winter temperatures in the interior of the grain bulks in the larger grain stores on farms, since *T. castaneum* cannot survive less than 2 °C for more than one month. He does not suggest any cold adaptation. Any adaptation of major grain infesting species to tolerance of lower temperatures could make more difficult and expensive the prevention of infestation by grain cooling.

Infestation by *Ephestia kuehniella* of American maize stored in a silo in Glasgow is currently being investigated because the population there appears to be developing at lower temperatures than laboratory experiments indicate is normal. *E. kuehniella* is cold-hardy, but development normally ceases at about 10 °C. There has not been any spontaneous heating in the bins and it may be that a race adapted to development at lower temperatures has appeared, although it is possible that a series of very mild winters has had some influence.

TOLERANCE OF CONTACT INSECTICIDES AND FUMIGANTS (RESISTANCE)

The virtual elimination of infestations of Australian wheat was achieved mainly by treatment with malathion when taken into store (which started in 1960), and subsequent further applications of malathion, or fumigation with methyl bromide or hydrogen cyanide. This freedom from infestation, an important commercial asset for exports, is threatened by the emergence of pest strains resistant to (or tolerant of) malathion and other insecticides. Such resistance is of course not confined to Australia but is now known from many parts of the world.

Dyte (1974) reported insecticide resistance in seventeen species of storage pests (twelve beetles and five moths) and at least one species of mite (*Acarus siro*), to all the main types of insecticide including juvenile hormone analogues (Table 5). Although some strains are resistant to only a single chemical or type of chemical, there are strains which are resistant to several. This may be cross resistance, where a single resistance mechanism confers protection to several compounds, or multiple resistance where a particular

Table 5. Occurrence of insecticide resistance in some stored product pests (Based on Dyte 1974).

Pest	Toxicant								
	A	B	C	D	E	F	G	H	I
Caryedon serratus				×					
Dermestes maculatus				×					
Oryzaephilus mercator	×			×					
Oryzaephilus surinamensis	×	×		×					
Rhyzopertha dominica	×			×			×	×	
Sitophilus granarius	×	×	×	×	×		×	×	×
Sitophilus oryzae	×		×	×	×				
Sitophilus zeamais	×			×					
Tribolium castaneum	×	×	×	×	×	×	×	×	×
Tribolium confusum	×		×	×			×	×	
Trogoderma granarium	×								
Tenebroides mauritanicus							×		
Ephestia cautella	×				×				
Plodia interpunctella	×				×				
Sitotroga cerealella				×					
Acarus siro				×					

× Indicates resistance
A Malathion
B Carbamates
C DDT
D Lindane
E Pyrethroids
F Juvenile hormone mimics
G Methyl bromide
H Phosphine
I Ethylene dibromide

strain has developed independent mechanisms for resisting several compounds.

Resistance to fumigants, demonstrated as possible by laboratory experiments, has occurred locally in Canada, especially to ethylene dibromide used for spot fumigation in flour mills for control of *Tribolium castaneum*. Resistance to methyl bromide and phostoxin has been found in a number of grain infesting beetles in various parts of the world, but so far no difficulties have been reported in achieving control, probably because existing fumigation schedules are adequate to compensate for the present resistance levels.

In addition to metabolic resistance to insecticidal compounds, behavioural resistance has been discovered in some storage insects. Pinniger (1974) has shown experimentally that in an insecticidal environment there is selection in favour of individuals of *Tribolium castaneum* which remain in refuges and which therefore are either not exposed at all or only for short periods to the insecticide. This phenomenon first became apparent in a maltings in Britain where malathion had been used for several years to control *Trogoderma granarium*. Whilst this species was much reduced, *Tribolium castaneum* greatly increased in numbers and tests showed that a resistant strain was present. The insects also tended to remain longer in cracks and to spend only short periods on treated surfaces (Green *et al.* 1975).

Resistant strains have been intercepted in imports into Great Britain (Dyte & Blackman 1970). In the United Kingdom most inland strains of *Tribolium castaneum*, *Oryzaephilus surinamensis*, *Sitophilus oryzae* and *S. zeamais* are resistant to lindane, and resistance has been found less commonly in other species. Most strains of *T. castaneum* are also resistant to malathion whilst resistant strains of *T. confusum* are widespread (Dyte 1974). A strain of *Acarus siro* L. resistant to lindane has been found in a cheese store (Wilkin 1973).

The principal danger at present is the introduction or appearance within the country of malathion-resistant strains of *O. surinamensis*, since this species is the principal pest of farm-stored grain and malathion the main grain protectant. So far, only one small outbreak has occurred on a British farm and that was eradicated. Despite monitoring and testing all *O. surinamensis* infestations detected in imports and control of any resistant infestations that are discovered, establishment of imported resistant strains is only a matter of time. There is always the risk, as for anti-coagulants and rodents, of resistant strains developing within the country.

A major difficulty in the detection of insecticide-resistant strains is that they are morphologically identical with susceptible strains, so that a quarantine against them is difficult to enforce except by fumigating all infested cargoes from countries where resistant strains are known to occur, or by holding up commodities for the time necessary to carry out tests.

At present pirimiphos methyl is the only effective alternative chemical which can be mixed with grain as a protectant; resistance to this may develop in time.

The development of resistance by a range of common storage and domestic pests will make difficult the control of residual insects and mites in structures, although for a time considerably increased dosages may have an effect. This is not normally practicable for application to food because of national and international restrictions on the kinds of pesticides used on food and the levels of residues. This is particularly serious with certain fumigants because the dosages and periods of exposure required to kill certain species already produce residues near the limits.

One cannot expect a steady flow of alternative pesticides to be introduced because the high cost of development and safety testing makes it uneconomic to produce an insecticide specially for the relatively small stored-product market; new stored-product insecticides are only likely to appear if they also have a much wider use.

The main effects of the appearance of resistance must therefore be to stimulate research on the mechanism of resistance and of the mode of action of insecticides and fumigants together with more study of non-chemical control measures. It is difficult however to envisage control measures so versatile and economic as contact pesticides and gaseous fumigants.

Discussion

New problems in stored products and domestic entomology, acarology and mycology originate partly through changes in industrial or agricultural technology (for example in flour milling and grain storage) and partly because of biological changes in the pests themselves (for example resistance to insecticides). Both may occur fairly rapidly, since new technologies such as pneumatic conveying in flour milling spread rapidly as does insecticide resistance, especially in those species of storage insects and mites carried in international trade.

Changes in methods of international transport (containerization), and in speed of movement (air transport), increase the risk of spread of pests to areas in which they may establish themselves. Changes in domestic practice (central heating and high-rise flats) provide new ecological niches which invaders can exploit, and may change the status of existing pests (the decline of the common clothes moth, *Tineola bisselliella*). Such changes occur less quickly than in industry or agriculture.

There are not many commodities moving in international trade or stored in Britain which remain entirely free from insect or mite pests for long.

Provided that the produce moves quickly from producer to processor and consumer, loss in weight or nutritive value may be small and infestation is only significant if the mere presence of insects or mites or evidence of their presence is important commercially. New problems usually occur when produce has to be stored for longer periods than is normal commercially, when new processes are introduced, or if a new pest or strain of an existing pest appears, as has happened for insecticide resistance to malathion in *O. surinamensis* and to lindane in *Acarus siro*.

It is difficult to forecast the exact nature of the problems which will occur and carry out the necessary research so that the solutions are available in advance, because so many result from sudden changes in commercial conditions or in government policies; from alterations to industrial processes carried out in secrecy behind closed doors; or because of unpredictably bountiful harvests.

For example, before the entry of the United Kingdom into the E.E.C. dried skimmed milk, a by-product of butter manufacture, was stored only for short periods by processors. After entry, virtually all existing stocks were offered to and were taken over by the Intervention Board for Agricultural Produce, which has acquired more subsequently. The total amount of dried skimmed milk stored in the E.E.C. was estimated at 365,000 t in January 1975 and had risen to 1,200,000 t by March 1975, of which some 2% was stored in Britain (Clayton 1976). It is known that in the United States of America, dried skimmed milk is susceptible to attack by certain Dermestid beetles, but in Britain, Psocids (especially *Lepinotus patruelis* Pearm.) show signs of becoming a pest, living on spilt powder outside the bags as well as penetrating into them. Apart from occasional heavy infestations on cereal imports (Broadhead 1954), Psocids have not merited much attention as stored product pests, but their potential as pests of dried milk justifies some work on their biology and ecology and on the physical properties of dried skimmed milk as they affect susceptibility to attack. The effects of fumigation also require study.

It remains to be seen whether the recent change of policy by the E.E.C. Commission, requiring compound feeding-stuffs manufacturers to incorporate dried skimmed milk into certain kinds of animal feeds, will result in shorter periods of storage. In that case the infestation problem will disappear, and with it the immediate need to study Psocids, unlike the continuing problems arising out of the storage of grain and of the processing of food and feed.

In normal commercial practice neither home produced nor imported grain is stored for long periods in this country, since virtually all the harvest of one year has been consumed or exported before the next one. Thus, in 1974/75 59% of the wheat crop and 39% of the barley crop were on the farm at the end of December and 10% and 5% at the end of June (Table 3). If however, bumper harvests occur in Britain and the price falls, sales into intervention may be necessary and wheat may remain in store for several years.

Although precautions have been taken based partly on experiences in storing government stocks of wheat and barley between 1938 and 1953 and on later research, the hazards of storing large bulks for several years are different from those of storing the normal 30–50 tons in the farm bin over several months during the cooler part of the year. Further research, especially in the grain stores, will be necessary, but this cannot be done until the grain is actually taken into store.

On a world scale similar unforeseen problems may arise in putting into effect FAO's proposals for the storage of reserve stocks of grain in various producer countries. Recent experiences of losses by infestation (especially by *Trogoderma granarium*) to famine stocks held in countries in the Sahelian zone, indicate that proper precautions must be taken at the beginning of storage if serious damage is to be prevented in areas of climate favourable to insect development. That this can be done successfully has been demonstrated in Australia.

Up to the present the prevention of attack and the control of infestation by insects and mites has depended mainly on the use of contact insecticides and acaricides and on fumigants, except in those countries such as central Canada, where temperatures in winter are low enough to provide natural protection, unless spontaneous heating occurs.

There are a number of developments which are increasing the difficulty of preventing or controlling infestation—especially of human food. On the one hand there is an increased demand in developed industrialized countries that food should be free from living or dead insects and mites and from evidence of damage done by them. This is reflected in grading standards, and in the requirements of food manufacturers, public health authorities and consumers. The need to prevent infestation of food in transit leads to demands for control of infestation in non-food commodities (e.g. bones), in order to avoid cross infestation.

Fulfilment of these requirements has hitherto been possible because competition between exporting countries to sell a surplus of primary agricultural products has brought about the application of minimum quality standards at time of export (e.g. China's standards in regard to Australian wheat). Such standards, whether enforced through official grading and inspection systems or by commercial contracts, often include provisions regarding insects and insect damage. They have been applied to cereals, cocoa beans, goundnuts, walnuts, dried fruits, etc.

The need to meet these requirements has been a strong incentive to carry out control measures in countries of origin, but continued control is threatened from two directions. The first is increased restrictions on the use of pesticides on or near food either directly or through lowering tolerance limits for their residues. The second is the spread of resistance to contact insecticides, especially malathion and lindane, and the possibility of similar resistance developing to fumigants.

A further factor which may increase infestation of imports is the likely disappearance of surpluses, especially of oilseeds and oilcakes mainly produced in countries in the tropics with rapid increases in human and livestock populations. In such circumstances whilst it will clearly be in the interest of producer countries to prevent loss due to infestation during storage of crops retained until the next harvest, there would appear to be little inducement to conserve export crops since there will be a ready market for them in deficit countries whatever their quality or state of infestation.

There would not appear to be any major important stored product pests in other parts of the world with a similar climate to our own which are not already here. The problems are: 1. The risk of importing insecticide-resistant strains (e.g. malathion-resistant saw-toothed grain beetle (*O. surinamensis*)). 2. The development within the country of insecticide-resistant or cold-hardy strains. 3. Changes in industrial or agricultural practices which will favour some of the imported species (e.g. the changes in flour milling practice which formerly favoured the Mediterranean flour moth, *Ephestia kuehniella*, and now favour the rust red flour beetle, *Tribolium castaneum*, as against the confused flour beetle, *Tribolium confusum*). Such changes are difficult to forecast because of industrial secrecy.

There would appear to be a greater chance of introduction and establishment of new domestic pests, especially with the extension of central heating and blocks of flats having ducted interconnected services through which insects can spread from one floor to another. There is however a considerable element of chance in the introduction and establishment of new pests and it may be many years before an invader finds exactly the right conditions at the right time for successful establishment.

These developments may well result in some increase in the incidence and level of infestation in food and feed both in imports and within the country, before practical and economical alternatives to chemical measures can be developed. The situation has not only encouraged more intensive research into the mode of action of insecticides and fumigants with special reference to resistance, but also into alternative chemicals, such as pheromones and hormones, as well as physical methods, especially the use of low temperatures and forms of radiation. There are already food manufacturers who store dried fruit and nuts in cool stores (12 °C or less) rather than in conventional warehouses with all the risks of infestation derived from the building or from other stocks, and the need therefore, for regular insecticidal spraying during the summer.

There is also the possibility of breeding into plants some resistance to attack in store, as for example by tight-fitting long sheaths round maize cobs, or by some change in the composition of seeds which, whilst not rendering them unfit for their purpose, will prevent insect development. Many pea and bean seeds for example, are resistant to attack by most species except Bruchid beetles.

When a new problem appears, especially if it is of economic significance or is a domestic nuisance, it is important that there should be sufficient basic information about the biology of the species concerned (along the lines described in the first part of this paper) and about their control, to allow a correct diagnosis and some measure of first-aid, since some research and development may be necessary before a fundamental solution can be obtained. It is too late to have to start investigations when the problem has declared itself, as already mentioned in regard to Psocids on dried skimmed milk and *Pyemotes* sp. in poultry houses. A long-term programme, with priorities derived from sound information regarding likely industrial, commercial, agricultural and domestic developments, is the main means of acquiring the necessary scientific capital which can be rapidly drawn upon when required.

References

AITKEN A.D. (1975) Insect Travellers Vol. 1, Coleoptera. *Tech. Bull. 31*. MAFF, HMSO, London.

ARMSTRONG M.T. & HOWE R.W. (1963) The saw-toothed grain beetle, *Oryzaephilus surinamensis* in home-grown grain. *J. agric. Engng. Res.* 8, 256–61.

BAILEY J. (1972) Problems of transport. Pest control in air transport. *Proc. 3rd British Pest Control Conference (Jersey, 1971)* 101–6.

BEZANT E.T. (1957a) The Australian carpet beetle, *Anthrenocerus australis* (Hope) (Col. Dermestidae) in Britain. *Ent. mon. Mag.* 93, 207.

BEZANT E.T. (1957b) Further records of the Australian carpet beetle *Anthrenocerus australis* (Hope) (Col. Dermestidae) in Britain. *Ent. mon. Mag.* 92, 401.

BEZANT E.T. (1963) The occurrence of *Dermestes peruvianus* La Porte and *Dermestes haemorrhoidalis* Küster (Col. Dermestidae) in Britain. *Ent. mon. Mag.* 99, 30–1.

BEZANT E.T. (1965) Food-handling premises in Britain as an ecological niche for *Dermestes* (Coleoptera, Dermestidae). *Proc. 12th Int. Congr. Ent. (London, 1964)* 647–8

BEZANT E.T. (1973) Infestation of premises by two species of hide beetles. *Pest Infestation Control 1970–73*, MAFF, HMSO, London, p. 18.

BILLS G.T. (1965) The occurrence of *Periplaneta brunnea* (Burm.) (Dictyoptera, Blattidae) in an international airport in Britain. *J. stored Prod. Res.* 1, 203–4

BILLS G.T. (1967) The second established colony of *Periplaneta brunnea* (Burm.) (Dictyoptera Blattidae) in an airport in Britain. *Ent. mon. Mag.* **102**, 130.

BRETT G.A. (1973) The dark flour beetle, a potential pest in the United Kingdom. *Pest Infestation Control 1968–70.* MAFF, HMSO, London, pp. 15–16.

BRETT G.A. & KENNEDY R. (1973) The lesser mealworm in poultry-house litter. *Pest Infestation Control 1968–70,* MAFF, HMSO, London, p. 22.

BROADHEAD E. (1954) The infestation of warehouses and ships' holds by Psocids in Britain. *Ent. mon. Mag.* **90**, 103–5.

BRYCE P.H. (1889) *Bull. 1. Provincial Bd. Health Ont.* Ontario Dept Agric.

BURGES H.D. (1959) Studies on the Dermestid beetle, *Trogoderma granarium* Everts. III. Ecology in malt stores. *Ann. appl. Biol.* **47**, 445–62.

BURGES H.D. & BURRELL N.J. (1964) Cooling bulk grain in the British climate to control storage insects and to improve keeping quality. *J. Sci. Fd Agric.* **1**, 32–50.

BURRELL N.J. & HAVERS S.J. (1976) The effects of cooling on mites in bulk grain. *Ann appl. Biol.* **82**, 192–7.

CHITTY A.J. (1904) *Ptinus tectus* and *Lathridius bergrothi* in Holborn. *Ent. mon. Mag.* **40**, 109.

CLAYTON H. (1976) 'No easy way of levelling the skim milk mountain'. *The Times,* London. March 4.

COX P.D. (1975) The suitability of dried fruits, almonds and carobs for the development of *Ephestia figulilella* Gregson, *E. calidella* (Guenée) and *E. cautella* (Walker) (Lepidoptera, Phycitidae). *J. stored Prod. Res.* **11**, 229–34.

CUNNINGTON A.M. (1965) Physical limits for complete development of the grain mite, *Acarus siro* L. (Acarina, Acaridae), in relation to its world distribution. *J. appl. Ecol.* **2**, 295–306.

CURTIS J. (1883) *Farm Insects.* Blackie, Edinburgh.

DENDY A. & ELKINGTON H.D. (1919) On the phenomenon known as 'webbing' in stored grain. *Royal Society Grain Pests (War) Committee Rept.* No. 4, 14–17.

DOBIE P. (1974) The laboratory assessment of the inherent susceptibility of maize varieties to post-harvest infestation by *Sitophilus zeamais* Motsch. (Coleoptera, Curculionidae). *J. stored Prod. Res.* **10**, 183–97.

DOBIE P. (1975) Susceptibility of different types of maize to post-harvest infestation. *Proc. 1st Int. Working Conf. on Stored Prod. Ent.* (Savannah, 1974) 93.

DYTE C.E. (1965) Studies on insect infestations in the machinery of three English flour mills in relation to seasonal temperature changes. *J. stored Prod. Res.* **1**, 129–44.

DYTE C.E. (1974) Problems arising from insecticide resistance in storage pests. *European and Mediterranean Plant Protection Organisation (EPPO) Bull.* **4**, 275–89.

DYTE C.E. & BLACKMAN D.G. (1970) The spread of insecticide resistance in *Tribolium castaneum* (Herbst) (Coleoptera, Tenebrionidae). *J. stored Prod. Res.* **6**, 255–61.

FLEURAT-LESSARD F. & ANGLADE P. (1973) Influence de la teneur en eau des grains stockés sur le développement des populations d'acariens. *Ann. Technol. agric.* **22**, 531–40.

FRAENKEL G. & BLEWETT M. (1943) The natural foods and the food requirements of several species of stored product insects. *Trans. R. ent. Soc. Lond.* **93**, 457–90.

FRAENKEL G. & BLEWETT M. (1944) The utilisation of metabolic water in insects. *Bull. ent. Res.* **35**, 127–39.

FREEMAN J.A. (1948) Stored products pests: A survey of the principal entomological problems in the United Kingdom. *Ann. appl. Biol.* **35**, 294–301.

FREEMAN J.A. (1950) Methods of spread of stored products and origin of infestation in stored products. *Proc. 8th Int. Congr. Ent.* (*Stockholm,* 1948), 815–25.

FREEMAN J.A. (1958). Infestation of stored products in Iran. Report of a survey carried out in October, November 1957 and January 1958. Central Treaty Organisation.

FREEMAN J.A. (1962) The influence of climate on insect populations of flour mills. *Proc. 11th Int. Congr. Ent. (Vienna, 1960)* II, 301–8.

FREEMAN J.A. (1967) Implications for infestation control of the use of freight containers in international trade. *Rept. Int. Conf. on the Protection of Stored Products (Lisbon Oeiras) 1967.* EPPO Pubns Ser. A No. 46E, 87–9.

FREEMAN J.A. (1974) A review of changes in the pattern of infestation in international trade. *European and Mediterranean Plant Protection Organisation (EPPO), Bull.* 4, 251–73.

FREEMAN J.A. (1976) Problems of stored product entomology in Britain arising out of the import of tropical products. *Ann. appl. Biol.* 84, 120–4.

GILES P.H. & ASHMAN F. (1971) A study of pre-harvest infestation of maize by *Sitophilus zeamais* Motsch. (Coleoptera, Curculionidae) in the Kenya Highlands. *J. stored Prod. Res.* 7, 69–83.

GRADIDGE J.M.G., WILLIAMS G.C., WAYMAN C. & FREEMAN J.A. (1969) Infestation in poultry houses. *Pest Infestation Control 1965–67.* MAFF, HMSO, London, pp. 19–20.

GREEN A.A., DYTE C.E., BILLS G.T., KANE M.J., TYLER P.S., WILKIN D.R., MAHON P.A. & FORD J.W. (1961) Control of dieldrin-resistant German cockroaches. *Pest Infest. Res. 1960,* ARC. HMSO, London, pp. 37–9.

GREEN A.A., DARLEY M.J., PINNIGER D.B. & BRADGATE N.C. (1975) The resistance of stored product insects and mites to pesticides—The rust-red flour beetle. *Pest Infestation Control 1971–73,* MAFF, HMSO, London, pp. 77–8.

GRIFFITHS D.A., WILKIN D.R., SOUTHGATE B.J. & LYNCH S.M. (1976) A survey of mites in bulk grain stored on farms in England and Wales. *Ann. appl. Biol.* 82, 180–5.

HALSTEAD D.G.H. (1975) Changes in the status of insect pests in storage and domestic habitats. *Proc. 1st Working Conf. on Stored Prod. Ent.* (Savannah, 1974) 142 53.

HALSTEAD D.G.H., RICHARDSON J. & WALKER A. (1975) *Tribolium destructor. Pest Infestation Control Laboratory Report 1971–73.* MAFF, HMSO, London, p. 10.

HOWE R.W. (1952) Entomological problems of food storage in Northern Nigeria. *Bull. ent. Res.* 43, 111–44.

HOWE R.W. (1953a) The rapid determination of the intrinsic rate of increase of an insect population. *Ann. appl. Biol.* 40, 134–51.

HOWE R.W. (1953b) Studies on beetles of the family Ptinidae. VIII. The intrinsic rate of increase of some Ptinid beetles. *Ann. appl. Biol.* 40, 121–33.

HOWE R.W. (1956) The biology of two common storage species of *Oryzaephilus* (Coleoptera, Cucujidae). *Ann. appl. Biol.* 44, 341–55.

HOWE R.W. (1957) A laboratory study of the cigarette beetle, *Lasioderma serricorne* (F.) (Col., Anobiidae) with a critical view of the literature on its biology. *Bull. ent. Res.* 48, 9–56.

HOWE R.W. (1958) A theoretical evaluation of the potential range and importance of *Trogoderma granarium* Everts in N. America. *Proc. 10th Int. Congr. Ent.* (Montreal 1956) 4, 23 8.

HOWE R.W. (1962) A study of the heating of stored grain caused by insects. *Ann. appl. Biol.* 50, 137–58.

HOWE R.W. (1963) The prediction of the status of a pest by means of laboratory experiments. *World Rev. Pest Contr.* 2, 2–12.

HOWE R.W. (1965) A summary of estimates of optimal and minimal conditions for population increase of some stored products insects. *J. stored Prod. Res.* 1, 177–84.

Howe R.W. (1971) A parameter for expressing the suitability of an environment for insect development. *J. stored Prod. Res.* **7**, 63–5.

Howe R.W. & Burges H.D. (1953) Studies on beetles of the family Ptinidae: IX. A laboratory study of the biology of *Ptinus tectus* Boield. *Bull. ent. Res.* **44**, 461–516.

Howe R.W. & Lindgren D.L. (1957) How much can the Khapra beetle spread in the USA? *J. econ. Ent.* **50**, 374–5.

Hunter F.A., Tulloch J.B.M. & Lambourne M.G. (1973) Insects and mites of maltings in the East Midlands of England. *J. stored Prod. Res.* **9**, 119–41.

Johnson W.G. (1895) On the mediterranean flour moth. *Appendix to 19th rept. State Ent. Illinois.*

Kennedy R. (1973) The rust-red flour beetle in flour mills. *Pest Infestation Control 1968–70.* MAFF, HMSO, London, p. 15.

Klein S.T. (1887) Appearance in London of *Ephestia kühniella* and the remedy provided by nature. *Trans. Middlesex nat. Hist. Soc.* 16–20.

Le Cato G.L. (1976) Yield, development and weight of *Cadra cautella* (Walker) and *Plodia interpunctella* (Hübner) on twenty-one diets derived from natural products. *J. stored Prod. Res.* **12**, 43–8.

Loschiavo S.R. (1975a) The detection of insects by traps in grain filled box-cars during transit. *Proc. 1st Int. Working Conf. on Stored Prod. Ent.* (*Savannah 1974*) 639–74.

Loschiavo S.R. (1975b) Field tests of devices to detect insects in different kinds of grain storages. *Canad. Ent.* **107**, 385–9.

Loschiavo S.R. & Atkinson J.M. (1973) An improved trap to detect beetles (Coleoptera) in stored grain. *Canad. Ent.* **105**, 437–40.

Mason F.A. (1921) The destruction of stored grain by *Trogoderma khapra*. *Bull. Bur. Biotechnol. Leeds* **2**, 27–38.

Miles H.W. (1928) Investigations on the control of the khapra beetle (*Trogoderma granarium* Everts) with calcium cyanide. *Bull. ent. Res.* **18**, 251–5.

Munro J.W. (1940) *Report on a Survey of the Infestation of Grain by Insects.* Dept. of Sci. Ind. Res. HMSO, London.

Munro J.W. (1966) *Pests of Stored Products.* Hutchinson, London.

Norris J.D. (1958) Observations on the control of mite infestations in stored wheat by *Cheyletus* spp. (Acarina, Cheyletidae). *Ann. appl. Biol.* **46**, 411–22.

Ormerod E.A. (1890) 13th Report of observations of injurious insects and common farm pests during the year 1889 with methods of prevention and remedy. London, pp. 49–54.

Osborne P.J. (1971) An insect fauna from the Roman site at Alcester, Warwicks. *Britannia* **2**, 156–65.

Osborne P.J. (1973) *Airaphilus elongatus* (Gyll.) Col. Cucujidae present in Britain in Roman times. *Ent. mon. Mag.* **109**, 219.

Oxley T.A. (1948) *The Scientific Principles of Grain Storage.* N. Publ. Co., Liverpool.

Oxley T.A. (1949) Survival and development of the leather beetle. *Pest Infestation Research 1948.* DSIR, HMSO, London, pp. 9–10.

Park T. (1954) Experimental studies of interspecies competition. II. Temperature, humidity and competition in two species of *Tribolium*. *Physiol. Zool.* **27**, 177–238.

Pinniger D.B. (1975) The use of bait traps for assessment of stored product insect populations. *U.S.D.A. Co-op. Econ. Ins. Rpt.* **25** (49–52), 907–9.

Pinniger D.B. (1974) A laboratory simulation of residual populations of stored products pests and an assessment of their susceptibility to a contact insecticide. *J. stored Prod. Res.* **10**, 217–23.

POTTER C. (1939) The occurrence of *Tribolium destructor* Uytt. in seeds in England. *Ent. mon. Mag.* **75**, 114–15.

PULPAN J. & VERNER P.H. (1965) Control of tyroglyphoid mites in stored grain by the predatory mite *Cheyletus eruditus* (Schrank). *Can. J. Zool.* **43**, 417–32.

RICHARDS O.W. & WALOFF N. (1946) The study of a population of *Ephestia elutella* Hübner (Lepidoptera: Phycitidae) living on bulk grain. *Trans. R. ent. Soc. Lond.* **97**, 253–98.

SMITH K.G. (1956) The occurrence and distribution of *Aphomia gularis* (Zell.) (Lep. Galleridae), a pest of stored products. *Bull. ent. Res.* **47**, 655–67.

SMITH K.G. (1960) Insect infestation associated with French shelled walnuts with particular reference to the occurrence of *Aphomia gularis* (Zell.) (Lep. Galleridae). *Bull. ent. Res.* **50**, 711–16.

SMITH K.G. (1961) Recent occurrences and a review of *Aphomia gularis* (Zell.) (Lep. Galleridae) as a storage pest in Britain. *Ent. mon. Mag.* **96**, 167–8.

SMITH K.G. (1965) Some aspects of the biology of *Paralipsa* (*Aphomia*) *gularis* (Zell.) (Lepidoptera) in relation to its distribution. *Proc. 12th Int. Congr. Ent.* (*London 1964*) 626.

SMITH K.G. (1969a) The Mediterranean flour moth. *Infestation Control 1965–67*. MAFF, HMSO, London, p. 12.

SMITH K.G. (1969b) Infestation by mites as a factor in the long term storage of flour. *Proc. 2nd Int. Congr. Acarology* (1967) 249–53.

SMITH K.G. (1974) Containerisation of cargo and its effects on the control of stored products pests in international trade. *Trop. stored Prod. Inf.* **27**, 31–6.

SMITH L.B. (1975) Role of low temperature to control stored food pests. *Proc. 1st Int. Working Conf. on Stored Prod. Ent.* (*Savannah 1974*) 418–30.

SOLOMON M.E. (1962) Ecology of the flour mite, *Acarus siro* L. (= *Tyroglyphus farinae* DeG.). *Ann. appl. Biol.* **50**, 178–84.

SOLOMON M.E. (1967) Experiments on predator–prey interactions of storage mites. *Acarologia* **11**, 484–503.

SOLOMON M.E. & ADAMSON B.E. (1955) The powers of survival of storage and domestic pests under winter conditions in Britain. *Bull. ent. Res.* **46**, 311–55.

SURTEES G. (1965) Ecological significance and practical implications of behaviour patterns determining the spatial structure of insect populations in stored grain. *Bull. ent. Res.* **56**, 201–13.

THOMAS I., JANSON J.W. & AITKEN A.D. (1968) Common names of British insects and other pests. *Tech. Bull. MAFF*, **6**, 2nd Edn.

WALKER J.J. (1917) Note on *Trogoderma khapra* Arrow, a recently described Dermestid granary pest. *Ent. mon. Mag.* **53**, 165.

WALOFF N. & RICHARDS O.W. (1946) Observations on the behaviour of *Ephestia elutella* (Lep., Phycitidae) breeding on bulk grain. *Trans. R. ent. Soc. Lond.* **97**, 299–335.

WARD N. & CALVERLEY D.J.B. (1972) A literature survey on the climate relationship in stores. *Trop. stored Prod. Inf* **23**, 35–44.

WAYMAN C. (1969) The control of insects in stores of home grown grain. The insect and mite pests. *Chem. Inds.* 1445–7.

WILKIN D.R. (1973) Resistance to lindane in *Acarus siro* from an English cheese store. *J. stored Prod. Res.* **9**, 101.

WOODROFFE G.E. (1953) An ecological study of the insects and mites in the nests of certain birds in Britain. *Bull. ent. Res.* **44**, 739–72.

WOODROFFE G.E. (1965) The sterol requirements of several species of Dermestes (Col. Dermestidae). *Proc. 12th Int. Congr. Ent. (London 1964)* 625.

WOODROFFE G.E. (1967) *Anthrenus sarnicus* Mroczkowski (Coleoptera, Dermestidae) in Britain. *J. stored Prod. Res.* 3, 263–5.

WOODROFFE G.E. & PARKIN E.A. (1957) The first household infestation of the Australian carpet beetle, *Anthrenocerus australis* (Hope) (Col. Dermestidae) in Britain. *Ent. mon. Mag.* 93, 144.

WOODROFFE G.E. & SOUTHGATE B.J. (1951) Birds' nests as a source of domestic pests. *Proc. zool. Soc. Lond.* 121, 55–62.

Present and future parasite problems in African game

R. SACHS *Tropeninstitut, Bernhard-Nocht-Strasse 74,
Hamburg, Germany*

It has been suggested that wild African game animals might be used as human food if reared under ranch conditions (Dasmann 1964), and pilot projects have been set up in various parts of the continent.

A potential danger in such projects is that the wild game may carry parasites pathogenic to man, or capable of being transmitted to domestic stock kept on the same site. This danger is more acute when we remember that indigenous hosts better adapted to resist or tolerate such pathogens may not exhibit disease symptoms, so that the infection may go unrecognized.

Traditional animal husbandry greatly alters parasite and disease complexes through changes to habitats and micro-environments, through hygiene and vector control and through therapeutic measures directed against specific pathogens. Whilst some parasites may become eradicated from domestic stock, others remain successful as the increasing cysticercosis/taeniasis problem shows. Even man is at risk when he is included in the chain of infection, either as a carrier, or as an epidemiologically unimportant dead end. The great progress made in recent years in combating these zoonoses has depended on the recognition of life cycles, and the realization that treatment of the human patient alone would not improve the disease situation.

If the familiar domestic species, grazed in the traditional monoculture, or few species mixed culture are to be replaced on any scale in Africa by parasitologically unfamiliar game species ranched extensively in mixed herds, we need much more information about the parasites they carry, their life cycles and population dynamics, if we are to have any chance of predicting and forestalling new pest problems.

In some of the developing African countries where National Parks and Game Reserves have been established it is still possible to study these game species together with their parasite faunas in a natural setting where life cycles can be disentangled from a knowledge of the existing food chains. This paper records some of my own observations made from carcass inspection in the Serengeti National Park, Tanzania, East Africa.

The parasite complex encountered

Thirteen large mammal species were included in the survey: giraffe (*Giraffa camelopardalis*), buffalo (*Syncerus caffer*), zebra (*Equus burchelli*), warthog (*Phacochoerus aethiopicus*), eland (*Taurotragus oryx*), waterbuck (*Kobus defassa*), wildebeest (*Connochaetes taurinus*), hartebeest (*Alcelaphus buselaphus cokei*), topi (*Damaliscus korrigum*), impala (*Aepyceros melampus*), Grant

Table 1. Parasitic infections* and their herbivorous hosts in the Serengeti National Park, Tanzania.

	Giraffe	Buffalo	Zebra	Warthog	Eland	Waterbuck	Wildebeest	Hartebeest	Topi	Impala	Grant	Thomson	Dik-dik
Muscle-cysts		3			3	3	1	1	2	3	1	4	1
Viscera-cysts						2	2	3		3	2	2	3
Sacrum-cysts							2	1	1		4		
Hydatid-cysts	4			3			3						
Spargana		3	3	3	3	3	3	3	3				
Filaria subcutis	3						3	3	3	3	3		
Nematodes in musculature		4	1										
Sarcocystis	3	1		3		2	3	3	3		3	3	3
Liver flukes		3			3		3		3				
Liver tapeworm					1	2					3		
Liver nematodes	3		1								2		
Pentastomids		1		3	3	3	2	1	2		3	3	3
Dictyocaulids			2				3	3	2				
Large protostrongylids							3		1				
Small protostrongylids							1	1	1		1	1	1
Syngamids		4				1							
Oestridae in nose	4		3				1	1	1				
Oestridae in brain							2						

* Excluding gastro-intestinal helminths and filarial setariid worms occurring free in the abdominal cavity.

Key   1 = almost all animals infected.
   2 = about two-thirds infected.
   3 = about one-third infected.
   4 = infrequent infection.

(*Gazella granti*), Thomson (*Gazella thomsoni*), dik-dik (*Rhynchotragus kirkii*).

The parasites studied were those of relevance to meat hygiene and so of public health importance, and the whole complex is illustrated in Table 1. The main parasitic conditions are enumerated below.

Muscle cysts

Cysticerci, the larval stages of tapeworms of the genus *Taenia*, were recovered from the muscular tissue of all antelope species and buffalo. The infection rate in some animals was high, indicating a very close relation between the herbivorous intermediate hosts and the carnivorous animals which prey on them. The adult taenias of predatory carnivores found to be responsible for the infection with muscle cysticerci were *Taenia gonyamai* of lion and *T. crocutae* of the spotted hyaena. A specific host/parasite relation does obviously exist between impala and wildebeest muscle cysts and the lion tapeworm, and between Grant and Thomson gazelle muscle cysts and the hyaena tapeworm. Differentiation was possible by using rostellar hooklets which bear the same morphological characteristics in both larval and adult parasites (Fig. 1). Of special interest is the heavy infestation of Grant in contrast to the low infestation of Thomson gazelles with muscle cysticerci, since both are animals of the same genus with similar behaviour and food preferences.

Viscera cysts

Cysticerci in the viscera attached to internal organs and serosal membranes, were recovered from all antelope species. Microscopic examination and comparison of the rostellar hooks revealed their difference from the muscle cysts. Those of the larger antelopes, wildebeest, topi and hartebeest were recognized as the larval stages of *Taenia regis* of the lion. Those of the smaller gazelles were different but are not yet identified.

Sacrum cysts

These resembled the viscera cysts in size but were located only inside the sacrum bone, and were obviously specifically adapted to antelopes of the subfamily Alcelaphinae (the hartebeest group). Because of their strange localization within the body they need a bone-eating animal to complete their life cycle, and the spotted hyaena was found to be the carrier of the adult tapeworm stage *Taenia olngojinei* (Dinnik & Sachs 1969).

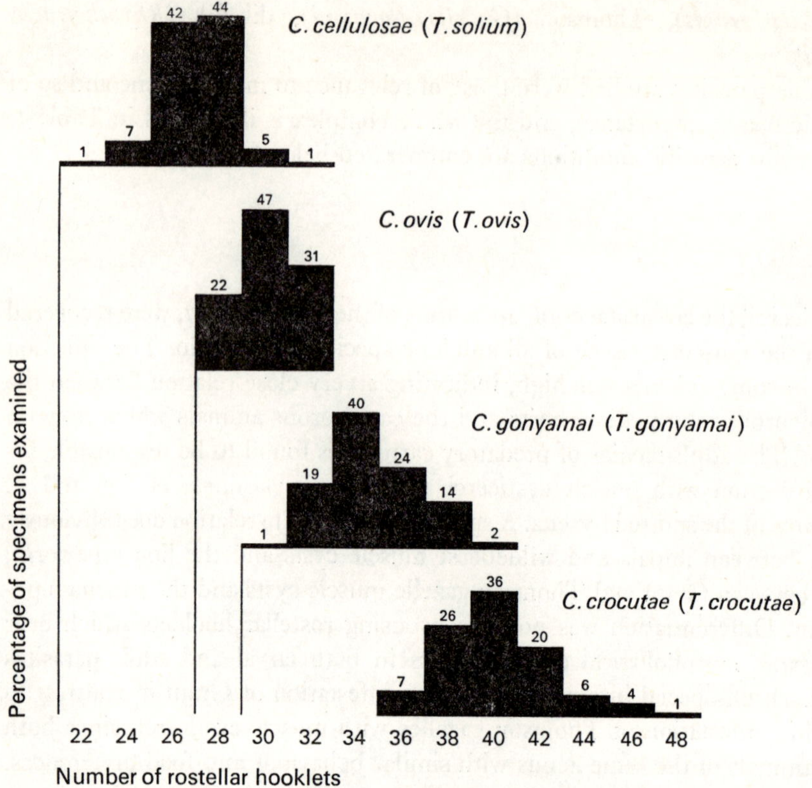

Figure 1. Number of rostellar hooks of muscle cysticerci as a possible means of identification. Note: *T. saginata* cysticerci (*Cysticercus bovis*) possess no rostellar hooks.

Hydatid cysts

These larval stages of *Echinococcus* tapeworms were found in the lungs of a few antelope, in the liver of giraffe and more frequently in the liver of warthog. Adult *Echinococcus granulosus felidis* were recovered from lions in the study area. Because the observations were made in a region free from domestic stock, a specific sylvatic lion/warthog cycle of echinococcosis seemingly occurs.

Spargana

These are larval stages of diphyllobotrid tapeworms, and were recovered from larger herbivores but not from the smaller antelopes. The preferred site of the parasite was the subcutaneous tissue especially in the region of the hock joints in herbivores; in warthog, the tongue musculature was predominantly infected. Tapeworms of the genus *Spirometra* found in various carnivorous animals in the area were responsible for the condition in the intermediate mammalian host.

Filaria in subcutis

Whitish filarial nematodes were found in subcutis and between the muscles in connective tissue. They were identified as *Dirofilaria* species. Onchocerciasis was not found in the game animals examined.

Nematodes in musculature

In zebra, the musculature of the abdominal wall frequently contained nematodes which were identified as juvenile strongyles specific to the horse family. Nodules collected from the muscles of buffalo contained small helminths not yet identified.

Sarcocystis

The whitish cysts of Miescher's tubes were recovered from the musculature of many animals. The sarcosporidia cysts varied in size from barely visible to the naked eye, to spindle-shaped and thumb-sized nodules. Buffalo carried the largest cysts and were frequently heavily infested, rendering such carcasses unfit for human consumption. The marked differences in shape and size suggest that the various game animals harbour different species of sarcocystis. Microscopic examination of smears made from the heart musculature showed that many animals not identified by naked-eye meat inspection were carriers of parasites.

Liver flukes

These were found infrequently in some of the free-living antelopes examined, and were identified as *Fasciola gigantica,* known also to infect cattle.

Liver tapeworms

Tapeworms of the genus *Stilesia* were found in the bile ducts of three taxonomically and ecologically unrelated antelope species. Domestic sheep and goats are known to be carriers of *Stilesia hepatica*, but game parasites have not yet been identified to species.

Liver nematodes

Impala carried the round worm *Cooperioides hepaticae* in the bile ducts, often associated with *Stilesia* infection, whilst zebra livers were often heavily parasitized by young *Delafondia vulgaris* parasites.

Pentastomids

Larvae of this tongue-shaped parasite were found in liver, kidney and lymph nodes of their intermediate hosts. After being ingested by a suitable carnivore (*Linguatula multiannulata* in hyaena, *L. serrata* in jackal and wild dog, and *Neolinguatula nuttalli* in lion and leopard) they develop in the nasal cavity. Although young stages of these three species are found in many herbivorous hosts, the adult stages are very specific, solely occurring in their host families—Hyaenidae, Canidae or Felidae.

Dictyocaulids

The long whitish lungworms *Dictyocaulus viviparus*, known also to infect domestic cattle, were found in the bronchi and trachea of the three Alcelaphinae antelope subfamilies. Zebra were infected with *D. arnfieldi*, a parasite of horses.

Large protostrongylids

Wildebeest, topi and hartebeest were carriers of a large, darkish roundworm which occurred in walnut-sized nodules within the lung tissue. The parasite was identified as *Protostrongylus africanus*, showing a marked host group specificity to the antelopes of the subfamily Alcelaphinae.

Small protostrongylids

These parasites, barely visible to the naked eye, produce significant lesions at the lung lobe margins, and were found in almost every animal of the hartebeest group, the two gazelle species, and impala. *Protostrongylus* (*P. gazellae* and *P. etoshai*), and *Pneumostrongylus* (*P. calcaratus* and *P. cornigerus*) were identified. A very close host/parasite adaptation exists as domestic animals were not found infected with the lungworm species identified from game in the study area.

Syngamids

Syngamid worms occurring in mammalian hosts belong to the genus *Mammomonogamus* and resemble the gape worm of domestic fowl (*Syngamus trachea*). *Mammomonogamus* worms were recorded from the trachea of elephant and hippo in Africa, and have now been recovered from the nasal cavities of waterbuck in the Serengeti. A specific host/parasite relation seemingly exists with the kob-family, as we found similar worms in the nasal cavity of puku (*Kobus vardoni*) and Uganda kob (*Kobus kob*).

Oestridae in the nose

The maggots of the nasal fly were frequently found in frontal sinuses and head cavities of the hartebeest group, giraffe and zebra. The latter harboured *Rhinoestrus usbekistanicus*, and *R. giraffae* was identified from giraffe. The three Alcelaphinae antelopes were almost always highly infected with *Oestrus aureoargentatus*, *O. variolosus*, *Kirkioestrus minutus*, *Gedoelstia cristata* and *G. haessleri*. Domestic sheep and goats in the study area were infected exclusively with *Oestrus ovis* which was never recovered from the wild antelopes.

Oestridae in the brain

First stage larvae of *Gedoelstia haessleri* were frequently found on the brain of wildebeest, associated with inflammation and tissue reaction.

It is clear that East African game carry a rich parasite fauna and that its transmission through the ecological food web is both complex and highly evolved. When man alters this ecology, he may well find himself involved in a parasitic chain and the origins of a new disease syndrome. Two examples will suffice to illustrate this.

The case of *Trichinella spiralis*

For a long time this parasite was not thought to occur in Africa south of the Sahara, this conclusion being based on the negative results from examining domestic pigs, which were obtained at various abattoirs in South Africa and Kenya (Cluver 1958, Ginsberg 1958). When the first human cases of trichinellosis in East Africa were diagnosed in 1959 (Forrester *et al.* 1961), they were attributed to the consumption of infected meat from wild suids such as the bush pig and giant forest hog (Forrester 1964). In West Africa, smoked ham prepared from warthog was the source of further human infections (Grétillat & Vassiliadès 1967).

Nelson *et al.* (1963) searching for the disease origin found a variety of wild East African carnivores in Kenya and Tanzania were natural carriers of the parasite. We undertook similar investigations in the Serengeti National Park, and found *T. spiralis* in lions, leopards and jackals, and in fifteen out of twenty spotted hyaenas (Sachs 1970). In the Kruger National Park in South Africa a similarly high percentage of the hyaena population was found to be infected with trichinellosis (Young & Kruger 1967).

These findings confirmed that *T. spiralis* is well established in Africa, and is neither restricted to moderate climate-zones nor to the pig/rat cycle. The main carrier in Africa is obviously the hyaena, and its omnivorous scavenging behaviour offers very favourable dispersal opportunities for the parasite. What is not clear is why the warthogs of the Serengeti were not infected. In areas where lions and hyaenas guard their kill, these omnivorous suids (which are themselves prey for the carnivores) rarely have access to the remains of an animal carcass. Consequently, they have less chance to become infected. Where hyaenas and other carnivorous predators are removed by man, as is probably the case in West Africa, and where wild omnivores take over the role as scavengers, warthogs become infected and open the route to man.

It is not improbable that the present situation in Africa reflects the original situation in Europe, when wild carnivores such as bears and wolves roamed the country. After their extermination by expanding human settlement, the dynamics of the disease changed. This assumption is supported by the report of Schoop & Schade (1939) who found that twelve out of thirteen outbreaks of trichinellosis in domestic pigs in Germany occurred after wild fox carcasses had been fed to domestic pigs. In the light of these data, the theory that trichinellosis was brought to Europe by domestic pigs imported from China at the beginning of the 19th century (Bartels 1968) is doubtful. I believe that in Europe its original and sylvatic cycle is an infection chain of carnivorous animals.

The existence of a focus of human pathogenic parasites in game animal populations is a continuous danger. Man can become infected directly, as with trichinellosis when animal species used as food (wild suids) have obtained the infection from a primary sylvatic cycle (wild carnivores), or when the carcass of the natural carrier (wild carnivore) is fed to a domestic food animal (domestic pig). In the latter case, the parasite gains access to the domestic animal/man cycle.

The case of *Paragonimus*

The West African lung fluke infection of man and animals which was recently epidemic in Nigeria (Voelker *et al.* 1975), is a food-borne disease caused by a trematode of the genus *Paragonimus*. The life cycle of the parasite involves two invertebrate intermediate hosts.

The natural host in Nigeria is *Viverra civetta*, the African civet (Voelker & Sachs 1974) although in neighbouring Cameroon, the mongoose *Crossarchus obscurus* was found to be the main natural reservoir (Vogel & Crewe 1965). Two *Paragonimus* species are involved, *P. africanus* being predominant in Cameroon, and *P. uterobilateralis* found in Nigeria and Liberia (Voelker & Vogel 1965).

The carnivores carry the adult parasites in the lung tissue and these produce eggs containing miracidia. These must find a suitable snail host for further development to the cercarial stage. The cercariae invade fresh-water crabs and develop to infectious metacercariae in the musculature of this second intermediate host. Any crab-eating mammal is a potential final host.

Infection of man depends on his food and his methods of preparing it. If crabs used for consumption are adequately boiled or roasted, the infectious stages within the crab's flesh are destroyed. In East Asia where the disease also occurs but is caused by other *Paragonimus* species, the parasite finds ideal conditions for its survival. Many dishes are prepared from raw crabs and the juice of fresh crabs is often added to improve the flavour. Africans however do not normally eat raw meat or raw crabs, and the epidemic in the Eastern province of Nigeria was the result of the Nigerian Civil War 1967/70. The starving human population in the war area used any available protein, including fresh-water crabs, and contrary to normal practice, these were often consumed raw or insufficiently cooked, partly through hunger, and partly in an attempt to avoid making fires which would have attracted the attention of approaching enemy troops. Man, by means of abnormal circumstances, was thus accidentally included in the life cycle of a parasite that was able to adapt itself to the new host.

The parasite fauna of African mammals still has the potential for creating further parasite problems both for man and his domestic animals and further

L

research is needed if we are to anticipate the consequences of changes in tropical ecosystems, and in man's diet.

References

BARTELS H. (1968) *Die Untersuchung der Schlachttiere und des Fleisches*. Paul Parey.

CLUVER E.H. (1958) *Public Health in South Africa*. Central News Agency Ltd, South Africa.

DASMANN R.F. (1964) *African Game Ranching*. Pergamon Press, London and Macmillan, New York.

DINNIK J.A. & SACHS R. (1969) Zystizerkose der Kreuzbeinwirbel bei Antilopen und *Taenia olngojinei* sp. nov. der Tüpfelhyaene. *Z. Parasitenk.* 31, 326–39.

FORRESTER A.T.T. (1964) Human trichinellosis in Kenya. *Proc. 1st Int. Congr. Parasit.*, Rome 2, 669–771.

FORRESTER A.T.T., NELSON G.S. & SANDER G. (1961) The first record of an outbreak of trichinosis in Africa south of the Sahara. *Trans. R. Soc. trop. Med. Hyg.* 55, 503–15.

GINSBERG A. (1958) Helminthic zoonoses in meat inspection. *Bull. epiz. Dis. Afr.* 6, 141–9.

GRÉTILLAT S. & VASSILIADÈS G. (1967) Présence de *Trichinella spiralis* (Owen, 1835) chez les carnivores et suidés sauvages de la région du delta du fleuve Sénégal. *C.R. Acad. Sci. (Paris)* 266, 1139–41.

NELSON G.S., GUGGISBERG C.W.A. & MUKUNDI J. (1963) Animal hosts of *Trichinella spiralis* in East Africa. *Ann. trop. Med. Parasit.* 57, 332–46.

SACHS R. (1970) Zur Epidemiologie der Trichinellose in Afrika. *Z. Tropenmed. Parasit.* 21, 117–26.

SCHOOP G. & SCHADE M. (1939) Der Fuchs als Verbreiter der Trichinosis. *Dtsch. Tierärztl. Wchschr.* 47, 553–61.

VOELKER J. & SACHS R. (1974) Observations on the life-cycle of *Paragonimus uterobilateralis*: the African civet (*Viverra civetta*) as natural reservoir in Nigeria. *Proc. 3rd Int. Congr. Parasit.*, München 1, 529–30.

VOELKER J., SACHS R., VOLKMER K.J. & BRABAND H. (1975) On the epidemiology of paragonimiasis in man and animals in Nigeria, West Africa. *Vet. Med. Rev. (Bayer)*, No. 1/2, 161–75.

VOELKER J. & VOGEL H. (1965) Zwei neue *Paragonimus*-Arten aus West-Afrika, *Paragonimus africanus* und *Paragonimus uterobilateralis* (Troglotrematidae, Trematoda). *Z. Tropenmed. Parasit.* 16, 125–48.

VOGEL H. & CREWE W. (1965) Beobachtungen über die Lungenegelinfektion in Kamerun (Westafrika). *Z. Tropenmed. Parasit.* 16, 109–25.

YOUNG E. & KRUGER S.P. (1967) *Trichinella spiralis* (Owen 1835) Railliet, 1895 infestation of wild carnivores and rodents in South Africa. *J. S. Afr. vet. med. Ass.* 38, 441–3.

Crop pest problems resulting from chemical control

T. H. COAKER *Department of Applied Biology, Pembroke Street, Cambridge*

Changes in the pest status of phytophagous arthropods can arise from a variety of causes ranging from changes in agronomical practices and weather to pesticide use. It is well known that many pest outbreaks following the use of various chemicals to control insects and fungi have on investigation been found to be due to the adverse effects of the pesticide on the natural enemies of the pest. In many cases however, this has been presumed; the outbreaks for example, may have arisen from direct effects of the pesticide on the fecundity of the pest, from the elimination of its competitors or from the improved nutritional status of the host-plant resulting from pesticide application. Some of these outbreaks have arisen from resurgences of treated populations of established major pests, others of species of secondary or potential importance. Outbreaks have also occurred following the development of strains resistant to pesticides. This paper will be mainly confined to the phenomenon of resurgence.

Resurgence can occur when the target species that was initially suppressed by the insecticidal treatment shows a rapid recovery after decline of the treatment effect. This is characteristic of pest situations where natural regulation and weather conditions provide only a partial suppression of the pest's numbers. Secondary or sporadic pests vary in importance and may create problems in localized areas or in certain years. Their numbers are usually under adequate regulation from biotic and abiotic factors which may occasionally break down allowing the pest to exceed its economic injury threshold. Potential pests unlike secondary pests may rarely cause economic injury unless some change in the agroecosystem occurs that raises them to secondary or major importance.

Resurgence is one of the more interesting phenomena associated with the use of insecticides, particularly since the introduction of synthetic compounds, and constitutes a major weakness of this method of insect control. It may be speculated as to whether or not this is the principle limitation that may lead to the obsolescence of chemical control, at least as far as the broad spectrum compounds are concerned. Certainly, the early expectations of synthetic insecticides being able to bring about a more or less permanent reduction in crop pest populations have not been widely realized. It is also generally true that pest control problems need to be faced repetitively each season with populations similar in size to the previous season regardless of the intensity of pesticide use.

Outbreaks following pesticide use

If insecticides are responsible for outbreaks of secondary pests and resurgence of treated ones, then evidence for this should be found from areas and crops that have been most heavily treated. In California, for example, out of the twenty-five major crop pests, twenty are resistant to one or more insecticides, twenty-two have been involved in resurgences or secondary pest outbreaks and nine, as a consequence of chemical treatments applied to control them, have produced outbreaks in other species. The following case histories illustrate the changes in pest status that have occurred where a complexity of phytophagous species was present in a heavily sprayed crop.

A substantial change in the pest status of the insects present on cotton in the southern United States of America occurred within four distinct periods; the pre-insecticide, the calcium arsenate, the organochlorine and the organo-phosphorus periods (Newsom 1974). The introduced boll weevil, *Anthonomus grandis* Boheman, and the cotton leaf worm, *Alabama argillacea* (Hübner), were the major pests prior to the introduction of calcium arsenate. During its period of use between 1924 and 1945, the bollworm, *Heliothis zea* (Boddie), and the cotton aphid, *Aphis gossypii* Glover, rose to major pest status; the bugs, *Lygus lineolaris* (P. de B.), and *Pseudatomoscelis seriatus* (Reuter) became more common and the thrips, *Frankliniella* spp., were recorded as a pest for the first time during 1930. After the organochlorines replaced calcium arsenate, the spider mites, *Tetranychus* spp., tobacco bud worm, *Heliothis virescens* Fab., cabbage looper, *Trichoplusia ni* (Hübner), and beet army worm, *Spodoptera exigua* (Hübner), also began to cause injury to the crop. Prior to this time none of these native species had caused concern as cotton pests in this region. In 1955, methyl parathion was added, to control the organochlorine-resistant boll weevil, to the spray programme which also included DDT and toxaphene, resulting in further changes. The tobacco bud worm became a major pest in 1964 and, on the other hand, the cotton leaf worm and the cotton aphid declined from major to minor pest status.

Pierce (1922) listed twenty phytophagous insect and mite species from cotton in the southern United States of America during the pre-insecticide period. There has been no change in the pest status of sixteen of these and therefore they could not be presumed to have been potential pests under the circumstances described although they have all been reported to feed on cotton. However, eight other species, not recorded by Pierce possibly because they were insignificant at the time, are currently major pests (Newsom 1974).

Another significant feature of this situation is that the pest status of the cotton leaf worm and the cotton aphid declined after the introduction of methyl parathion. Both the organochlorine and organophosphorus compounds used could be expected to be toxic to the natural enemies of these species, so the reason for the decline in their status is that parathion effectively maintains their numbers below economic threshold, a situation unlikely to change until resistance develops (Newsom & Brazzel 1968). The increased pest status of the banded white fly, *Trialeurodes abutilonea* (Haldeman) provides some evidence to support this view. This species first attracted attention after the organochlorines were applied to cotton, but its numbers were not serious until methyl parathion had been used for ten years, when resistance to this insecticide was confirmed. Where organophosphorus compounds have not been used, e.g. on soybean, the whitefly is still maintained below pest status by its natural enemies even where soybean is grown in rotation with cotton heavily treated with insecticides. Where cotton is not treated intensively it is still an unimportant species on that crop (Watve 1971).

Although these events suggest that insecticides have been responsible for the changes that have occurred on cotton over the past fifty years, possibly through their suppression of natural enemy populations, there is little information to substantiate this view, no doubt because of the difficulty in studying such complex relationships.

Similar changes in the pest complex in apple orchards have occurred in the south Tyrol of Italy following the intensive use of insecticides (Kolbe & Kremer 1969, Kremer 1971). Initially five of the twenty phytophagous arthropod species present were considered worthy of insecticide control. By 1971 this number had increased in commercial orchards to six but now includes only one of the original five pests, *Psylla piri* (L.), the remainder were not included in the original list. The reason for this change was thought to be due to the eradication of those species with a low reproductive potential and the survival of those with a high one. Consequently, in this area some of the common apple pests such as codling moth, *Cydia pomonella* (L.), apple blossom weevil, *Anthonomus pomorum* (L.), winter moth, *Operophtera brumata* (L.), saw flies, *Holocampa* spp., and apple sucker, *Psylla mali* (Schm.) are no longer found in commercial orchards.

These examples illustrate the kind of changes to the pest status of phytophagous insects and mites that have taken place since the introduction of pesticides. It is also indicated from these examples, although perhaps a fact that is often ignored, that changes have occurred in both directions, i.e. some species have risen from secondary to major pest status while others have been eradicated from both classes.

The causes of pest outbreaks

The toxicity of a pesticide to an organism is dependent on its dose and its effects are of two general kinds, direct lethal and sublethal on the organism itself and indirect on the other species present such as natural enemies, competitors and prey (Dempster 1975).

Unquestionably, many of the so-called short-term effects, that is those immediately following pesticide application, have probably arisen from the disruptive effects of pesticides upon the natural enemies (Ripper 1956); but this is not always so, there being a definite tendency to associate changes in pest status with natural enemy suppression. In consequence, little attention has been given to the cumulative or long-term effects of non-selective pesticides on insect populations, with the exception of the effect on the segregation of resistant strains (Georghiou 1972).

Correlation between natural enemy elimination either by pesticides, hand removal or other exclusion procedures with the subsequent increase of many kinds of pests has provided evidence that natural enemy destruction is a reasonable cause-effect explanation for most pesticide-induced outbreaks of crop pests. This explanation does not require the assumption of undemonstrated and complex mechanisms of translocation of the pesticide through the host plant or of direct contact effects by the pesticide to produce reproductive stimulation on the range of species involved (Putman 1963). Support for this latter view has, however, arisen due to the dissatisfaction with the general tendency to ascribe observed effects as being caused by natural enemy destruction, particularly when by inference it is assumed that low incipient pest densities are maintained at that level by natural enemies alone. It is, therefore, not without justification that the contention that correlations of pesticide destruction of the natural enemies and pest outbreaks frequently do not prove the cause-effect relationship and that population increase may well arise for example, from plant nutritional changes unrelated to pesticide use (Rodriguez 1958, Post 1963, Chaboussou 1965, Storms 1965). Most of these studies were, however, short-term ones and carried out under artificial conditions thereby ignoring the ecological consequences. For example, increased fecundity does not automatically mean increased density, at least not to outbreak status, it ignores compensatory mortality (i.e. of a physical, physiological or ecological nature) and dispersion (van de Vrie 1974).

OUTBREAKS DUE TO INDIRECT EFFECTS

Ripper (1956) listed fifty arthropod species whose populations had increased following pesticide application, many of which were claimed to be the consequence of an indirect effect on the natural enemies. There is also considerable evidence to show that most synthetic insecticides exhibit broad spectrum activity on beneficial arthropods such as phytoseiid mites and other mite predators, predaceous coccinellids and carabids, although some are relatively non-toxic (Steiner 1959, Bartlett 1963, 1964, Colborn & Asquith 1971, Lingren & Ridgway 1967, Mowat & Coaker 1967, Edwards & Thompson 1975). It has also been shown that various predatory species can be poisoned by feeding on prey that have been feeding on plants treated with systemic insecticides (Ahmed *et al.* 1964, McClanahan 1967). While there is ample evidence of destructive effects on natural enemies, the criteria required to justify the conclusion that the removal of parasites and predators is responsible for the resurgence of pest populations are more tenuous and most cases, at best, are usually based on negative correlation between the number of natural enemies and those of the pest.

With a fuller understanding of the role of natural enemies in the regulation of pest species, however, the interaction of a pesticide on the pest/natural enemy complex can be more positively interpreted.

The number of cabbage rootfly, *Erioischia brassicae* (Bouché), is determined in Britain by egg loss due to predation by ground-dwelling coleoptera (Hughes & Mitchell 1960, Benson 1973). About sixty species of Carabidae and Staphylinidae are known to prey on the immature stages (Coaker & Williams 1963) and the survival of the pest is negatively correlated with a predatory index (sum of the number × relative predatory value for each species) for the predators within the crop (Coaker 1965) and, also with the more effective and numerically abundant predator, *Bembidion lampros* (Herbert) (Wright, Hughes & Worrall 1960). Soil residues of aldrin and dieldrin at concentrations not directly toxic to cabbage root fly eggs and larvae are toxic to some of the predatory species, due to the greater mobility of the predators and their continual contact with the residues in the soil. *Aleochara* spp. are the most susceptible being killed at concentrations of less than 0·1 ppm. *Bembidion lampros* is also killed by equally low concentrations after an initial phase of greater activity which increased the numbers trapped on plots containing less than 0·5 ppm in the soil. Other predatory species were less susceptible.

The reduction of predators implied by the trapping records increased the survival of cabbage root fly eggs and larvae and thus augmented damage to the infested brassicae crop which resulted in reductions in yield up to 70% (Coaker 1966).

Similar resurgence occurred with the small cabbage white butterfly *Pieris rapae* (L.) following the suppression of its natural enemies by DDT applied as a foliar spray and from the residues, arising from the spray, in the soil (Dempster 1968a). As with cabbage root fly the principle predators are ground-dwelling arthropods which, by climbing up the plant, feed on 50–60% of the young caterpillars (Dempster 1967). Again not all the predators were susceptible to the insecticide concentrations present in the soil and, *Nebria brevicollis* (F.) and *Trechus quadristriatus* (Schrank) were found in larger numbers on the DDT-treated plots than on the untreated ones. In this experiment DDT also killed the mesostigmatid mite predators of Collembola causing the Collembola to increase in numbers and these beetles to aggregate on the treated plots. Together with these disturbances, there was an increase of cabbage aphid *Brevicoryne brassicae* (L.) and the slug *Agriolimax reticulatus* (Muell.) on the treated plots (Dempster 1968b).

Although it is difficult to separate direct from indirect effects of the insecticides without considerable ecological knowledge of the species, in these examples the upsurgence of the two principle pest species being studied probably arose from the same cause, the destruction of their natural enemies.

Without equivalent quantitative knowledge of the mortality factors, Pimentel (1961) also concluded brassica pest resurgence to be mostly due to insecticide suppression of the natural enemies whereas Root & Skelsey (1969) thought the reasons to be more complex and to include reduction in interspecific competition since aphid outbreaks occurred when the density of other phytophagous species was reduced, and declined after the insecticide (carbaryl) was withdrawn.

Agricultural pesticides, especially insecticides, acaricides and fungicides can have a marked effect on predators of mites and the literature on this topic has been reviewed by Huffaker, van de Vrie & McMurtry (1970) and McMurtry, Huffaker & van de Vrie (1970). Most of the published accounts deal with the Tetranychidae particularly in association with orchard and citrus crops and one of the best documented examples of this is the fruit tree red spider mite *Panonychus ulmi* (Koch) which in most commercial orchards before 1925 in England, was mainly found in low numbers (Massee 1958), in fact it was even once thought to be a beneficial insect. In southern England at the beginning of the century about sixty insect species were recognized as potential apple pests and the demand for high quality produce led to the introduction of caustic soda winter wash, grease banding and whitewashing of the trunks to control the apple sucker, winter moth caterpillars, apple blossom weevil, apple saw fly and codling moth, the major pests at the time. This period, 1909–21, marked the beginning of a pest control programme in orchards, albeit an ineffective one as large pest populations existed during the spring and summer, nevertheless, predatory mirids and coccinellids were present amongst the beneficial insects. After this time the number of red spider and the apple capsid, *Plesiocoris rugicollis* (Fall), also previously a potential pest, increased following the use of tar oil winter wash which was not toxic to the overwintering red spider eggs but was lethal to the eggs and hibernating adults of many insect predators of the mite. Their destruction accounted for the rapid build-up of the mite population in the spring after hatching from the overwintering eggs, to persist unchecked through its several generations in the summer. In 1931–1932 DNOC was introduced to destroy the eggs of the red spider and capsid and, although it eventually replaced the tar oil washes, the capsid remained a major pest. DNOC did, however, succeed in eliminating the important predators that overwintered on the trees with the exception of the black-kneed capsid, *Blepharidopterus angulatus* (Fall.) which survived all the winter washes in general use. A further impetus was given to the red spider populations following the introduction of DDT and BHC in 1946 to control the apple capsid, apple blossom weevil, apple saw fly and winter moth caterpillars. The red spider then needed to be controlled several times annually, a practice that led to its development of resistance to most acaricides.

Similar outbreaks of various tetranychid mites induced by DDT and other broad spectrum pesticides are well known from many other parts of the world. In Nova Scotia, for example, orchards treated with sulphur sprays also suffered infestations of the red spider as well as increased damage by codling moth, where previously when treated with copper fungicides, red spider mite damage was unknown. The replacement of lead arsenate with DDT for codling moth control after resistance had developed to it in 1949 led to outbreaks of red spider greater than had been experienced before (Pickett & Patterson 1953).

In southern England, few of the forty species of predaceous insects and mites that feed on red spider survive the sprays applied to commercial orchards, particularly when broad spectrum compounds are used (Collyer 1953, Morris 1968). The effect of any one chemical may, however, be complicated, as well as being toxic to predators it may also stimulate the mite's reproduction.

Most pesticides used today have a more or less broad spectrum effect although some are relatively innocuous to certain natural enemies, or in particular situations. For example, the organochlorines, particularly DDT, may be less toxic to anthocorid and coccinellid predators than organophosphorus materials, but more toxic to mirids and predatory mites (Gratwick 1965); equally they may be more or less toxic to the immature stages of the predators than to the adult stages (Lingren & Ridgway 1967). Long-term effects may also be more detrimental than the immediate or direct effects. The predator may die out from the lack of food if the pesticide annihilates the prey species or another species providing an alternative food so that the pest is released from the predatory control and is then free to increase (Oatman 1965). Treatments may even increase certain predators by reducing their competitors (Cone 1963). They may also inhibit fungi, lichens and mosses furnishing alternative foods, cover or niches for predators. Low toxicity materials or sublethal doses may even reduce predator reproduction (van de Vrie 1962) and several years may have to elapse after the curtailment of the treatments before effective predation is restored.

As well as pesticides destroying parasites, predators and competitors they can also be inhibitory to the growth of entomopathogenic fungi. Included amongst the chemicals with inhibitory effects are not only fungicides (Wilding 1972) but also insecticides and acaricides that can be inhibitory to certain organisms at one-tenth of their recommended doses (Olmert & Kenneth 1974). The disruption of the fungi would likewise reduce their contribution to the natural regulation of the pest species (Ignoffo *et al.* 1975).

OUTBREAKS DUE TO DIRECT EFFECTS

The observed population increase of mites and aphids following pesticide application often seems best fitted to the pest-stimulation explanation. The delayed responses following pesticide application may be explained by the stimulation occurring through the plant rather than as a direct sublethal effect on the pest, due to the period necessary for absorption and translocation of the pesticide or its metabolites in the plant, before production of the stimulatory effect. Variations in the response would also depend on the dosage of the pesticide.

The sporadic nature of mite outbreaks caused by certain pesticides also provides some support for the stimulation view. Both DDT and carbaryl, while providing similar effects on mite predators, do not produce outbreaks with similar frequency, the effect of carbaryl being more consistent than DDT. Also some other materials that have low toxicity to the natural enemies are also difficult to reconcile with the opinion that resurgence is entirely due to natural enemy suppression (Bartlett 1968). Although a considerable amount of work has been done on this topic (van de Vrie,

McMurty & Huffaker (1972) quote over 120 references) no generally accepted hypothesis has yet been reached as to the cause.

Much of the early work was done with *Panonychus ulmi* on apple and it is characteristic that whenever field experiments with *P. ulmi* were made a population increase supposedly due to pesticide treatment was observed. When exact laboratory investigations were carried out, however, an increased fertility due to the direct contact with the insecticide residues on the leaves was not always obtained.

Hueck *et al.* (1952) showed in both field and laboratory experiments that DDT increased the fecundity of *P. ulmi* on apple by 50% in contrast to mites on untreated surfaces but these experiments were not designed to distinguish whether the stimulatory factor acted directly on the mite or through the leaves. Pielou (1960) did not observe such an effect and questioned the validity of Hueck's results because of overcrowding, low longevity and fecundity of *P. ulmi* in his experiments. Löcher (1958) using the more easily cultured mite *Tetranychus urticae* (Koch) on beans, obtained over a 70% increase in fecundity and claimed a direct effect from DDT since positive or negative results were obtained depending on whether juvenile or adult stages were used, he also found that the F_1 generation was stimulated following the treated one. Some of these findings were confirmed by Seifert (1961) who, from histological studies, observed increased oogenesis. These results generated much similar work but Attiah & Boudreaux (1954) amongst others, failed to show that DDT stimulated *Tetranychus urticae* and *T. cinnabarinus* (Boisduval).

Huffaker (1948) and Huffaker & Spitzer (1950) were the first to induce increased fecundity in *T. urticae* on DDT-treated beans and claimed the effect to be of a nutritional or hormonal nature within the host-plant. Stimulation by this indirect means would appear to be delicately balanced with the concentration of the insecticide, as Fleschner (1958), Saini & Cutkomp (1966) induced more rapid reproduction from mites placed on untreated terminal leaves above DDT-treated basal leaves, only at certain concentrations.

Chaboussou (1966) obtained an increase of protein N and K on leaves treated with DDT, carbaryl, diazinon, parathion and certain fungicides which in turn were correlated with mite fecundity. He showed that insecticides can affect the Ca:K ratio in leaves and that reduction of this ratio is inversely correlated with amino acid and reducing sugar content, a factor considered to be the primary reason for *Eotetranychus carpini* (Duds.) increase following the application of carbaryl and the reduction of the K:Ca ratio in grape leaves. Pielou (1962) and Cone (1963) on the other hand were unable to obtain a response with *Tetranychus urticae* and *Panonychus ulmi* from several insecticides including DDT and carbaryl.

More recent field results from Egypt that indicated a strong stimulation of *Tetranychus cinnabarinus* and *T. arabicus* Attiah from DDT and carbaryl treatments on cotton encouraged Dittrich, Streibert & Bathe (1974) to reinvestigate this phenomenon. Parent generations of *T. urtica* were kept on DDT and carbaryl residues on bean leaves at concentrations likely to produce hormoligosis, a sublethal effect of the stressing-agent causing stimulation to an organism and providing it with increased sensitivity to respond to changes in the environment (Luckey 1968). The mites were placed on freshly prepared detached leaves every fourth day to obtain maximum stimulation and to avoid secondary effects through the plant. Crowding was also avoided in the parent and F_1 generations. These procedures were not fully adopted in the previous studies and may have contributed to their inconsistencies. Both insecticides increased fecundity and also shifted the $\male:\female$ ratio in favour of the females in the F_1, suggesting a stimulatory effect on both sexes by increasing sperm production in the parent male and fecundity in both female generations. The authors concluded that the stimulation was a direct one and not via the plant and that both the higher number of eggs produced and the increased proportion of females could, therefore, contribute to resurgence of the mite in the field after DDT and carbaryl treatments.

DDT-treated plants also appear to cause mites to disperse because of irritation and repellancy (Davies 1952, Pielou 1960, Attiah & Boudreaux 1964) and may result in mites moving to the leaves uninhabited by natural enemies. Attiah & Boubreaux (1964) showed, however, that only relatively high DDT residues appreciably affected dispersion in *Tetranychus urticae* and *T. cinnabarinus* and that the advantages gained were offset by the toxic effects of the DDT on the immature stages.

Change in pest status following herbicide use

Herbicides have permitted crop husbandry practices to change dramatically over recent years and in doing so have improved the opportunities for certain crop pests. The new situations may provide less diversity of plant species from the time of crop emergence, facilitated by pre-emergence herbicides, or greater continuity, in that habitats are not so drastically disturbed by minimal cultivation techniques as with ploughing; in both cases an increase in pest problems has occurred.

Recent changes in sugar beet husbandry and the introduction of 'planting-to-a-stand' appears to have increased the pest status of the springtail *Onychiurus armatus* (Tull.) (Heijbroek 1971). Although regarded as an important pest of sugar beet seedlings because it feeds on the young roots, it has been shown to cause a greater loss of plant stand following application of selective pre-emergence herbicides. The herbicides by removing the alternative food plants of the springtail causes them to aggregate around the beet seedlings and at high densities have reduced the seedling population by 50% compared with crops treated with a post-emergence herbicide.

The discovery of the bipyridyl herbicides in the late 1950's lifted one of the constraints limiting the technique of direct drilling. This practice provides a very different habitat from ploughed soil for a number of soil-dwelling arthropods. Edwards (1975) found that not only were beneficial organisms (litter feeders) more numerous following direct drilling but that some pest species were also encouraged by the system. Slug attacks were found to be much greater particularly in mild, wet winters and when oilseed rape occurred in the rotation. Wireworms and dipterous stem borers can also be a danger to direct drilled crops following grass or even sometimes in continuous cereals. Other pests such as leatherjackets, chafers, cereal and root aphids, cutworms and army worms may also produce greater problems on direct drilled crops (Phillips & Young 1973, Edwards 1975). The numbers of carabid beetles that prey on some of these pests were also found to be fewer on direct drilled crops (Edwards 1975), a factor which might favour these pests.

Plant nematodes are also affected by this system and invasion of oat roots by *Heterodera avenae* (Woll.) was almost doubled by direct drilling as compared with ploughing (A. Spaull personal communication).

Oat seedling infestation of both resistant and susceptible cultivars by *Ditylenchus dipsaci* (Kuhn) has also been shown to increase following the application of 2,4-D before tillering and below recommended rates. The herbicide caused plant cell hypertrophy and proliferation which is a condition favourable to nematode development (Webster 1967).

Conclusion

Crop pest problems resulting from chemical control take many forms amongst which the most important is the development of resistance to insecticides. The widespread nature and acceleration in the development of resistant strains is such that it has been suggested that unless effective alternative control measures are developed some agricultural land could conceivably be taken out of production for certain crops.

Resurgence of crop pests treated with pesticides has also caused many problems. The mechanisms involved in this phenomenon are clearly not fully understood in most cases, although correlation evidence strongly supports the opinion that natural enemy suppression is the cause. Resurgence is characterized by an abnormally rapid return of the pest population to an economic injury threshold following insecticide treatment. This type of population increase is common when natural enemies do not have the capacity to provide adequate regulation of the pests' numbers, either naturally or because their numbers have been reduced by pesticide treatment.

Resurgence can be easily overlooked in its initial stages but as natural enemy destruction accumulates there is an increasing need for more frequent pesticide treatments. Equally resurgence could easily be mistaken for the onset of resistance development but conversely, resistance could be more rapidly produced as a consequence of increased insecticide use following resurgence. It is also misleading to assume that it is only the pest species than can resurge following pesticide application as it can also occur amongst predators. Morgan & Anderson (1958) found that following the virtual elimination of *Typhlodromus* spp. on apple trees by parathion sprays, the number of *T. occidentalis* Nesbitt increased to be similar to, if not greater than, those on unsprayed trees. Newsom (1974) has speculated that such resurgence could be caused by similar phenomena responsible for outbreaks of pest species since predators may also be subjected to the same direct and indirect effects from pesticides.

The interaction of a pesticide on the complex of natural enemies and the pest species is complicated enough, but in most cases where resurgence is claimed to be the consequence of natural enemy destruction due to pesticides, there is a lack of data on the level of natural mortality and particularly of the key predators or parasites involved in regulating the pest population, if they exist. We do not fully understand what mechanisms are responsible for maintaining some pest populations at high densities when they are subjected to repetitive application of broad spectrum insecticides any more than why others decline instead of resurging following pesticide treatment.

Associated with these effects is reproductive stimulation from the pesticide either directly on the pest or via the plant. It would be unrealistic to attribute the explosive development of red spider populations, for example, solely to biochemical changes within the nutritional status of the host-plant since some natural enemies can prevent mite outbreaks even in the presence of factors that promote fecundity (Collyer 1964, Huffaker & Flaherty 1966, van de Vrie 1970).

In addition there are many documented cases of other causes producing pest outbreaks similar to the sort often ascribed to pesticides, for example, improved soil fertility and new crop varieties, changes in cropping practice permitted by irrigation and large-scale cultivation of particular crops.

Factors causing outbreaks are consequently complex and more knowledge is required on the interactions existing in agroecosystems. That other factors as well as the destruction of natural enemies can be involved in changes in pest status suggests that some caution should, therefore, be applied before ascribing all such outbreaks to the disruptive effects of pesticides.

References

AHMED M.K., NEWSON L.D., EMERSON R.B. & ROUSSEL J.S. (1964) The effect of systox on some common predators of the cotton aphid. *J. econ. Ent.* **47**, 445–59.

ATTIAH H.H. & BOUDREAUX H.B. (1964) Influence of DDT on egg-laying in the spider mites. *J. econ. Ent.* **57**, 53–7.

BARTLETT B.R. (1963) The contact toxicity of some pesticide residues to hymenopterous parasites and coccinellid predators. *J. econ. Ent.* **56**, 694–8.

BARTLETT B.R. (1964) The toxicity of some pesticide residues to adult *Amblyseius hibisci* with a compilation of the effects of pesticides upon phytoseuid mites. *J. econ. Ent.* **57**, 559–63.

BARTLETT B.R. (1968) Outbreaks of two-spotted spider mites and cotton aphids following pesticide treatment. I. Pest stimulation *vs.* natural enemy destruction as the cause of outbreaks. *J. econ. Ent.* **61**, 297–308.

BENSON J.F. (1973) Population dynamics of cabbage root fly in Canada and England. *J. appl. Ecol.* **10**, 437–46.

CHABOUSSOU F. (1965) La multiplication voi trophique des tetranyques à la suite des traitements pesticides. *Bull. Zool. Agric. Bachic. (Série II)* **7**, 143–84.

CHABOUSSOU F. (1966) Die Vermehrung der Milben als Folge der Verwendung von Pflanzenschutzmitteln und die biochemischen Veränderungen die diese auf die Pflanze ausüben. *Z. angew. Zool.* **53**, 257–76.

COAKER T.H. (1965) Further experiments on the effect of beetle predators on the numbers of cabbage root fly, *Erioischia brassicae* (Bouché), attacking brassica crops. *Ann. appl. Biol.* **56**, 7–20.

COAKER T.H. (1966) The effect of soil insecticides on the predators and parasites of the cabbage root fly (*Erioischia brassicae* (Bouché)) and on the subsequent damage caused by the pest. *Ann. appl. Biol.* **59**, 339–47.

COAKER T.H. & WILLIAMS D.A. (1963) The importance of some Carabidae and Staphylinidae as predators of the cabbage root fly, *Erioischia brassicae* (Bouché). *Ent. exp. appl.* **6**, 156–64.

COLBORN R. & ASQUITH D. (1971) Tolerance of the stages of *Stethorus puncrum* to selected insecticides and miticides. *J. econ. Ent.* **64**, 1072–3.

COLLYER E. (1953) Biology of some predatory insects and mites associated with the Fruit Tree Red Spider Mite (*Metatetranychus ulmi* (Koch)). IV. The predator–mite relationship. *J. hort. Sci.* **28**, 246–59.

COLLYER E. (1964) A summary of experiments to demonstrate the role of *Typhlodromus pyri* Schent. in the control of *Panonychus ulmi* (Koch) in England. *C.r. 1er Congrès Int. d'Acarologie, Fort Collins, Colo., U.S.A.* (1963), 363–71.

CONE W.W. (1963) Insecticides as a factor in population fluctuations of mites on alfalfa. *Wash. Agr. Expt. Sta. Tech. Bull.* **41**, pp 55.

DAVIES D.W. (1952) Some effects of DDT on spider mites *J. econ. Ent.* **45**, 1011–9.

DEMPSTER J.P. (1967) The control of *Pieris rapae* with DDT. I. The natural mortality of the young stages of *Pieris. J. appl. Ecol.* **4**, 485–500.

DEMPSTER J.P. (1968a) The control of *Pieris rapae* with DDT. II. Survival of the young stages of *Pieris* after spraying. *J. appl. Ecol.* **5**, 451–62.

DEMPSTER J.P. (1968b) The control of *Pieris rapae* with DDT. III. Some changes in the crop fauna. *J. appl. Ecol.* **5**, 463–75.

DEMPSTER J.P. (1975) Effects of organochlorine insecticides on animal populations. In *Organochlorine Insecticides* (Ed. by F. Moriarty), pp. 231–48. Academic Press, London.

DITTRICH V., STREIBERT P. & BATHE P.A. (1974) An old case reopened: Mite stimulation by insecticide residues. *Env. Ent.* **3**, 534–40.

EDWARDS C.A. (1975) Effects of direct drilling on the soil fauna. *Outl. Agric.* **8**, 243–4.

EDWARDS C.A. & THOMPSON A.R. (1975) Some effects of insecticides on predatory beetles. *Ann. appl. Biol.* **80**, 132–5.

FLESCHNER C.A. (1958) Field approach to population studies of tetranychid mites on citrus and avocado in California. *Proc. 10th Intern. Congr. Ent. Montreal (1956)* **2**, 669–74.

GEORGHIOU G.P. (1972) The evolution of resistance to pesticides. *A. Rev. Ecol. Syst.* **3**, 133–67.

GRATWICK M. (1965) Laboratory studies of the relative toxicities of orchard insecticides to predatory insects. *Rep. E. Malling Res. Sta. for 1964*, 171–6.

HEIJBROEK W. (1971) De mogelijkheden voor de bestrijding van de belangrijkste voor jaarsplagen. III. De springstaart (*Onychrurus armatus* Tullb.) *Meded. Inst. rat. Suikprod.* **38**, 1–48.

HUECK H.J., KUENEN D.J., DENBOER P.J. & JAEGER-DRAAFSEL E. (1952) The increase of egg production of the fruit tree red spider mite under influence of DDT. *Phyciologia comp. Oecol.* **2**, 371–7.

HUFFAKER C.B. (1948) A technique for translocation of DDT in plants. *J. econ. Ent.* **41**, 650–1.

HUFFAKER C.B. & FLAHERTY D.L. (1966) Potential of biological control of 2-spotted spider mite on strawberries in California. *J. econ. Ent.* **59**, 786–92.

HUFFAKER C.B. & SPITZER C. (1950) Some factors affecting red spider populations on pears in California. *J. econ. Ent.* **43**, 819–31.

HUFFAKER C.B., VAN DE VRIE N. & McMURTY J.A. (1970) Ecology of tetranychid mites and their natural enemies. II. Tetranychid populations and their possible control by predators: an evaluation. *Hilgardia* **40**, 391–458.

HUGHES R.D. & MITCHELL B. (1960) The natural mortality of *Erioischia brassicae* (Bouché) (Dipt. Anthomyiidae): life tables and their interpretation. *J. anim. Ecol.* **78**, 359–74.

IGNOFFO C.M., HORTELLER D.L., GAREIA C. & PINNELL R.E. (1975) Sensitivity of the entomopathogenic fungus *Nomuraea rileyi* to chemical pesticides used on soybeans. *Env. Ent.* **4**, 765–8.

KOLBE W. & KREMER F.W. (1969) Untersuchungen über den Einfluss einiger Arthropoden auf Raubmilben (Acari). *Z. angew. Zool.* **48**, 257–311.

KREMER F.W. (1971) Change of species dominance among pests on intensively protected pome fruit crops in Italy. *Pflanzenschutz-Nachrichten* **24**, 232–8.

LINGREN P.D. & RIDGWAY R.L. (1967) Toxicity of five insecticides to several insect predators. *J. econ. Ent.* **60**, 1639–41.

LUCKEY T.D. (1968) Insecticide hormoligosis. *J. econ. Ent.* **61**, 7–12.

LÖCHER F.J. (1958) Der Einfluss von Dichloriddiphenyltrichloromethylmethan (DDT) auf einige Tetranychiden (Acari, Tetranychidae). *Z. angew. Zool.* **45**, 201–48.

MASSEE A.M. (1958) The effect of the balance of arthropod populations in orchards arising from unrestricted use of chemicals. *Proc. 10th Intern. Congr. Ent. Montreal (1956)* **3**, 163–8.

McCLANAHAN H.J. (1967) Food chain toxicity of systemic acaricides to predaceous mites. *Nature* 215, 1001.

McMURTRY J.A., HUFFAKER C.B. & VAN DE VRIE M. (1970) Ecology of tetranychid mites and their natural enemies. II. Tetranychid enemies: Their biological characters and the impact of spray practices. *Hilgardia* 40, 331–90.

MORGAN C.V.G. & ANDERSON N.H. (1958) Notes on parathion-resistant strains of two phytophagous mites and a predacious mite in British Columbia. *Can. Ent.* 90, 92–7.

MORRIS M.G. (1968) The effect of sprays on the fauna of apple trees. V. DDT/BHC and lead arsenate/nicotine applied at the green cluster stage. *J. appl. Ecol.* 5, 409–29.

MOWAT D.J. & COAKER T.H. (1967) The toxicity of some soil insecticides to carabid predators of the cabbage root fly (*Erioischia brassicae* (Bouché)). *Ann. appl. Biol.* 59, 349–54.

NEWSOM L.D. (1974) Outbreaks of secondary pests and resurgence of treated populations. *Proc. F.A.O. Conf. in Ecol. in Relation to Plant Pest Control*, Rome, 145–60.

NEWSOM L.D. & BRAZZEL J.R. (1968) Pests and their control. In *Advances in Production and Utilization of Quality Cotton: Principles and Practices* (Ed. by F.C. Eliot, M. Hoover & W.K. Porter), pp. 367–405. The Iowa State Univ. Press.

OATMAN E.R. (1965) Effects of preblossom miticides and subsequent insecticide applications on mite populations in apple in Wisconsin. *J. econ. Ent.* 58, 335–43.

OLMERT I. & KENNETH R.G. (1974) Sensitivity of entomopathogenic fungi, *Beauveria bassiana*, *Verticillium lecani* and *Verticillium* spp. to fungicides and insecticides. *Env. Ent.* 3, 33–8.

PHILLIPS S.H. & YOUNG H.M. (1973) *No-Tillage Farming*. Reiman Associates, Wisconsin pp. 224.

PICKETT A.D. & PATTERSON N.A. (1953) The influence of spray programs on the fauna of apple orchards in Nova Scotia. IV. A review. *Can. Ent.* 85, 472–8.

PIELOU D.P. (1960) The effect of DDT on oviposition and behaviour in the European red mite *Panomychus ulmi* Koch. *Can. J. Zool.* 38, 1147–51.

PIELOU D.P. (1962) The ineffectiveness of Sevin in increasing oviposition in the spider mite *Tetranychus telarius* (L.). *Can. J. Zool.* 40, 9–11.

PIERCE W.D. (1922) How insects affect the cotton plant and means of combating them. *U.S. Dept. Agri. Farm Bull.* 890.

PIMENTEL D. (1961) An ecological approach to the insecticidal problem. *J. econ. Ent.* 54, 108–14.

POST A. (1963) The effect of cultural measures on the population density of harmful and beneficial organisms in orchards. *Entomophaga* 7, 257–62.

PUTMAN W.L. (1963) Lack of effect of DDT on the fecundity and dispersion of the European red mite, *Panomychus ulmi* Koch in peach orchards. *Can. J. Zool.* 41, 603–10.

RIPPER W.E. (1956) Effect of pesticides on the balance of arthropod populations. *A. Rev. Ent.* 1, 403–38.

RODRIGUEZ J.G. (1958) The comparative NPK nutrition of *Panomychus ulmi* (Koch) and *Tetranychus telarius* (L.) on apple trees. *J. econ. Ent.* 51, 369–73.

ROOT R.B. & SKELSEY J.J. (1969) Biotic factors involved in crucifer aphid outbreaks following insecticide application. *J. econ. Ent.* 62, 223–33.

SAINI R.S. & CUTKOMP L.K. (1966) The effects of DDT and sublethal doses of Dicofol on reproduction of the two-spotted spider mite. *J. econ. Ent.* 59, 249–53.

SEIFERT G. (1961) Der Einfluss von DDT auf die Eiproduktion von *Metatetranychus ulmi* Koch (Acari, Tetranychidae). *Z. angew. Zool.* 48, 441–52.

STEINER H. (1959) Verschiebungen in der Miridenfauna der Apfelbaumes durch Behandlung mit Insektiziden, Akaraziden und Fungiziden (Heteroptera, Miridae). *Verh. IV. Intern. Pflanzensch. Kongr. Hamburg* (1957) 1, 971–4.

STORMS J.J.H. (1965) Rearing methods for studying the effect of the physiological condition of the host plant on the population development of *Panomychus ulmi* Koch. *Bull. Zool. Agric. Bachic. (Serie II)* 7, 119–30.

VAN DE VRIE M. (1962) The influence of spray chemicals on predatory and phytophagous mites on apple trees in laboratory and field trials in the Netherlands. *Entomophaga* 7, 243–50.

VAN DE VRIE M. (1970) The influence of the predaceous mite *Typhlodromus* (A.) *potentillae* (Garman) on the population development of *Panomychus ulmi* (Koch) on apple grown under various nitrogen conditions. *Entomophaga* 15, 291–304.

VAN DE VRIE M. (1974) Studies on prey/predator interactions between *Panomychus ulmi* and *Typhlodromus* (A.) *potentillae* (Acarina; Tetranychidae, Phytoseiidae) on apple in the Netherlands. *Proc. F.A.O. Conf. on Ecol. in relation to Plant Pest Control, Rome* (1973), 145–60.

VAN DE VRIE M., McMURTRY J.A. & HUFFAKER C.B. (1972) Ecology of Tetranychid mites and their natural enemies. *Hilgardia* 41, 343–432.

WATVE C.M. (1971) *Biology and control of the banded whitefly* Trialeurodes abutilonea *(Haldeman) on cotton in Louisiana.* Ph.D. thesis, Louisiana State Univ.

WEBSTER J.M. (1967) Some effects of 2,4-dichlorophenoxyacetic acid herbicides on nematode-infested cereals. *Pl. Path.* 16, 23–6.

WILDING N. (1972) The effect of systemic fungicides on the aphid pathogen, *Cephalosporium aphidicola*. *Pl. Path.* 21, 137–9.

WRIGHT D.W., HUGHES R.D. & WORRALL J. (1960) The effect of certain predators on the number of cabbage root fly (*Erioischia brassicae* (Bouché)) and the subsequent damage caused by the pest. *Ann. appl. Biol.* 48, 756–63.

Human pest and disease problems: contrasts between developing and developed countries

DAVID J. BRADLEY *Ross Institute of Tropical Hygiene,*
London School of Hygiene and Tropical Medicine, Keppel St.,
Gower St., London

It is increasingly recognized that disease is an aspect of ecology and also, conversely, that an understanding of ecology is a necessary basis for any attempt to understand disease in the human community. Because of the impossibly broad field to be covered in this paper, much selection has been necessary and the account of many topics is inevitably superficial. This is not to imply that the problems or their solutions are simple. On the contrary, one aim of this paper is to show that the relations between man and his diseases, particularly his infections, are very complex. Simplistic views by those directing economic development may do much damage. Human 'pests' are here interpreted as disease vectors, though lice, fleas and mosquitoes are unpleasant and annoying even when they are not 'vecting'.

The differences between more and less developed countries

THE SIZE OF THE DIFFERENCE

What is the difference in disease problems between developing (or less developed) and industrial countries? On any reckoning it is very great. The crudest measure of disease, but the only one for which comparable data for many areas are to be had, is the death rate in relation to age. This can only tell us about fatal disease, but its message is very clear. If this symposium were being held in Malawi or Niger about half of those attending would not be there: they would be dead! The basic survivorship curves for two model countries are given in Fig. 1. The lower line represents the percentage of live-born children surviving to various ages in a little-developed country while the upper line shows survivorship at the other extreme, as is seen in Scandinavia and other highly developed countries. The obliquely shaded area between these curves is my topic.

There are three main differences between the two survivorship curves. In the first year of life there is an appreciable death rate, mainly from congenital defects, in the industrial country but a *very* much greater one in the less developed nation, with two out of every ten babies failing to reach their second, and often first, birthday. Between the ages of 1 and 5 years

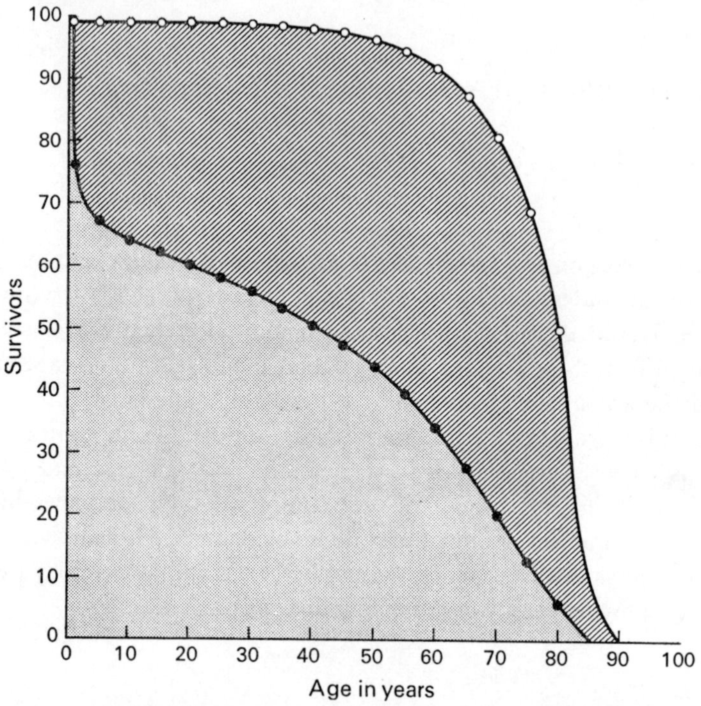

Figure 1. Survivorship curves for 100 live-born babies in a prosperous industrial country (open circles) and a poor developing country (solid circles).

there is a further loss of up to 10% or more of children in the developing land whilst mortality in the other country is negligible. Thirdly, death rates remain higher throughout life in the less developed country. The result is that less than half the people get appreciably beyond their fortieth birthday, and life expectation at birth may be half that in the developed country. The age-specific mortalities are explicitly compared in Fig. 2, where sex differences are brought out. Two effects are due to trauma. The dip in the late teens is due to road accidents to young Swedes, whilst the second female peak results from obstetric deaths in Sri Lanka (a country with relatively good data and better health than many developing countries). Otherwise, the differences result from disease.

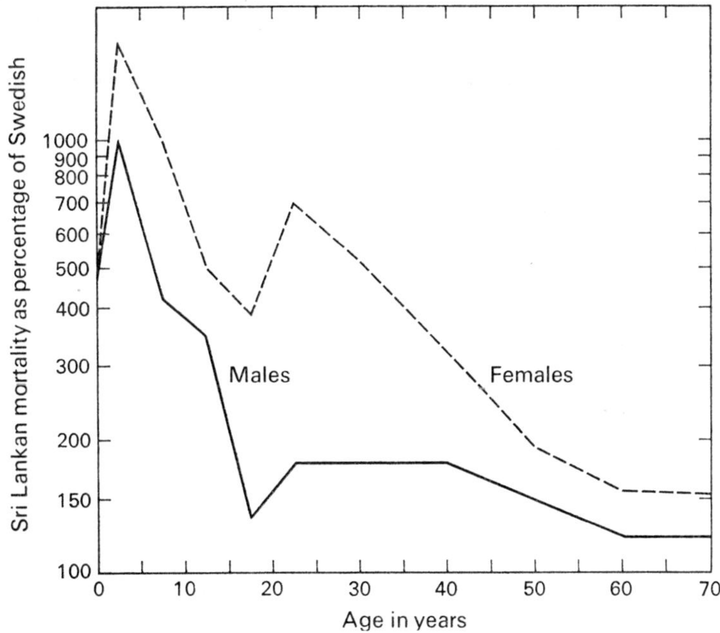

Figure 2. The ratio of age-specific mortality in Sri Lanka to that of Sweden. The Sri Lanka mortality is given as a percentage of the Swedish. (From Myrdal, Gunnar, *Asian Drama: An Enquiry into the Poverty of Nations*, Vol II. p. 1412 © 1968 by the Twentieth Century Fund, New York. (Issued in England by Allen Lane, The Penguin Press, London, 1968.))

THE CAUSES OF THE DIFFERENCES

Why do these differences occur? The majority of the 'excess' deaths are due to infections. More than 58% of deaths in a Nigerian children's hospital (Table 1), were due to infections and the figure is as high in most developing countries. They may be attributed to two basic causes: poverty and climate.

Many illnesses of less developed countries are diseases of poverty. The disease pattern resembles that seen in the poor of Britain last century. Poverty operates in at least four ways: through poor nutrition, an unhealthy physical environment in the narrow sanitary sense, ignorance with its inter-related social and cultural effects on disease, and lack of access to health care facilities. Even if the facilities exist, they may be too far away, financially out of reach of the sick people, or provide such poor quality care that the illness remains unaltered. The physical environment and nutrition need further mention here.

The bare minima of a sanitary human environment are adequate supplies of unpolluted water, safe removal of human and other wastes, and adequate shelter. These all cost money, require some appreciation of the microbial theory of disease if they are to be properly used, and need maintenance which requires a suitable infrastructure. Few countries remove rubbish adequately, while sewerage in urban areas poses many problems. A quarter to half of the serious infections in the tropics are due to the access of faeces to mouths, directly or indirectly.

Table 1.   Major causes of death
in a Nigerian children's hospital, 1963
(Morley 1963).

Diarrhoea	12%
Pneumonia	12%
Malnutrition	12%
Malaria	8%
Pertussis	8%
Measles	8%
Tuberculosis	5%
Smallpox	5%
All others	30%

Nutrition is the other great problem of less developed countries. Whether the world's food is adequate to feed its population can be argued. What is certain is that much food is in the wrong places to feed the hungry, which is one reason for considering malnutrition as a disease of poverty. It occurs on a great scale, with Bengoa's (1974) careful figures suggesting around a hundred million severe cases in children and therefore an immense milder and subclinical amount in addition. The child may be short of all major nutrients, giving rise to marasmus, or the lack of protein may predominate, giving the swollen limbs, skin changes, and reddish hair colour of kwashiorkor. It is gradually being realized how far malnutrition and infection interact (Scrimshaw *et al.* 1968). A malnourished child's immune responses may be impaired and the clinical severity of illness may be much increased. Measles will send a well-fed child to bed, but a badly nourished child may die. Conversely, the catabolic effects of infection and the way it stops a child eating may precipitate it into overt malnutrition.

So in large part the disease pattern of a less developed country is due to poverty. But also, since most such countries are in the tropics, there is an important group of diseases, especially parasitic and viral, which depend on the warm temperature (Macdonald 1965). These are caused by disease agents which have a temperature-dependent developmental stage outside the human body. For example, malaria parasites taken up by a suitable anopheline mosquito have to undergo a development cycle lasting some ten days before stages infective to man reach the salivary glands. This extrinsic cycle is prolonged at low temperatures and exceeds the adult mosquito life-span in colder climates. Temperature-dependent development of this type is found in a range of vector-borne diseases such as malaria, schistosomiasis and onchocerciasis as well as in some parasitic helminths whose eggs or larvae develop in the soil, such as *Ascaris* the human roundworm, and the hookworms. Some of these will tend to get rarer as poverty is reduced. This will happen with the worms whose eggs reach the soil via the faeces. But some of this category of warm climate diseases are relatively independent of poverty directly and will only be reduced by specific measures aimed at their control.

So the differences between industrial and developing countries may be summarised by concluding that the overall position is much better in industrial countries. They are richer, and so have better nutrition and a more sanitary environment, and they are mostly in colder climates. The best data, discussed above, is on deaths. The differences in illnesses (morbidity) between the two categories of country are even greater. This is simply illustrated by comparing pre-school children (toddlers), in rural Uganda and Britain. Few British children in the community in a single random survey will have skin disease. Forty-seven per cent of Ankole children on any day will be suffering from scabies, skin ulcers, or fungus infections of their skin (Cook 1967). The level would be much higher still in New Guinea (Feachem 1973).

We may take this reasoning a step further, in that rates of infection with parasites are even higher than for illnesses. In non-mountainous areas of sub-Saharan Africa four out of every five children between the ages of 5 and 9 years will have malarial parasites in a single blood smear. Over 95% of children in northern Tanzania will have urinary schistosomiasis by the age of 12 years. The 'average' developing country child will have not one but several infections at any one time.

Origins of the disease problems

Many studies of agricultural pest and disease problems look back wistfully at the Utopian natural ecosystem when such difficulties had not emerged. If we look back for the 'natural man' there is some difficulty in finding him. If there was an idyllic primordial state it was a long way back. The nearest approach that can be made in practice is to study communities of hunter-gatherers. They can be found living at low density in several tropical areas. The Hadza (Bennett *et al.* 1970) have remained a fairly isolated group. They show many of the parasites found in other neighbouring tribes, though infections were often relatively light. They were certainly more ill and had a shorter life expectancy than the people of an industrial country, though their low population density reduced the impact of infections on them by comparison with their neighbours.

At the other extreme, a high and rising population density raises problems of resource depletion. But what of the human equivalent to monoculture, urbanization? The industrial countries had very major disease problems in the 19th century as their cities grew. During the industrial revolution in Britain the urban death rates were higher than in the rural areas. This experience has not so far been repeated in the tropics where the available evidence suggests that it is safer, in terms of survival, to live in the city than the rural areas. This is rather surprising to someone who has visited cities in tropical areas, but urban squalor is generally more obvious than rural.

So far, illness and death have been considered in general terms. The remainder of the paper will be confined to infections because, apart from malnutrition, they are responsible for most of the health differences between industrial and developing countries.

EPIDEMIOLOGICAL BASES FOR THE ORIGIN OF DISEASE PROBLEMS

To understand origins and outbreaks, some consideration of the epidemiology of infective diseases is needed. Two extreme types may be recognized: epidemic and endemic infections. In considering disease ecology there is both the human or other host (micro-habitat) and the host community and environment (macro-habitat).

In classical epidemic infectious disease of man the microorganism or parasite gains access to a host and multiplies rapidly. This continues until either the host dies or is treated, or an immune response is mounted against the microbe. This destroys the microbial population and the host recovers, immune to further attacks of the disease. Spread is dependent on some of the microbes passing from the infected to a susceptible host before death or the immune response supervenes. Examples of such diseases are measles, smallpox and cholera. Transmission depends on rapid multiplication of the microbe in man and on a supply of susceptibles being maintained. Where the population is very large, as in a city of over half a million people, for measles the supply of susceptibles is maintained by birth. Persistence in smaller communities is only maintained where there are many partly isolated sub-populations. In practice, real infections of this sort (but not smallpox or measles) persist also by giving rise to a few 'carrier' hosts who may be clinically well but continue to maintain and spread the microbe.

At the other extreme is the endemic infection. Here the infecting microbe is able to maintain itself in the host in spite of immune responses. The evolutionary pressure towards rapid multiplication is less but a microbe or helminth that does not reach the numbers or stage necessary for transmission in the time the host takes to mount an immune response, must evolve a means of avoiding it. Since very efficient transmission will not in this case carry with it a risk of microbial extinction there tends to be a variable, but often high, rate of potential transmission of very persistent infections. Several examples may be given, in which the persistent infection of hosts is achieved in different ways. Leprosy and tuberculosis are highly persistent endemic infections (organisms persist intracellularly inside macrophages and other cells or in chronic walled-off inflammatory lesions to the interior of which the protective mechanisms do not readily reach). In schistosomiasis there is evidence that adult worms may be able to avoid the immune response they have induced by 'disguising' themselves with host red cell antigen while young worms, from further exposure to infection are destroyed before this can occur (Smithers *et al.* 1969). The sleeping sickness trypanosomes avoid host responses by undergoing antigenic variation in a controlled manner just before host antibodies destroy the parasites (Gray 1965).

Of course these are the extreme examples and most real diseases have components of both endemicity and epidemic behaviour. In particular, any infection newly in an area will tend towards epidemic behaviour while such infections as malaria will follow one or other pattern depending on the transmission level, which will in turn depend on the biology of the local vector mosquitoes (Macdonald 1957) and the ambient temperature.

The relation between epidemic and endemic infections and the natural regulation of parasite populations is developed elsewhere (Bradley 1972, 1974a). The epidemic pattern of disease can sometimes be a usually sporadic infection which has encountered unusually favourable environmental conditions for its rapid spread, corresponding to an 'unregulated' population of the type discussed fully by Andrewartha and Birch (1955). Often however, it will be an infection ended by total immunity in the individual and thus regulated by the pattern of immunity in the population.

It is of interest to compare the situation with that which obtains in island biogeography. Each human host stands in relation to the disease organism as a discrete habitat, an island, discontinuous from other hosts. The behaviour of an epidemic microbe corresponds to that of an r-selected species in island ecology (MacArthur & Wilson 1967). It operates on a strategy of rapid invasion and multiplication at the price of being rapidly eliminated from the host. On the other hand the endemic infection corresponds to the K-selected island species. It is adapted to long-term persistence in the host and is relatively insensitive to changes in the ease of transmission. The

K-selected parasite creates a persistent or chronic problem without dramatic effects and its public health significance tends to be underrated. The r-type parasite, by its more flamboyant epidemic behaviour, manages to draw attention to itself and is much more of a problem in the eyes of most people. Southwood (1977) has considered the concept of r- and K-selection in relation to pest types. There is a difference between its application to parasite and islands in that the host responds directly by an immune response to the parasite, and competition with other colonists is not usually a major issue in limiting the parasite population. Also, reducing the invasion rate of a K-type species on an island may simply render it unimportant. An endemic infection in a new area or one whose transmission is greatly reduced may behave as an epidemic infection and this important origin of disease problems will be considered below in some detail.

Another difference between men and islands is that people are mobile, and many disease outbreaks result from human movement. We now relate the general picture set out here to outbreaks of disease in the real world, considering initially the consequences of human mobility.

Movement as a cause of disease outbreaks

Human disease outbreaks frequently arise either from the introduction of infection into a community by people or conversely, by the migration of susceptible people into a focus of transmission. Movement frequencies and ranges tend to increase in the modern world and disease problems of this sort are important.

The first introduction of a pathogen into a human community may have disastrous effects. Well-known examples are the introduction of measles to Fiji and of tuberculosis to the North American Indians. The course of infection was far more rapid than in the countries where the infections had been present for a long time, and the initial mortality was very high. The level of intercommunication in the world now leaves rather few opportunities for this type of happening in crude form. There are a few notable gaps (yellow fever does not appear to have reached India as yet) but they are sufficiently remarkable that there may be some form of heterologous immunity preventing the introduction. An interesting example of spread across Africa at the beginning of this century is provided by sleeping sickness which was probably carried to Uganda from the west of Africa by Hausa bearers accompanying the explorer Stanley. The infection spread widely throughout the Lake Victoria shore in Uganda and produced a mortality of several hundred thousand people so that the shoreline was depopulated by the effects of the disease combined with the administrative attempts to stop transmission. Over seventy years since the great epidemic, the natural history of the infection appears to have changed and become more chronic.

A more frequent occurrence is the introduction by migrants of disease to areas from which it has temporarily been removed. Air travel generally, and such mass movements as the pilgrimage to Mecca have led to outbreaks of cholera which are far harder to control in developing countries with poor water supplies and sanitation than in industrial nations.

The second way in which human mobility leads to disease outbreaks is where susceptibles move into transmission foci. They may then become infected either from other people or from vertebrates other than man.

Infections to which both man and other vertebrates are susceptible may be maintained in nature at sites uninhabited by man. Thus strains of *Leishmania* are transmitted between small rodents and sandflies in the Brazilian jungle. In the southern U.S.S.R. colonial rodents, especially gerbils, may also maintain *Leishmania*, while further north in the coniferous forests encephalitis viruses may be transmitted between ticks and wild animals. Similarly with Rocky Mountain spotted fever in the United States. Disease affects man when he disturbs these 'natural foci' as they have been called by Pavlovsky (1966). In the case of solitary hunters in the Rockies, only sporadic infections occur, but in Brazil the construction of a highway through the forest was accompanied by a leishmaniasis epidemic among construction workers. The settlement of Siberia led to regular outbreaks of the zoonoses. These were sufficiently predictable to the Russian research workers who knew the habitats of the different vectors and reservoirs that a whole subject of 'landscape epidemiology' was created.

More prosaic than the zoonoses are the outbreaks which follow movement of migrant labour from the overpopulated hill areas of many tropical developing countries down to the plains. Malaria outbreaks among such people are so regular an occurrence as to have acquired the term 'malaria of the tropical aggregation of labour'. A lesser problem may follow migration between malarious areas since there is some degree of strain-specificity in acquired immunity to malaria.

Shorter distance movements may be responsible for disease outbreaks, as when villagers have to fetch water from a wooded stream beside which tsetse flies rest and bite, giving rise to sleeping sickness. This can be avoided by clearing the trees and shrubs from the area around the water point.

Many more complex disease outbreaks result from a mixture of the above happenings with social patterns, vector biology and particular local circumstances. An illustration from southern Uganda (Robertson 1963) shows the complexity. The great sleeping sickness epidemic ended when the administration ordered evacuation of the shores of Lake Victoria. More recently it was decided to resettle that area. By now the risk of Gambian sleeping sickness spread by riverine tsetse was less than that of the more acute Rhodesian type of disease transmitted by a thicket tsetse. It was proposed to resettle at a population density adequate to clear the forest and thus drive out the tsetse. Unfortunately, the peoples' memories exceeded those of the administration and they insisted on returning to their ancestral plots from before the 1908 clearance. Consequently one had a scattering of settlers in tiny clearings in the forest, providing veritable fly restaurants for the tsetse, and the predicted outbreak of Rhodesian sleeping sickness occurred. Sleeping sickness also raises the fascinating problem of how the disease, especially the Rhodesian type, arose in the first place. Is it a mutation from the salivarian trypanosome of wild game *Trypanosoma brucei*; has it arisen once and spread, or does it emerge recurrently?

Genetic aspects of the origin of disease problems

Ideally, this would be the first and most biologically interesting section of this paper, but it has to be too speculative to be so placed. A totally new disease is fascinating but rarely seen. Too often it proves to be, like America before Columbus, present but unstudied. For example, the notorious Lassa fever which recently reached the headlines seems to have been present in small rodents in West Africa and to have produced human cases long before recent cases in expatriates.

Influenza periodically causes major world-wide epidemics, that in 1918 causing well over a million deaths, and these are due to 'new' antigenic types which are unaffected by antibodies in the population. It has been generally considered that these new variants arise by mutation but there is now considerable uncertainty as to how far these derive from pre-existing strains in non-human vertebrates.

So genetic change causing disease is little understood. The genetic modifications that occur following introduction of a parasite are much better known, especially for viral myxomatosis of the European rabbit in Australia (Fenner & Ratcliffe 1965). When the disease, originally causing mild lesions in a different genus of American rabbit, was introduced into Australia the mortality was extremely high. After ten years it was possible to show that there had been progressive attenuation of the virus strain as tested in the original lines of rabbit and also an increase in rabbit resistance when tested with the original virus. Thus there was evolution of both host and parasite towards a more benign and transmissible infection. The vector mosquitoes had a better chance of transmitting from chronic rather than rapidly fatal lesions. By contrast, in Britain where transmission was more by rabbit fleas which especially leave a dead host and thus maintain transmission of lethal virus strains, attentuation of the myxomatosis virus was less than in Australia.

Host genetics are an important determinant of infection problems. The heterozygous carrier of the sickling gene which affects human haemoglobin, is relatively protected against some lethal consequences of falciparum malaria, and is widespread in malarious areas of the tropics, especially Africa. It creates a secondary disease problem, especially noticeable when malaria is controlled, in that the homozygous sickle cell person has a haemolytic anaemia, many other pathological problems, and a poor expectation of life.

Other genetic features determine the susceptibility of populations to infection. It has been known for some time that susceptibility of mice to virus infections is sometimes under single gene control (Bang & Warwick 1960) and more recently that a single gene largely affects susceptibility to visceral leishmaniasis in mice (Bradley 1974b). It has now been shown that the reason why vivax malaria is not a problem in West Africa is that the Duffy blood group antigen, which is very rare in West Africans, acts as the parasite receptor on the red blood cell (Miller *et al.* 1975). The genetics of the immune response has recently emerged as a most active research area and there is no doubt in my mind that it will contribute greatly to the understanding of infective disease problems, though it is too early to point to clear examples.

No such lack exists for two other genetic areas: the genetically based resistance of disease microbes to chemotherapeutic medicines and resistance of vectors to insecticides. These have led to the recurrence of disease problems in very intractable form after control attempts have been made. Initially, concern was mainly over resistance of microbes and vectors to the toxic substances directed at them. Now it is realized that vectors are becoming resistant as a result of incidental exposure to pesticides directed at other insects, usually of agricultural importance. Resistance in bacteria can be acquired directly or the genetic elements responsible may be transferred to a pathogen from another bacterium exposed to the selection pressure from the drug concerned. Thus the use of tetracycline to promote growth of pigs and other stock encourages the emergence of resistant commensal intestinal organisms which may then transfer the drug resistance to human pathogens of the typhoid group.

The problem of microbial resistance to antibodies and other chemotherapeutic agents is world-wide. In industrial countries a greater variety of expensive drugs is available and the common pathogenic bacteria are resistant to more of them. The range of resistance is less in developing countries but the problems are as great or greater since these nations do not have the funds to purchase more expensive alternatives. The relative lack of control over medicines in many developing countries tends to increase their incomplete and indiscriminate use, thus promoting selection for resistance.

Insecticide resistance has been recorded from all but two of the medically important genera of vectors and cross-resistance between members of a chemical group of pesticides means that the problems of resistance are of great practical relevance and we shall consider them further in relation to malaria.

The problems of almost successful control

EFFECTS ON SOME DISEASES OF POVERTY

Increasingly, disease outbreak problems are in part the consequences of the partial or attempted control of infections, whether incidentally to other aspects of development, or deliberately.

The standard of environmental hygiene in industrial countries rose fairly steadily until recently. Now, some viruses transmitted through the faeces have very efficient modes of spread. One such group is the poliovirus group, which in developing countries is effectively spread to everyone in childhood. The consequence is a number of cases of persisting 'infantile paralysis' while most people become immune without becoming ill from the virus. Shortly before polio vaccine was successfully introduced, the pattern had been different in some industrial countries, including Britain. Healthy young men had suddenly contracted severe polio with not only limb paralysis but also respiratory paralysis from brain-stem involvement. Here sanitary improvements were delaying the time of first exposure to infection until later when clinically the effects of the virus were more serious. A new disease problem, but due to reducing transmission of a well-known microbe. This seems to be happening with other viruses to some extent, particularly infective hepatitis and possibly Eb virus which appears aetiologically related to glandular fever. Essentially one is moving from an endemic to an epidemic situation.

The same sort of phenomenon results from reducing the transmission of malaria in a place where it is highly endemic due to very efficient vector mosquitoes. The regulation of infection in man will be mainly due to human immunity, but the parasite is capable of rapid multiplication within the host. The highly endemic malaria may not appear to do much harm though in fact it is likely to be killing 5–10% of all children born. It does not cause obvious massive illness beyond childhood. A consequence of moderately successful malaria control may be to reduce transmission so that immunity and infection both decrease greatly, only to leave the whole community susceptible to the return of infection and illness. This is a real problem, but it has most often occurred from the breakdown of control in an area of already epidemic unstable malaria, and this for several reasons.

OUTBREAK PROBLEMS AND MALARIA ERADICATION

A particularly clear example of human disease problems arising from the failure of attempts at control is human malaria. Both the disease itself and the biological and political issues involved in recent eradication attempts are of immense complexity, but the course of events raises many points of general interest.

If almost everyone gets malaria there is much disease but there may be less of a 'problem' because people complain less about a perennial than a novel ill. Efforts in the first two decades of this century have removed the threat of malaria from many countries and greatly reduced the prevalence of infection in others. For many reasons, that eradication effort has diminished recently, leading to disease problems affecting some millions of people in several countries, especially in southern Asia. The causes merit discussion, since they could equally well apply to other vector-borne diseases.

There are four main origins of the recrudescent malaria problem: two biological and two social.

The most prominent social reason is economic. The attack on malaria in the past two decades has been aimed at eradication: to remove the last indigenous case of malaria from a country by first stopping transmission using residual insecticides on the walls of houses and then going on to clear up the reservoir of infection by treating any residual cases during the 'consolidation' phase, when complete regular coverage of the population is needed. Eradication appears financially as an expensive capital project with a clear aim and time scale. As such, the project begins with immense positive feed-back. If a lot of money has been spent and only a little more is needed to achieve success, then there is great incentive to provide it. If neighbouring countries have succeeded in eradicating malaria, they will be only too delighted for you to do the same, since it will reduce the hazard of reinfection to them. The results early on are dramatic, with a great fall in the amount of malaria. Unless success is fairly well assured, there is later on great danger of a negative feedback. Once confidence in the project fails, no government wishes to give money to a capital expenditure scheme which will lead to nothing. Getting rid of the last few cases is undramatic and very demanding on the health services and where the infrastructure of dispensaries and staff is inadequate, as in so many developing countries, it may prove insuperably difficult while morale and public cooperation are falling. This is the second social reason. But, returning to economic difficulties, the health budget for many developing countries is between 50p and £1 per head per year, including all hospitals, disease control schemes, and rural health facilities, so that the cost of the malaria schemes was crucial. Inflation has increased all costs, and insecticide is a major one. Manufacture depends on oil products the costs of which have gone up particularly. The price of DDT, the main insecticide for malaria work, depended on its being a small part of the U.S. agricultural production. Now that the United States of America has stopped using large amounts of DDT on its crops, the economies of scale have gone and the insecticide is still more expensive. So economics, the psychology of eradication economics, and health infrastructure have been social blocks to preventing malaria eradication.

The biological origins of recrudescent malaria are insecticide and drug resistance. By 1975 a total of forty-two anopheline mosquito species had developed insecticide resistance (World Health Organization 1976). A dramatic illustration is from Turkey where *Anopheles saccharovi* is resistant to all four major insecticide groups: DDT, dieldrin/hexachlorobenzene, organophosphates and carbamates. Resistance of the malaria parasites themselves to the major relevant medicines has also arisen. The most important has been chloroquine resistance in South-East Asia and the Americas, which has meant that the most useful drug for malaria treatment is no longer of use in some areas.

Over the last few years malaria has increased for all these reasons from extremely low levels in Sri Lanka and low ones in India to become a massive and much more intractable problem. Malaria control is one among many attempts to raise living standards, and quite a number have, as side effects, given rise to disease problems.

Side effects of attempts to raise living standards

There are many ways in which these can be subdivided and I have selected four categories, but they have one feature in common—failure to look widely enough. A particular objective is in mind and, so often, action is taken to achieve that objective without adequate thought about other consequences.

DEVELOPMENT TOWARDS A CASH ECONOMY

In an agricultural community, the chief opportunity to raise income is by planting cash crops. This takes up time and land which may previously have produced food for the family to subsist on. Under these pressures there is in many developing countries a tendency to move towards nutritionally inadequate food crops, and it is not rare to see family income rising while their nutritional state and their children's health may fall. If the cash income rises, it may be spent on things apparently valued by the elite of that society, which may be disastrous for health. Tinned baby milk has acquired prestige value, and too often it is made up with polluted water. There is no refrigerator so that it is left around for days to grow bacteria and it is often over-diluted. The infant death rate rises.

Disease problems of industrialization resemble in type the problems of the industrial revolution in England though in many countries these have been forestalled and standards of occupational health in a number of otherwise poor countries are good.

MAJOR WATER DEVELOPMENT PROJECTS

The disease problems of man-made lakes and irrigation schemes in the tropics have been repeatedly discussed in recent years (Stanley & Alpers 1975) and although most ecologists must be well aware of them, those concerned with planning the projects often appear unaware. It is necessary to be realistic. It may not be possible at a reasonable cost to prevent schistosomiasis entering an irrigation project, but it should be possible to control it, as has been shown financially profitable in Tanzania (Fenwick & Figenschou 1972).

Man-made lakes raise large disease problems for those who live at the site. It is unfortunately true that in every case of a large tropical man-made lake the local inhabitants have been made worse off and often needlessly so. The costs in social terms of hydroelectric dams and their impoundments fall on the local people. The benefits chiefly accrue to urban dwellers some distance away.

PROBLEMS FROM ATTEMPTED HEALTH IMPROVEMENTS

There are rather few generally applicable solutions to health problems. The large hospital has much to offer at a certain stage of development. Earlier on, it absorbs most of the health budget and prevents far more cost-effective disease control measures being taken.

So simple a measure as the pit latrine is also not universally applicable. In rural areas it is meant to reduce the level of faecal-oral diseases and hookworm. If it is dug at a site with a high water table, the bottom becomes

M

flooded. This makes an ideal breeding site for the mosquito *Culex pipiens fatigans*, an important vector of filariasis over much of the world, so that what reduces one disease may be the origin of another.

MISPLACED ECOLOGICAL ADVICE

There are other illustrations that solutions to ecological problems are not general. The plant *Lantana camara* (a nuisance in Mexico) was successfully used to help drain land in Malaya, thereby reducing malarial mosquito breeding. It was introduced to Kenya in the Alego area to replace *Euphorbia* hedges around the paddocks and houses at the suggestion of an agricultural worker. It was little better than the *Euphorbia*, but it maintained so high a relative humidity in the thickets it formed that the stream-haunting tsetse *Glossina fuscipes* was able to leave the streams, live in the *Lantana* thicket and take up a peri-domestic habitat. This resulted in a very sharp epidemic of Rhodesian sleeping sickness transmitted from man to man by the 'wrong' tsetse fly (Willett 1965).

Conclusions

Since this symposium is concerned with the origins of disease problems, the presentation has had to concentrate on the sombre aspects of development. This should not cloud our overall view. In terms of reducing premature death, disability and misery, economic development is highly desirable. Disease patterns are extremely varied and complex. They raise lots of fascinating ecological and genetic problems. Because of this complexity rapid development can easily make some problems worse but often the difficulties can be avoided if they are foreseen. There is no substitute for thinking broadly and ecologically about the possible consequences of any changes we plan. If we do so, we can avoid being the origin of many of the human pest and disease problems of our complex world.

References

ANDREWARTHA H.G. & BIRCH L.C. (1955) *The Distribution and Abundance of Animals.* Chicago Univ. Press.

BANG F.B. & WARWICK A. (1960) Mouse macrophages as host cells for the mouse hepatitis virus and the genetic basis of their susceptibility. *Proc. natn. Acad. Sci. U.S.A.* 46, 1065–75.

BENGOA J.M. (1974) The problem of malnutrition. *WHO Chronicle* 28, 3–7.

BENNETT F.J., KAGAN I.G., BARNICOT N.A. & WOODBURN J.C. (1970) Helminth and protozoal parasites of the Hadza of Tanzania. *Trans. R. Soc. trop. Med. Hyg.* 64, 857–80.

BRADLEY D.J. (1972) Regulation of parasite populations. A general theory of the epidemiology and control of parasitic infections. *Trans. R. Soc. trop. Med. Hyg.* 66, 697–708.

BRADLEY D.J. (1974a) Stability in host–parasite systems. In *Ecological Stability* (Ed. by M.B. Usher & M.H. Williamson), pp. 71–87. Chapman & Hall, London.

BRADLEY D.J. (1974b) Genetic control of natural resistance to *Leishmania donovani*. *Nature, Lond.* 250, 353–4.

COOK R. (1967) The Ankole pre-school protection programme. Mimeographed. Kampala, Uganda.

FEACHEM R.G. (1973) Environment and health in a New Guinea highlands community. Unpublished Ph.D. thesis, University of New South Wales.

FENNER F. & RATCLIFFE F.N. (1965) *Myxomatosis*. Cambridge University Press, London.

FENWICK A. & FIGENSCHOU B.H. (1972) The effect of *Schistosoma mansoni* infection on the productivity of cane cutters on a sugar estate in Tanzania. *Bull. Wld Hlth Org.* 47, 567–72.

GRAY A.R. (1965) Antigenic variation in clones of *Trypanosoma brucei*. *Ann. trop. Med. Parasit.* 59, 27–36.

MACARTHUR R.H. & WILSON E.O. (1967) *The Theory of Island Biogeography*. University Press, Princeton.

MACDONALD G. (1957) *The Epidemiology and Control of Malaria*. Oxford University Press, London.

MACDONALD G. (1965) On the scientific basis of tropical hygiene. *Trans. R. Soc. trop. Med. Hyg.* 59, 611–20.

MILLER L.H., MASON S.J., DVORAK J.A., McGINNISS M.H. & ROTHMAN I.K. (1975) Erythrocyte receptors for (*Plasmodium knowlesi*) malaria: Duffy blood group determinants. *Science* 189, 561–3.

MORLEY D. (1963) A medical service for children under five years of age in West Africa. *Trans. R. Soc. trop. Med. Hyg.* 57, 79–94.

PAVLOVSKY E.N. (1966) *Natural Nidality of Transmissible Diseases*. University of Illinois Press, Urbana.

ROBERTSON D.H.H. (1963) Human trypanosomiasis in south-east Uganda. A further study of the epidemiology of the disease among fishermen and peasant cultivators. *Bull. Wld Hlth Org.* 28, 627–43.

SCRIMSHAW N.S., TAYLOR C.E. & GORDON J.E. (1968) *The Interactions of Nutrition and Infection*. WHO, Geneva.

SMITHERS S.R., TERRY R.J. & HOCKLEY D.J. (1969) Host antigens in schistosomiasis. *Proc. R. Soc. Ser. B.* 171, 483–94.

SOUTHWOOD T.R.E. (1977). The relevance of population dynamic theory to pest status. In *Origins of Pest, Parasite, Disease and Weed Problems* (Ed. by J.M. Cherrett & G.R. Sagar), pp. 35–54. Blackwell Scientific Publications, Oxford.

STANLEY N.F. & ALPERS M.P. (1975) (Eds). *Man-made Lakes and Human Health*. Academic Press, London, New York and San Francisco.

WILLETT K.C. (1965) Some observations on the recent epidemiology of sleeping sickness in Nyanza region, Kenya, and its relation to the general epidemiology of Gambian and Rhodesian sleeping sickness in Africa. *Trans. R. Soc. trop. Med. Hyg.* 59, 374–94.

WORLD HEALTH ORGANISATION (1976) Resistance of vectors and reservoirs of disease to pesticides. *Technical Report Series* 585. WHO, Geneva.

The prediction of pest and disease problems in man

GORDON SURTEES *c/o Ministry of Defence,*
Whitehall, London

I was invited in my particular contribution to show how our knowledge of pathogens and their vectors, coupled with predictions about social developments in man—especially in developing countries—may enable us to forecast pest and disease problems in coming years. It is a daunting enough task to be asked to reduce a field in such a way as to produce a digestible product. As I have been removed from active ecological research, especially in the tropics, for some time now, my view will to some extent be a long-distance one. In this presentation I apply myself largely to problems associated with impoundment, irrigation and agricultural development. We have to recognize that problems associated with urban development have grown up over the past one hundred years or more largely divorced from any ecological or other scientific influence. With further inevitable rapid and largely unplanned urban expansion in the future, this will continue to be the case. During my professional career I was much involved with problems connected with the epidemiology of arthropod-borne virus diseases and much of the following discussion is from this standpoint. However, the horizon is broadened wherever possible.

Looking back over the past twenty years or so in the field of vector-borne human disease, I have come to the conclusion that the most significant word in my terms of reference is the word 'may'. This conclusion could be based on a number of lines of reasoning. Some may feel that the word should be 'will'—that is, we know, therefore we can predict. Or perhaps it should be 'may, after further research'. Pessimistically, some may feel that the word should be deleted and some more cautious expression put in its place. The scenario I explore adopts none of these starting points. I argue that in this field in 1976, the word is largely irrelevant. Furthermore, that over the past twenty years little of any real weight has been added to our understanding of these problems, that there has been a sad lack of rapport between scientist

"

and administrator, and that schemes where ecologists may become involved can be more motivated by politics than by either science or economics.

Extent of our knowledge

My first proposition is that in 1955 our knowledge of vectors and pathogens was sufficient (not all-inclusive—but sufficient) to enable us to predict what would be the major public health consequences of any social or agricultural change in the tropics. In particular, the essential ecological facts about breeding behaviour, distribution and abundance of the main species of medical importance were in our possession.

Much research since then, and this applies not only to ecology, has seen the specialist absorbed in minutiae of interest mainly to the cognoscenti. We have witnessed superb examples of academic research of the profoundest nature, executed in elegant isolation at the expense of the taxpayer, but no doubt earning the eternal gratitude of the rustic cowherd of the upper Zambesi.

It is not my intention to give an extensive bibliography here, as many aspects of this problem have been very fully reviewed elsewhere (Surtees 1970a, 1970b, 1971, 1975). In 1955 and in some cases a decade or more earlier, we were aware of the following facts.

Anopheles gambiae, the major vector of malaria in tropical Africa, Central and Southern America and in the southern United States, was intimately associated with irrigation schemes and in particular, the extension of rice-fields. *Culex fatigans*, the urban vector of bancroftian filariasis throughout tropical Asia, was typical of large, densely populated urban centres with inadequate sewerage systems. *Anopheles stephensi*, an important vector of malaria in India, was also known to be urban but restricted to wells and man-made water containers. *Aedes aegypti*, the classical vector of urban yellow fever in Africa and America, and of dengue fever in tropical Asia, was clearly identified with urban and peri-urban situations where it bred in man-made water containers, typically close to human dwellings, while in the Ugandan forests there were wild populations breeding in tree-holes. *Aedes simpsoni*, a vector of yellow fever in East Africa, was common in plantations of species with water-retaining axils such as banana and pineapple. *Anopheles gambiae* was recognized as a vector of filariasis in East Africa as well as transmitting malaria in that region. Both trypanosomiasis and oncerceriasis arose consequent upon large-scale impoundment and development. Apart from these broader issues, there were other well-documented facts of considerable predictive value to hand. Epidemic malaria in Venezuela was associated with seasonal breeding of *Anopheles albimanus* in ricefields, while

Anopheles gambiae was a major product of banana plantations in Liberia. In East Africa, *Anopheles gambiae* and *Anopheles funestus* bred prolifically in new ricefields, while the species succession in this ecosystem was clearly understood. Irrigated cotton in Egypt, banana plantations in Africa and date ranches in California were all known to be producing species of medical importance. The extension of rice growing in Portugal had caused a resurgence of malaria in that country, while *Culex tarsalis*, transmitting to man a number of encephalitis-causing viruses, was produced in vast numbers annually from numerous irrigation schemes in the United States. Extensive seepage areas around the Jebul Aliyah dam in Egypt supported populations of *Anopheles* mosquitoes with the result that malaria increased in the surrounding population. Grading of land in preparation for the construction of roads interfered with natural drainage and allowed pools to form which then supported breeding populations of mosquitoes. New disease syndromes were recognized as emerging in dense urban foci in tropical Asia where the seasonal populations of *Aedes aegypti* were very high. *A. aegypti* had followed man and all his works. *A. albopictus* travelled to Hawaii by ship and this species with *A. scutellaris* travelled to the United States by the same means. Man had introduced *Anopheles gambiae* to South America with disastrous consequences. Already in 1943 air transport was recognized as a hazard in moving species of medical importance around the world.

This record could be extended considerably, while the time frame could be pushed back into the 1930's and 1920's. During the ensuing twenty years we have embellished, embroidered, polished and perfected, but the essential canon of ecological law in these fields has remained substantially unchanged. What then have we achieved?

It would be unjust to say that we have achieved nothing; there have been many worthwhile advances. My main concern is this; has what we have done had any real effect on the overall standard of health of people generally in the tropics, compared with efforts in other fields such as urban improvement, nutrition and child health, drainage and some areas of agricultural research? I would argue more positively that resources and effort in vector-borne human disease control could have been allocated in more practical and effective ways. This is of course a *post hoc* judgement. I think it could have been forward-looking policy.

Instead of allowing graduates, each representing a considerable quantum of national investment, to go off to tropical research units to pursue well-worn paths and to confirm for the *n*th time the findings of workers twenty or thirty years earlier, with only minor variations or slight geographical changes of emphasis, a more positive management stance should have been adopted by the controlling organizations and agencies, directing resources towards health and welfare improvement and not solely to basic research. This approach is now being forced on many of us in the present economic climate and we are at last beginning to question in more than a superficial way the all too common attitude of 'it is there, therefore we will climb it'.

I realize that this approach is still unpopular. In the eyes of some it is frankly 'unscientific'. I am afraid it is a fact of life. With restricted resources for research, a critical assessment of capabilities and capacity, aims and objective, is truly scientific. It is efficient, and the whole basis of our philosophy as natural scientists is that efficiency in a given environment is a prerequisite for survival.

The use of our knowledge

My first proposition was that at least twenty years ago we had in our possession all essential predictive data. My second proposition therefore is: why, despite the existence of these data with all their biomedical implications, did large-scale and drastic changes in land-use continue to be made? Who did not take any notice, and why?

We have already considered some of the data and their implications, but simply to ask why was no notice taken of them is to pose a one-sided question, which can only produce a distorted answer. Let us examine three possible reasons. First, possible motives behind some schemes; second, problems of scientific communication; and third, some local attitudes to health and welfare.

MOTIVATION FOR DEVELOPMENT SCHEMES

Stanley & Alpers (1975) comment that large-scale projects are frequently initiated by political expedience or engineering ambition and invariably constructed in ignorance of biomedical changes set in train. You will notice that they imply that economic reasons are of less importance. Thus it could be that if you become involved in research ostensibly in support of a development scheme, you could also be contributing to a political manoeuvre quite divorced from community health or human welfare. Development schemes so called, may look good at election time albeit having little regard to health, local economics, or even increased food production. Outside influence may also play a part. For instance the Aswan High Dam was financed by the Soviet Union to the extent of about 240 million dollars, so providing a considerable sphere of influence. Internal politics may also be important—Kariba was the lynch pin of a forlorn dream of Central African federation.

In such a context what sort of impact can you make as ecologists? You could answer that politics is not your business, the 'ours not to reason why' approach. You could add that involvement gives the opportunity for research, and research leads to the publication of data. You could also add that publication leads to communication! Which brings me to the second point for examination; the communication of scientific results.

SCIENTIFIC COMMUNICATION

Much ecology and indeed other scientific work when published, is directed solely to acolytes of the wisdom. Seldom are scientific results presented in a language and format which will communicate to the non-specialist—often the very one who could, if properly briefed, have an impact on policy and planning. When such a one does ask a specific question of a specialist, the kind of answer he may well get is a three-year research programme and a grant application. This is not simply game playing—it is unrealistic and even dishonest. Scientific stature should not be measured in kilograms of paper, but by the ability to play a significant role in society by contributing in a responsible and economic way to problem solving and decision making, as required by that society.

LOCAL ATTITUDES

The third point for examination is this: developing countries that invite expatriate research workers and doctors to carry out programmes of work, may well not be facing up to their own domestic problems as they should. When qualified, many doctors and scientists in developing countries opt for a desk in the ministry rather than go out to the rural areas where the real health problems are, and where the solutions must be put into practice.

I would sum up at this stage by saying that even though the predictive data existed, they were obscured by bad communication, not seen as relevant, or simply ignored along with the intrinsic local problem.

The likely impact of our knowledge if applied

Now to my third, and in some way consequential proposition. Data existed and major land changes still took place. Would the local populations have been subject to less of a health hazard if, as a result of advice about the biomedical implications, these changes had not taken place?

As a generalization, over most of tropical Africa and Asia, nearly everybody is infected with something. Disease is largely endemic and its expression subclinical, except for occasional flare-ups. Epidemics do occur, but these are more likely in populations of immigrant workers than in the indigenous people. In many parts of Africa, malaria is endemic or hyperendemic, so that any extension of say ricefields or other irrigated land favouring *Anopheles* mosquitoes, will do little to exacerbate the situation. Throughout tropical Asia a very large proportion of the population over the age of twelve, carry antibodies indicating previous infection with group B encephalitic arboviruses. Ricefield extension or the growth of fish farming, will again have a negligible effect. Development is more likely to cause a change in the spectrum of infectious agents than an increase in the overall rate of infection.

The danger of a quantitative, as well as a qualitative change in this pattern is more real in arid or semi-arid areas where, due to impoundment or extensive irrigation, a vector population at a previously very low level may explode in numbers. Even so, this of itself will not necessarily constitute a danger to health. This will only arise if hosts of infectious agents actually in the infective stage, are imported, and even then human and other biologically suitable host populations must continue at a sufficiently high level to ensure susceptible hosts at a density that will maintain a disease focus.

Turning for a moment from agricultural development, probably the supreme example of a disease problem with severe epidemic episodes which would not have emerged if development had not taken place, is the unrestricted and rapid growth of towns in the tropics with totally inadequate sewerage systems. Having said this of course, towns did, and will continue to grow in a largely unplanned way. This uncontrolled growth has been directly responsible for the massive problem of filariasis in tropical Asia and latterly in Africa as well. This urban explosion may also have contributed to the emergence of new viral syndromes in Asia.

Thus, although there are long-standing problems associated with urban centres throughout tropical Africa and Asia, the overall level of disease in man does not change significantly in the long term solely as a result of local or even regional agricultural development and irrigation schemes. We have reached the point where even though the data existed but were ignored, it would have made little difference in these macro terms had they been heeded and the subsequent course of events different.

Constraints

The stark fact remains that population growth, combined with a strong tide of expectation of higher standards of living, demand that emergent countries increase food production considerably. But again this is only one side and a very small aspect of development in the real sense. Consideration of this larger problem does help us to forecast some future localized pest and disease problems in man. However, right at the outset history and economics warn us that perhaps the prerequisites for such broad spectrum development do not exist in the countries under consideration. It is worth noting that we have yet to see really massive agricultural schemes, but I suggest that we have sufficient predictive data were these to occur.

Britain in the mid-eighteenth century experienced the conception and rapid maturation of the industrial revolution. Without going into this too deeply we should take note of some of its essential components. Immediately before 1760 we had had an agrarian revolution that raised agricultural efficiency and productivity by orders of magnitude. At about this time, the population rose very sharply with an overall improvement in community health and in particular, a fall in child mortality and adult morbidity. This rise in the population, together with fundamental legislative moves made the factory system, with its economies of scale and division of labour, possible. This came about in advance of the motive power essential for later industrial development. Alongside these events were economic catalysts such as massive private investment in public works and the growth of the entrepreneur. In addition to all this, the world was our market and we exploited it to the full.

As a result, the classical take-off point was attained. From this point a self-sustaining and accelerating economic system emerged. It needs little extension of this argument to see that, apart from an increasing population and some decline in child mortality these parameters do not exist in any great measure in the emergent countries.

We must now ask, what schemes are likely to have the desired effect in producing economic take-off? Consider the following scenario. Hydro-electric schemes and major impoundments in support of urban and industrial development are increasingly common. Hydroelectric schemes will provide the power for industry into which twentieth-century technology can be injected. With such an injection and outside investment, the 1760–1970's time jump may be made. Such industry in its turn will boost the export trade, particularly if rich natural resources are to hand. A growing trading balance will in its turn attract more outside investment. Collateral effects can then be foreseen in terms of greater social investment in areas such as child health, hospitals and schools. This sequence of events would lead us to argue that in emergent countries, industrial development must precede really large agricultural extension.

This is a plausible argument. It has many attractions and seems to answer many questions, but it is, I fear, totally unrealistic. Changes in orders of magnitude and technological jumps require more than planning, they demand massive proportional investment. This is difficult to conceive at the present time, at least on the required scale. Western industrialized societies cannot afford it, and the emergent countries themselves do not have it to afford.

This is a bleak enough prospect, but let me for a moment go back to one of my earlier propositions and consider some possible consequences. Supposing the scientific and medical resources of Europe and America had been allocated in such a way as to support less self-generated research, but more direct improvement in living standards, health and welfare. Suppose also that due to prophylaxis, improved drainage, controlled instead of continuous irrigation, coupled with sounder land management, we had greatly improved community health on a pan-tropical scale. Suppose malaria no longer followed in the wake of ricefield development in Africa, and also suppose that elegant drainage systems had eliminated filariasis from tropical Asia. One could see as a consequence, an increased population coupled with enhanced longevity of truly epidemic proportions. As such a population would almost certainly outstrip its food supply in the same time scale, this rapid growth would probably be followed by malnutrition and increased susceptibility to community diseases. This could have pandemic sequelae.

Predictions

I predict four main areas where there will continue to be pest and disease problems for man. These are, changes in land use associated with impoundment, irrigation and agricultural extension; the movement of people, pathogens and vectors in some cases on a global scale; intrusion into hitherto undisturbed ecosystems and evolution of new disease syndromes; and finally, collateral effects in non–human vertebrate hosts of disease agents.

PROBLEMS FROM CHANGES IN LAND USE

Changes in land use will generally continue on their present scale but in order to see the implications for man we must understand the environmental changes involved, the vector species and disease agents likely to be favoured and areas at particular risk.

Any area to be irrigated and used for agricultural extension which is not totally arid will contain a certain amount of tree and shrub growth with available free water distributed as discrete water holes or river courses subject to a seasonal pattern of inundation and drying. Within this mosaic the distribution and abundance of animals will reflect their particular ecological requirements.

A large-scale irrigation scheme will initially disrupt and simplify this habitat resulting in an immediate reduction in the variety and number of plants and animals. A species complex, characteristic of formal agriculture will later replace this. In Africa south of the Sahara, loss of natural water holes containing such plants as *Pistia stratiotes* and their replacement by an extensive uniformly irrigated area, will significantly change species composition. Further simplification will be reflected in shaded habitats being replaced by sunlit ones, loss of water-retaining leaf axils and rot holes in trees and the disappearance of broad-leaved plants found in natural bodies of water that extend marginal conditions out into deeper water.

Simplification of the habitat will also be reflected in the loss of plant communities and natural successions and their replacement by sharp transitions between crop and field boundary. While irrigated areas will be independent of seasonal fluctuations in water supply, which will in turn tend to minimize fluctuations in numbers of breeding mosquitoes, changes in numbers, and species succession will occur as a crop grows and matures.

Irrigation schemes directly increase the acreage of above-ground water beyond a main inpoundment. A general raising of the water table as a result will lead to marshy areas and pool formation. If seepage from the impoundment is extensive, sedge swamps will build up and old water courses will refill. A raised water table will also favour the growth of dense vegetation peripheral to the controlled area. All of these changes directly affect the distribution and abundance of mosquito species, natural vertebrate hosts of disease agents, and as a consequence the health of man.

Having thus indicated some major physical changes expected as a result of dam construction and irrigation, we can note some of the mosquitoes of medical importance that will be favoured and whether they are important in the transmission of malaria (M), filariasis (F) or arboviruses (A) to man.

Three types of dam are commonly found in East Africa. True dams built across main drainage channels hold water all the year round and have well-established vegetation. When the water level drops, however, isolated pools are formed which support breeding populations of *Anopheles gambiae* (M, F, A) and *A. funestus* (M, F, A). Other species may also be found from time to time including *A. rivulorum*, *A. squamosus*, *Culex annulioris*, *C. tigripes*, *Mansonia africana* (A) and *M. uniformis* (A).

Tank dams are usually constructed in open country on smooth gentle gradients, drawing water from furrows dug down the slope. These dry out seasonally and are only sporadic sources of mosquitoes. Hafir dams however, are smaller versions of tank dams in which the water level fluctuates widely, and are usually prolific sources of *A. gambiae*.

The physical structure of reservoirs usually influences the distribution and abundance of breeding populations. Production is low where the shoreline is steep but high where this slopes away gradually. Wave action in reservoirs where the water is maintained at a constant level produces sand bars which enclose small pools. In Africa these constitute ideal breeding conditions for *A. gambiae*. Marginal vegetation favours breeding while floating vegetation extends marginal conditions out into deeper water, thus increasing mosquito production. In Tennessee reservoirs, marginal vegetation is colonized by *Anopheles quadrimaculatus* (M, A), *Culex erraticus* and *C. territans*, while in the Nile Valley schemes, floating grass rafts support large breeding populations of *Anopheles pharoensis* (M.) In India, growth of *Pistia*, *Spirogyra* and *Ceratophyllum* spp. favours colonization by *Anopheles hyrcanus* (M), *A. barbirostris* (M), *A. pallidus* and *A. vagus*. In Surinam *A. darlingi* (M) breeds along reservoir margins overgrown with grass and annual weeds.

Outside the main impoundment with its own problems of mosquito production, nearly every crop requiring irrigation supports mosquito species of medical importance. For instance, irrigated pastures in Nebraska are sources of *Culex tarsalis* (A) and *Culiseta inornata* (A), while about half of the farm acreage of California supports mosquito populations. Faulty land construction, inadequate maintenance and unnecessarily excessive irrigation all exacerbate this problem. Irrigation channels in Wyoming produce *Aedes dorsalis* (A) while the same habitat in central Europe produces *A. vexans* (A). Irrigation channels in cotton plantations in Egypt support *Anopheles gambiae*, drainage channels on banana plantations in West Africa produce *A. coustani*, *A. hancocki*, and *A. hargrevesi*, while irrigation canals on Californian date ranches produce *Psorophora confinnis* (A).

Rice is the major crop associated with irrigation. In almost every area, species of medical importance are found; *Culex tritaeniorrhynchus* (A) throughout tropical Asia, India and Japan, *P. confinnis* (A), *Aedes dorsalis* (A) and *Culex tarsalis* (A) in the United States, *Anopheles freeborni* in Central America, *C. modestus* (A) in Mediterranean Europe and *Anopheles maculipennis* (M) in central Europe. In East Africa, this man-made system produces *Anopheles gambiae* (M, F, A), *A. funestus* (M, F, A), *A. coustani*, *Culex univittatus* (A), *C. antennatus* (A), *C. annulioris*, *Mansonia uniformis* (A), and *M. africana* (A).

Some examples will show the effect of these man-made environmental changes and their consequent production of arthropod vectors on the health of man.

O'nyong-nyong virus has been isolated from *Anopheles gambiae* and *A. funestus*, both species being encouraged by impoundment and crop irrigation. This virus causes a febrile illness, generally non-fatal, characterized by sudden onset of totally crippling joint pains, and recovery is slow. There was an epidemic of ONN in East Africa between 1959 and 1962 involving over two million people, in the worst hit areas over 70% of the population being incapacitated. Village populations in Tanzania, living near ricefields that produced large numbers of *A. gambiae* and *A. funestus* have been found to have higher microfilarial rates than village populations in non-ricegrowing localities, where the numbers of *Anopheles* mosquitoes were lower. Western encephalitis is widespread in the United States where it can cause large epidemics. The virus has been isolated repeatedly from *Culex tarsalis* in irrigated areas. Onset is characterized by headache and fever followed by neurologic signs: paralysis occurs in 15% of those infected. The principal host of Japanese encephalitis virus is *Culex tritaeniorhynchus* the dominant ricefield species in tropical Asia and the disease occurs from the far-eastern U.S.S.R., south to Borneo and the Indian sub-continent. Epidemics in Korea, China and Japan are often annual and involve thousands of cases, mainly of children. Up to 10% of those infected exhibit some neurologic or psychotic changes.

On the basis of this evidence, certain predictions may be made. Simplification of the habitat south of the Sahara will reduce the number of naturally occurring water holes containing *Pistia* and other aerenchymatous plants with a consequent reduction in *Mansonia* species. Impoundment and irrigation will favour species such as *Anopheles gambiae*. This change will result in increased transmission of malaria, bancroftian filariasis and such arboviruses as o'nyong nyong, Bwamba, Nyando and Sindbis with reduced transmission of Nduma, Wesselsbron, Spondweni and Pongola viruses. In tropical Australia this initial environmental change could be followed by reduction of the shaded swamp breeder *A. bancrofti* and its replacement by *A. amictus* although both are vectors of malaria. Loss of brackish water would eliminate the vector of Ross River virus, *Aedes vigilax*.

Increased acreage of above ground water in Australia will favour *Culex annulirostris*, host of Murray Valley encephalitis and Sindbis viruses, *C. bitaeniorhynchus* another host of MVE and *Anopheles farauti*, a vector of both malaria and Bancroftian filariasis. In the United States, mosquito hosts of Western, Californian and St Louis encephalitis viruses will all increase in abundance. A longer term consequence of impoundment and irrigation is the growth of dense peripheral vegetation and shaded breeding sites. This will encourage *Anopheles annulipes* and *Aedes normenensis* in Australia, both hosts of Sindbis and MVE viruses, as well as *Anopheles bancrofti* vector of malaria. In the United States, *Culex tarsalis* and *Aedes dorsalis* will be favoured.

Mosquito production in major impoundments will depend to some extent on the physical structure and the growth of marginal vegetation. Species to be expected include *Anopheles gambiae* (Africa), *Culex tarsalis* and *Anopheles quadrimaculatus* (U.S.A.), *Anopheles hyrcanus* (Asia), *Anopheles darlingi* (S. America), and *Aedes nigrithorax* and *A. nivalis* (Australia).

PROBLEMS FROM THE MOVEMENT OF PEOPLE

The movement of people, pathogens and arthropod vectors raises very considerable epidemiological problems particularly when the consequences of those actions have to be predicted. However certain fundamental principles are recognized and may be considered alongside documented events.

Disease does not recognize political boundaries and its spread is facilitated by the greater speed of mass movement now possible. Depending on where they come from, persons entering a locality may be susceptible to a local disease or may arrive in an infective state and so introduce an exotic pathogen. Development requires labour forces and work sites will continue to provide a plethora of breeding sites for mosquitoes, including pockets of run-off water, containers, wheel tracks, machinery and temporary sewerage arrangements. In such circumstances, not only do mosquitoes breed in close proximity to man, but man himself lives in crowded conditions. Both conditions favour disease transmission. Over five million people are involved each year in migrant labour movements in Africa south of the Sahara.

Three parameters are important in considering the role of susceptibles, namely, an indigenous focus of infection, an efficient vector species and sufficient numbers of susceptible individuals. Outbreaks of disease in susceptible populations may not necessarily affect the normal local population, but the significance of such outbreaks will depend very much on the size of the immigrant population. Certainly in the past immigrant labour forces have been depleted by malaria and arbovirus infections and these dangers will remain in the future.

Movement into an area by infective persons will continue to constitute a major hazard, as they may introduce a pathogen against which the local population is immunologically unprotected. Malaria and filariasis are more likely to be introduced in this way than are arbovirus diseases. In the former two, a person may remain in, or return to, the infective state long after initial infection. Malaria parasites may reappear in peripheral blood for up to twenty years, while filarial worms may circulate in the blood stream for over ten years. Movement of infectives into an area may be on a large scale if a labour force is involved, or may be associated with more casual events, such as a visit by itinerant traders. The movement of workers has caused outbreaks of malaria in Nigeria, Venezuela and parts of India. Bancroftian filariasis was introduced to Taiwan by Chinese from the mainland and into Sri Lanka by students coming from an endemic area in India.

With the opening up of new areas and particularly with transcontinental communications crossing different ecological zones, problems of this nature will continue to arise. Further, with rapid global travel on an ever-increasing scale, these problems will not remain restricted to the tropics, as has been dramatically evidenced in recent years by Marburg disease and Lassa fever.

Communications and travel, as an adjunct of development, favour the spread of mosquitoes to all parts of the world. *Aedea aegypti* owes its world-wide distribution to its ability to travel with man. *Culex quinquifasciatus* was introduced into Australia by sailing ship in the late nineteenth century, while *Aedes aegypti* and *A. albopictus* entered Hawaii by the same means. There are two distinct aspects of this problem, the transport of larvae and the transport of adults. Larvae, if introduced to a new area, and if the subsequent adults survive, must feed on an infected host before they can transmit a disease agent. Adults however, can arrive at their destination already infected and transmit an exotic disease agent at once. The danger then is that the pathogen may be maintained in an indigenous mosquito species. Larvae of *Aedes aegypti* and *A. albopictus* were carried to the United States by ship from the Pacific area in fresh water trapped inside motor tyres and within the United States, movement of old tyres was responsible for the re-introduction of *Aedes aegypti* to the Rio Grande Valley after this region had been free of the species for five years.

Adult mosquitoes are carried considerable distances on trains. Between 1958 and 1960, 173 goods trains from the interior of Mexico were inspected at Brownsville, Texas (an important international air, land and sea junction) and over 3,000 adult mosquitoes collected; only a fraction of the likely population. Nine species from six different genera were recorded, all known vectors of malaria or arboviruses. Air travel is an increasing hazard in this respect. *Aedes aegypti* is widespread in the United States and foci of yellow fever are only a few jet-hours away. Recently a survey at Nairobi airport recorded thirteen species of mosquito from incoming aircraft, and one, *Aedes sollicitans* an American species, was collected from a plane which started at Rome. The transport of exotic species from one area to another poses problems for their survival. *Aedes aegypti* can survive, bite and breed during the summer in southern England (Surtees *et al.* 1971), so incoming species would have no difficulty in transmitting disease agents to a susceptible human population. There is an added possibility that an arbovirus, once introduced to England could become established, as both *Culex modestus* and *Aedes vexans* act as hosts of arboviruses on the continent.

PROBLEMS FROM INTRUSION INTO UNDISTURBED ECOSYSTEMS

The results of intrusion have been touched upon in considering trans-continental communications. When this happens, a hitherto animal-centred system may involve man. In the course of evolution, balanced vertebrate/

microorganism systems (zoonoses) have arisen which in some cases require an arthropod host or vector, for their continued maintenance. Typically, these microorganisms do not cause any clinical illness or cytopathogenic effect in their maintenance hosts. Zoonoses are usually associated with a particular locality or biocoenose and are essentially enzootic. If a strange host, such as man, enters this naturally balanced system, he may become infected. This may give rise to some clinical manifestation such as fever or encephalitis, and death may result. Such a disturbed system will produce epidemics if there is a sufficiently high level of transmission to susceptibles. An example of the consequences of human intrusion into a zoonosis is the epidemics of tick-borne Russian spring/summer encephalitis during the clearing, development and settling of Siberian forests. Even today, foresters and campers in central Europe still fall victim to the disease. Continued intrusion, particularly in the tropics, constitutes a severe danger in the context of land development.

PROBLEMS FROM THE EVOLUTION OF NEW DISEASE SYNDROMES

Evidence for the evolution of new disease syndromes comes mainly from urban situations but the underlying epidemiological principles apply to any situation where dense mosquito populations are favoured and where the human population is largely in a susceptible state. In tropical Asia during the 19th century, the growth of large towns was followed by urban epidemics of dengue fever. *Aedes aegypti* invaded this region during the latter half of the century and because of its ability to colonize domestic water containers and the widespread practice of water storage, large populations built up intimately associated with man. Dengue was endemic but essentially a rural disease transmitted to man by indigenous *Aedes* species, mainly *Aedes albopictus*. Urban dwellers were little exposed to this disease and therefore remained susceptible. When large populations of *Aedes aegypti* built up however, they transmitted this virus under urban conditions and there were severe epidemics in man. Following this phase the virus was always maintained at a high level and the human population was constantly in an immune state. As a consequence the problem of classical dengue epidemics receded, but in recent years a new syndrome has emerged with mortalities in children of up to 50%. Several strains of the causative virus closely related to known dengue virus but classified as distinct serotypes, were isolated from sick patients and from *Aedes aegypti*. The disease is characterized by a severe haemorrhagic syndrome. Why this syndrome emerged is not clear, but it has been suggested that due to repeated infection with dengue virus, the severe response is preceded by a phase of sensitization.

If superinfection/sensitization is the biological basis for this new syndrome the same events could be repeated in other areas, including rural ones.

PROBLEMS FROM COLLATERAL EFFECTS IN NON-HUMAN VERTEBRATE HOSTS

Habitats modified by man may bring into close juxtaposition mosquito species of medical importance and non-human vertebrate hosts. The latter may then act as maintenance or amplifying hosts of pathogens to the subsequent detriment of man's own health. Ricefields in East Africa are an example of such a biological complex and the hazards which may be involved. Several of the mosquito species known to transmit malaria and filariasis as well as arboviruses to man (*Anopheles gambiae*, *A. funestus*, *Mansonia uniformis* and *M. Africana*), also feed on birds. At least a dozen arboviruses have been isolated from birds. Vegetation peripheral to the ricefields will not only provide nesting sites for birds, mainly herons and egrets, but also provide ideal microclimate for resting mosquitoes. Coincidence of the production of large numbers of mosquitoes from early stages of the rice crop, and the appearance of susceptible nestlings, will constitute a potential epidemic situation for the local human population.

The urban environment also facilitates the maintenance of viruses and their spill-over into the human population. Domestic animals closely associated with man are particularly important here and their presence is a continual potential health risk. With the growth of towns, this risk will also grow. While urbanization is a major factor in reducing wild vertebrate species, it favours others and these may be domestic animals such as dogs, chickens and pigs, or birds common to towns.

While Venezuelan encephalitis virus is isolated mostly from equids, antibodies indicative of past infection are also found in dogs, while dogs are also implicated in the epidemic of Japanese encephalitis in Sarawak. In the United States, St Louis encephalitis is typical of dense urban environments and the virus has been repeatedly isolated from house sparrows, jays, pigeons and swifts. *Culex tarsalis* the main vector, bites both birds and man. A close correlation has been shown between weekly infection rates with western encephalitis virus in sparrows and in *Culex tarsalis*, leading up to the involvement of man.

As development encroaches into natural habitats the elimination of large vertebrates may result in a change in biting behaviour of mosquitoes, so that man becomes their main source of food. After the eradication of *Anopheles darlingi* and malaria from one locality in Guyana, there was an increase in the human population, loss of grazing land displaced the many cattle that had been found there previously, and the mechanization of farming techniques eliminated horses and donkeys. *Anopheles aquasalis* was the major species feeding on cattle and horses, but when these declined, the mosquito turned to man as its principal source of blood. Subsequent to this succession of events, immigrant workers entered the area from a malarious locality and the result was that malaria was re-established with *A. aquasalis* as the vector.

Conclusions

I would like to draw four conclusions from this survey.

1. As all necessary predictive data were to hand at least twenty years ago, and many important gaps have been filled in since, it has long been time to turn resources away from self-initiated research to direct environmental and health improvement. Ideally, this application of results should be a major consideration at the planning stage. This will require ecologists to leave active research and move into the administrative field.

2. In the past much data have not been put to good use. This has been due to a number of reasons but the one that must concern us most is communication on the part of the scientists. This is still a problem that is being ducked by many. It is too easy to hide behind gobbledegook, or in the absence of real understanding, to give something a longer and more impressive-sounding name. Editors of scientific journals must shoulder a large part of the blame for this. If it is not clear, send it back. Similarly, supervisors of research students must do their share. Unfortunately lack of communication is all too often a symptom of ignorance.

3. Lessons of economics and history tell us that we are unlikely to see any really massive changes in land use. For such changes as are proposed, we have predictive data to hand on which to base sound advice.

4. There are problems ahead, some of which could have implications for territories outside the tropics. The major problem areas of the future are further unplanned urban growth, the opening up of hitherto undisturbed ecosystems, the transport of pathogens and collateral effects in non-human hosts of disease agents.

Nevertheless, I am prepared to predict that whatever changes are planned, we have the data to hand to provide a sound basis for surveillance, planning and prevention programmes.

References

STANLEY N.F. & ALPERS M.P. (1975) *Man-made Lakes and Human Health*. Academic Press, London, New York and San Francisco.

SURTEES G. (1970a) Effects of irrigation on mosquito populations and mosquito-borne diseases in man, with particular reference to ricefield extension. *Int. J. Environ. Stud.* 1, 35–42.

SURTEES G. (1970b) Large-scale irrigation and arbovirus epidemiology, Kano Plain, Kenya. *J. Med. Entomol.* 7, 509–17.

SURTEES G. (1971) Urbanization and the epidemiology of mosquito-borne disease. *Abs. Hyg.* 46, 121–34.

SURTEES G. (1975) Mosquitoes, arboviruses and vertebrates. In *Man-made Lakes and Human Health* (Ed. by N.F. Stanley & M.P. Alpers), pp. 21–34. Academic Press, London, New York and San Francisco.

SURTEES G., HILL M.N. & BROADFOOT J. (1971) Survival and development of a tropical mosquito, *Aedes aegypti*, in southern England. *Bull. Wld Hlth Org.* 44, 707–9.

Discussions on ways of preventing problems arising

Ways of preventing problems arising by ecosystem management

The discussion centred on three main issues; the justification for pest control through ecosystem management, methodologies and future developments in research, advisory and extension work.

Most members of the group felt that the time had come for the developed agricultural technologies to have a much closer look at pest control through management methods. The most pressing reasons for this were those reviewed in the Kennedy report, namely increasing cost of agrochemicals in monetary and particularly energy terms for decreasing returns and the need to develop alternative pest control strategies to overcome pesticide resistance.

It was pointed out that the development of the agrochemical industry during the past thirty years had inhibited research on the biological and management control of pests and that the relevant research that was started before the Second World War had been discontinued. During the past thirty years yields per unit area had increased dramatically but it was perhaps significant that despite the use of pesticides long-term studies of orchard crops in the United Kingdom have shown no reduction in the percentage of the apple crop lost to pests and similar studies in the United States of America had shown a doubling (from 6% to 12%) of the proportion of the corn crop lost to insect pests. At the same time, despite the increase in yields farmers were no better off in relation to other professions in terms of return per unit of capital employed, and from the purely economic point of view the whole farming system could tolerate an increased loss to pests as there was spare capacity; moreover a reduction in crop yield on a national scale coupled with the inevitable increase in the price of the crop could result in the economic yield per unit area remaining stable or even increasing. Farmers on the whole were happy to try biological or management control of pests provided these did not involve obvious yield losses; thus they were more happy to try trap cropping for nematodes than delayed planting or sowing as the latter would involve an inevitable loss in yield through reducing the growth period of the crop. Unfortunately there seems to be little agreement or even information on the kinds of management methods of pest control that can be recommended or even suggested to the farming community.

The discussion on methodologies of pest control by management inevitably covered a wide range of possible methods and a surprisingly narrow range of examples. Apart from crop rotation and biological control most attention was focused on the effects of changing the spatial pattern of crops at the different scales ranging from the landscape scale of strip farming as in some areas in Europe and the United States of America, through strip farming at the within-field scale to the effects of intercropping and crop mixtures. Other possibilities mentioned involved manipulation of management variables such as timing of cultivation and planting or sowing and irrigation and varying plant population and fertilizer levels.

The discussion brought two points out very clearly. Firstly that there is already sufficient evidence to indicate that much of the present use of insecticides could be avoided by crop management methods. For instance the cotton and corn crops in the United States of America receive 65% of the insecticides applied to all crops in that country. The major pests of corn can be reduced by crop rotation and those of cotton by a variety of methods, e.g. either alfalfa strips or sowing 20% of the cotton fields with alfalfa; irrigation control; using strips of early planted cotton as baits for the boll weevil; intercropping with maize.

Secondly there has been far too little research done on management methods of crop control. What, for instance, is the effect on pest levels of varying the pattern of crops (a) from a landscape level (are pest levels in the strip farming systems similar to those of field systems?), (b) at the between-plant level (is the pest damage in oats or barley under modern managements different from that of the traditional 'mixed corn' where these crops were grown with field beans?).

The group recognized that the scientific community simply did not have enough information to give to the farming community; the techniques of researching the subject were open to investigation, and some felt that even the general strategy of research was open to dispute. The majority felt that research should be conducted within the present economic and techno-logical constraints of the agricultural industry. One of the problems of research was that of scale. Economics and labour resources demanded that the plots or units of experimentation be kept as small as possible but where one was concerned with mobile pests the plots needed to be large enough to prevent between-plot interference, and many kinds of investigation would necessarily involve large plot or whole farm experiments.

It was recognized that if management solutions to pest problems became available it may be difficult to put them across to the farmers as the very successes of the last twenty-five years had resulted in the advisory and extension services being committed to the use of chemical methods and there would inevitably be some resistance from the agrochemical industry which had a much larger number of representatives in the field than the 'unbiased' government advisers.

It therefore seems unlikely that in the short term pest problems will be significantly reduced by management methods but taking the long-term view there was good reason to suppose that most of the serious pest problems could be reduced to an acceptable level by management methods that did not involve chemicals. It should be emphasized that much of the necessary research will be expensive and time consuming but this research should be undertaken as the changes in agriculture can be dramatic and sudden and scientific research must build up the necessary corpus of knowledge that can be called upon if present methods of crop protection have to be discontinued.

Ways of preventing problems arising by crop and stock siting

Pest, disease and weed problems are often most difficult where a crop has been grown in one area for many years. The separation of a crop from major concentrations of its pests (to include pests, diseases and weeds) may thus significantly reduce damage. Separation can be achieved geographically, locally and in time.

Introducing a crop to a new geographical area may allow a period of diminished risk from attack especially from specialized pests. Such transfers may involve removal within the same climatic zone or culture of the crop near the margins of its climatic range where pest pressure may be less. Successful examples of such transfer include rubber, coffee, potato, *Pinus radiata*, *Eucalyptus globulus*, the northwards movement of maize culture in the United States of America and cotton growing outside the tropics. It is, however, difficult to separate the effects of changed location from improved management practice. Thus the success of rubber in Malaysia was as a new plantation crop.

Despite successes, many examples can be cited where the introduction of a crop to a new area has not led to decreased pest problems, e.g. rice into Kenya, and cocoa into West Africa. Problems arise from the invasion of the new crop by indigenous, mainly unspecialized pests and by the chance introduction of pests from the original area. Quarantine measures can greatly reduce the chance of the latter but many major pests eventually follow the crop, e.g. coffee rust into Brazil after about one hundred years of successful exclusion. Introduced pests are, however, often most amenable to biological control by the introduction of alien predators which are thereby themselves released from hyperparasitism.

Moving crops to new areas may be useful if pest pressure in existing areas of culture is very high but yields in the new sites must be good and this will involve an adequate agricultural infrastructure. Success is likely to be enhanced if good quarantine operates. A period of minimized pest loss may be achieved, but the length of this is unpredictable. Techniques of 'seed' production in areas of little or no pest attack, e.g. seed potatoes in 'aphid-free' areas, are likely to be of greater value for their potential in pest control is good given adequate international phytosanitary certification.

Techniques which alter the temporal relations of crop and pest such as timing of sowing to avoid the coincidence of maximum crop susceptibility with the period of maximum pest pressure may be of permanent value in diminishing pest damage. They are, however, of limited application and rotation or the use of breaks in cropping regimes have been more widely used. In many situations continuous cropping has long been practised, e.g. in sugarcane, pineapple, perennial crops and more recently in annual crops where economic pressures have made rotation less viable, e.g. in cereals. The advantages of rotations to pest control are lost, increases in pest attack occur but in many annual crops the increases are less than predicted. Breaks between successive crops are frequently as effective as full rotations and in some crops enforcement of compulsory breaks has greatly aided pest control. For example, in the United States of America cotton, tobacco and sugar beet may be harvested over large areas at fixed dates thereby interrupting pest cycles and in the Philippines a compulsory one-year break crop, after wheat, in rice culture, if enforced effectively, controls leaf and stem borer.

Legal enforcement is difficult and not readily acceptable. In some cases, e.g. in wheat, the lack of suitable, profitable 'break crops' is also a limiting factor. Development of this approach seems desirable in particular the introduction of new rotations of crop varieties of different susceptibilities to disease; but much depends on convincing growers of their value and economic viability. If this can be done legislative enforcement may be less useful than grower co-operation as shown in the use of the Sr6 gene for black stem rust resistance in winter wheat in areas of the northern United States of America and its restriction to spring wheats in adjacent Canada. Such schemes could be valuable parts of new strategic approaches to pest control in areas of intensive monoculture where other control methods are under severe pressure.

At field level the separation of crops, spatially or with barriers, from sources of pests and the enforcement of measures to prevent introductions are widely used for pest–crop combinations where other controls are ineffective. Since their efficacy depends on the pattern of pest dispersal such methods are most suitable for relatively immobile pests and largely ineffective for those readily transmitted by air movements.

Field size can influence pest losses; large fields diminish risks for some crop–pest combinations, e.g. in vector-borne viruses of cucurbits, but in others, particularly with polycyclic pathogens, a mosaic of small fields may decrease losses. Generalizations are unsafe and much depends on the sources of infection of the pests concerned.

Some fields or sites within a large area of crop are known to be regularly at greater risk from attack, e.g. in aphid infestation of field beans. Factors such as proximity to overwintering pest sources, micro-climatic differences and topographical features affecting dispersal patterns may all be involved. The identification and prediction of 'high risk fields' would aid control both by avoidance or allowing special measures to be concentrated in areas of highest risk.

Predictions of potential pest attacks are likely to be important to strategies which do not rely on routine use of a single method. Such strategies will assume increasing importance as other methods such as pesticide, or conventional plant breeding for disease resistance become stretched technically or economically. Cultural methods of control including the siting of crops and stock may then also play a greater role than at present where they are largely limited to situations where other control is ineffectual or inappropriate. However experience suggests that generalization about the use of cultural control are to be avoided.

The understanding of the biology and ecology of pests implicit in much of cultural control may be of increasing value in minimizing risks of pest attack as agricultural systems change and may inhibit change without thought of the pest problems which may arise.

Ways of preventing problems arising by plant breeding

Many problems that arise through the deliberate breeding of crop plants were caused by the uniformity of modern cultivars which have a uniform susceptibility to pests and diseases. The uniformity is most pronounced within varieties of modern inbred crops where legislation concerning registration of varieties has meant that great care is taken to bulk a single homozygous genotype. Moreover uniformity also extends between varieties. Modern British varieties of wheat, for example, have an extremely narrow genetic base and Cappelle Desprez is in the parentage of most of them. A widening of the genetic base is desirable and more diverse material from other countries is now being introduced into breeding programmes. A system of legislation to avoid the planting of a high acreage of one variety and imposing two- or three-year cycles of varieties was, although based on sound principles, considered to be an impractical approach. It was concluded that in the case of tree breeding where mistakes are expensive and long term the avoidance of uniformity was more important than in annual crops.

The manner in which resistance is introduced into crops was also discussed. Although there are many mechanisms of resistance possible, the least effective is the method of introducing major genes controlling resistance (R-genes) one at once. One way to improve effectiveness is to introduce several R-genes simultaneously into a crop either in a single genotype or in the form of multilines. Suitable sources of variability are not available for all crops and doubts were raised about the value of multilines. Introducing single R-genes is commercially sound, a new variety with one new R-gene can be produced more rapidly than one with several.

Alternatively, a more stable resistance than that imparted by resistance genes which give plants immunity should be used. It is possible to select plants that are susceptible at the seedling stage but display adult plant resistance, and such resistance may be more durable although consumer resistance may be higher. Other approaches were also considered such as crop 'avoidance' or 'escape', e.g. breeding fast-maturing varieties that can be planted later to avoid smut.

An interesting point to emerge from the discussion was that resistance to disease was not the whole problem, a difficulty lay in combining it with the highest possible yield. Thus Maris Widgeon, a PBI wheat variety, has a stable disease resistance even when widely grown, which has persisted to the present time. However, despite its good quality it has been rendered obsolete by more modern higher-yielding varieties.

It was recognized that breeding had created problems over pest resistance, e.g. post-harvest predation of maize having soft endosperm. This was the result of the breeder being concerned only with yield and not with post-harvest keeping quality. The solution is obviously to make resistance to pests the breeders' concern although to do so would increase the expense of breeding programmes. Pests are much more difficult than diseases to control by breeding and in a high technology agriculture the use of chemicals is inevitable.

It has to be concluded that legislation and commercial considerations are in conflict with scientific principles, and it is not until science overrides the others that plant breeding is likely to cease to give rise to pest and disease problems.

Ways of preventing problems arising by encouraging realistic consumer demand

It is usual in open discussions for there to be arguments about the meanings of words. This meeting was no exception but the group finally agreed that consumer demand was the demand for quality although it was accepted that for some crops yield is more important—especially if man is not the consumer! The word realistic caused more argument.

Many factors affect the demand by consumers for quality and it is impossible to reach many grand generalizations. If supplies are short and demand sustained, expectation of quality can be lower; where supply outstrips demand, quality may become important simply to sell the product; where there is a balance between supply and demand, higher quality can mean a higher price. Against this background the grower lacking a fixed price guaranteed market must gamble; often the pesticide bill is an expensive insurance policy.

Other forces also act to keep quality high. Public health authorities, with the backing of the law and enthusiasm from the public encourage the trend towards higher quality. Advertising, particularly in processed food areas, and the 'brand image' set standards which lead to increases in the quality demanded by the purchaser. For crops where the highest quality standards are set, the results of those standards are increased costs of production, greater use of pesticides, bigger risk of residues, higher wastage and naturally in many cases increased cost to the consumer. The individual consumer's present quality demands may have been set deliberately or inadvertently more by the system than by the consumer. Are present demands realistic?

The group in discussion reached one point of agreement that applied to crops sold direct to the housewife and to crops grown for feeding animals other than man or for conversion processing. Pesticide use sometimes, perhaps often, exceeds real need. Apart from the use for purely cosmetic purposes, the main cause of wastefulness was the apparent guarantee that the pest, parasite, disease or weed problem would not occur if the chemical was used. Frequently, of course, the problem would not have arisen even if the pesticide had not been used! The group identified only two ways in which pest, parasite, disease and weed problems might be reduced by encouraging realistic consumer demand.

1. The problems have led to the use of pesticides, the consequences of which are explored elsewhere in this volume. Some of the consequences are new problems and whether or not they would disappear if pesticide use was reduced remained unresolved. Education of the consumer to allow an end to cosmetic use of pesticides would be necessary and some changes in the Pure Food Act ('the greatest barrier to encouraging realistic consumer demand') essential!

2. The withdrawal of pest-, parasite- and disease-prone crops and foods.

Much discussion was devoted to the reasons why farmers and growers used pesticides wastefully. The probability of maximum yield increases if pesticides are used; sometimes it is cheaper to spray than it is to receive advice on the necessity of spraying; but most important is the general inability to predict the arrival and impact of a disease, a pest or a weed. Accurate forecasting systems would allow reductions in the frequency of pesticide use and secondarily some release from many of the pressures which chemical control measures have created. Perhaps the most realistic consumer demand ought to be for the development of accurate forecasting of the probability of pest, disease, parasite and weed problems.

Author Index

Figures in italics refer to pages where full references appear

o

Subject Index

Related references appear in brackets